Health of People, Health of Planet and Our Responsibility

Wael K. Al-Delaimy
Veerabhadran Ramanathan
Marcelo Sánchez Sorondo
Editors

Health of People, Health of Planet and Our Responsibility

Climate Change, Air Pollution and Health

 Springer Open

Editors
Wael K. Al-Delaimy
University of California San Diego
La Jolla, CA, USA

Veerabhadran Ramanathan
Scripps Institution of Oceanography
University of California at San Diego
La Jolla, CA, USA

Marcelo Sánchez Sorondo
Bishop-Chancellor of the Pontifical
Academy of Sciences
Vatican City, Holy See

ISBN 978-3-030-31127-8   ISBN 978-3-030-31125-4   (eBook)
https://doi.org/10.1007/978-3-030-31125-4

This Springer imprint is published by the registered company Springer Nature Switzerland AG
The registered company address is: Gewerbestrasse 11, 6330 Cham, Switzerland

Message from Pope Francis

Some forms of pollution are part of people's daily experience. Exposure to atmospheric pollutants produces a broad spectrum of health hazards, especially for the poor, and causes millions of premature deaths. People become sick, for example, from breathing high levels of smoke from fuels used in cooking or heating. There is also pollution that affects everyone, caused by transport, industrial fumes, substances that contribute to the acidification of soil and water, fertilizers, insecticides, fungicides, herbicides, and agrotoxins in general. Technology—which linked to business interests, is presented as the only way of solving these problems—in fact proves incapable of seeing the mysterious network of relations between things and so sometimes solves one problem only to create others.

O God of the poor,
Help us to rescue the abandoned
and forgotten of this earth,
So precious in your eyes.
Bring healing to our lives,
That we may protect the world
and not prey on it,
That we may sow beauty, not
pollution and destruction.

—Pope Francis, Laudato Si'

Foreword

The notion of the Anthropocene—the age of humans—comes from the observation that humanity is contesting nature in shaping our planet by profoundly changing the atmosphere, biosphere, land surface, and oceans. In the Anthropocene, climate change is tightly linked to many of the other grand challenges that we face—including the improvement of air quality, public health, sustainable development, and human well-being.

For example, the consumption of limited resources, as well as global warming and shifting precipitation patterns, can affect agriculture and jeopardize food security. Global warming and environmental pollution are also challenging issues of justice because the poor and future generations are the ones most affected. Global environmental change and societal disparity can lead to migration, territorial conflicts, and pressure on resources like water and arable land, which in turn can endanger international stability and peace. Thus, the well-being of humans and the integrity of our environment are closely coupled, as expressed by the term "planetary health."

Both the challenges and the opportunities of a safe and prosperous Anthropocene are reflected in this book, including perspectives of sciences, engineering, medicine, humanities, politics, and religions. I hope and wish that the integration of scientific knowledge with societal considerations will help to solve the challenges and grasp the opportunities of the Anthropocene for the well-being of humanity, now and for future generations.

Paul Crutzen[1]

Nobel Laureate in Chemistry 1995 (shared with Mario J. Molina and F. Sherwood Rowland).

Paul J. Crutzen
Max Planck Institute for Chemistry
Mainz, Germany

---

[1] I thank Uli Pöschl, Mark Lawrence, Jos Lelieveld, and Veerabhadran Ramanathan for helpful discussions.

Preface

This book was almost 8 years in the making at the Pontifical Academy of Sciences (PAS). It perhaps originated at a 2011 meeting that we organized on the topic of mountain glaciers, which was held inside the Vatican at the headquarters of PAS. The glacier meeting was followed by a historic meeting in 2014, again organized by PAS at the Vatican, in which we joined hands with the Pontifical Academy of Social Sciences (PASS). The joint PAS/PASS meeting was ambitious in scope and was titled *Sustainable Humanity, Sustainable Nature, Our Responsibility*. On the last day, we briefed Pope Francis on the primary findings of this meeting, including the moral and ethical issue of climate change, by pointing out the following: more than 50% of climate pollution is being emitted by the wealthiest one billion people, while the poorest three billion, who suffer the worst consequences of climate change, contribute about 5% or less.

A year later, Pope Francis published his climate encyclical *Laudato Si'*, with the best espousal to date of climate justice. Among other things, it clearly states that "The climate is a common good. ... A very solid scientific consensus indicates that we are presently witnessing a disturbing warming of the climatic system. ... A number of scientific studies indicate that most global warming in recent decades is due to the great concentration of greenhouse gases (carbon dioxide, methane, nitrogen oxides and others) released mainly as a result of human activity. As these gases build up in the atmosphere, they hamper the escape of heat produced by sunlight at the Earth's surface. The problem is aggravated by a model of development based on the *intensive use of fossil fuels*, which is at the heart of the worldwide energy system. Another determining factor has been an increase in changed uses of the soil, principally deforestation for agricultural purposes" (*Laudato Si'*).

Recent meetings at PAS on our responsibility vis-à-vis our "Common Home" have included, in addition to PAS academicians, a number of natural scientists, social scientists, philosophers, theologians, and faith leaders, including, at times, the Pope himself. A fourth of the PAS's 80 members are Nobel Laureates in physics, chemistry, and medicine. At the 2014 meeting on *Sustainable Humanity, Sustainable Nature, Our Responsibility*, the combination of thought leaders from around the world concluded that solving the climate change problem would require a transformational change in people's attitudes toward nature and that an alliance between science, policy, and religion could make such a transformational change happen.

The Pope's climate encyclical was published in 2015. A few months later at PAS, we helped the Holy Father organize two summits on climate change: one that consisted of mayors from around the world, and a second that comprised judges from our global community.

During the historic December 2015 UN Climate Conference in Paris, which gave birth to the Paris Agreement to limit warming to *well below 2 °C*, we both joined the Vatican delegation as science advisors. We had the unique privilege of witnessing firsthand the alliance between science, policy, and religion.

Around the time of the 2015 Paris Conference, there was another major momentous development in the science of climate change. An international group of climate and public health experts, called the Lancet Commission, released a report that concluded that climate change would pose unacceptable and catastrophic health risks.

In 2017, the American Meteorological Society (AMS) declared for the first time that "We are experiencing new weather extremes because we have made a new climate." Most weather forecasters, meteorologists, and climate scientists in the USA are members of AMS. The AMS statement is significant because this link between climate warming and extreme weather is directly relevant to public health, since weather extremes such as heat waves, droughts, and floods cause most of the fatalities attributed to natural factors. Numerous studies by individual scientists and institutions have suggested that by 2050, over two billion people could be exposed to extreme heat waves, major droughts, and severe floods.

These studies, along with the Lancet Commission's conclusion and the findings of the American Meteorological Society, have elevated the health impacts of climate change as a major emerging topic of climate change. In response, PAS organized a meeting of scientific leaders, faith leaders, and political leaders from around the world on November 2–4, 2017. The meeting was titled *Health of People, Health of Planet and Our Responsibility: Climate Change, Air Pollution and Health*. It was co-sponsored by the World Health Organization (through Dr. María Neira), the Institute of Public Health at the University of California at San Diego (through its then head, Dr. Bess Marcus; now at Brown University), and the Miraglo Foundation based in San Diego (through its directors Dr. E. Guarneri and Rauni King).

The November 2017 meeting on health produced a declaration signed by several Nobel Laureates, mayors, the governor of California, and faith leaders. This declaration is reproduced in this book. The papers submitted at this meeting provided the background materials for all of the chapters in this book. There are numerous reports and books that are being published on the topic of climate change and health, but this book is unique in one important aspect: the authors not only represent the natural and social sciences but also include faith leaders, political leaders, and practitioners of medicine. As such, this book is a product of an alliance between leaders in science, policy, medicine, and religion, and it is such alliances that can catalyze the sort of transformational change we need to solve the climate change problem.

The funding for the open access of this book was provided by the Center for Development Research, Bonn University (through Professor Joachim von Braun), and the Max Planck Institute of Chemie, Mainz (through Dr. Jos Lelieveld).

La Jolla, CA, USA   Veerabhadran Ramanathan
Vatican City, Holy See   Marcelo Sánchez Sorondo

Declaration: Our Planet, Our Health, Our Responsibility

Veerabhadran Ramanathan (PAS & UCSD), **Monsignor Marcelo Sánchez Sorondo** (PAS Chancellor).

Partha Dasgupta (PASS & CU), **Peter Raven** (PAS & MBC), **Joachim von Braun** (PAS President & UOB), **Jeffrey Sachs** (UN SDSN).

PAS: Pontifical Academy of Sciences; UCSD: University of California at San Diego; CU: Cambridge University; MBC: Missouri Botanical Garden; UOB: University of Bonn; UN SDSN: United Nations Sustainable Development Solutions Network.

This declaration is based on the data and concepts presented at the workshop:

Health of People, Health of Planet and Our Responsibility: Climate Change, Air Pollution and Health.

Organized by the Pontifical Academy of Sciences, Casina Pio IV, Vatican City, 2–4 November 2017.

Statement of the Problem

With unchecked climate change and air pollution, the very fabric of life on Earth, including that of humans, is at grave risk. We propose scalable solutions to avoid such catastrophic changes. There is less than a decade to put these solutions in place to preserve our quality of life for generations to come. The time to act is now.

We human beings are creating a new and dangerous phase of Earth's history that has been termed the Anthropocene. The term refers to the immense effects of human activity on all aspects of the Earth's physical systems and on life on the planet. We are dangerously warming the planet, leaving behind the climate in which civilization developed. With accelerating climate change, we put ourselves at grave risk of massive crop failures, new and re-emerging infectious diseases, heat extremes, droughts, mega-storms, floods, and sharply rising sea levels. The economic activities that contribute to global warming are also wreaking other profound damage, including air and water pollution, deforestation, and massive land degradation, causing a rate of species extinction unprecedented for the past 65 million years and a dire threat to human health through increases in heart disease, stroke, pulmonary disease, mental illness, infections, and cancer. Climate change threatens to exacerbate the current unprecedented flow of displacement of people and add to human misery by stoking violence and conflict.

The poorest people of the planet, who are still relying on nineteenth-century technologies to meet basic needs such as cooking and heating, are bearing a heavy brunt of the damage caused by the economic activities of the rich. The rich too are bearing heavy costs of increased flooding, mega-storms, heat extremes, droughts, and major forest fires. Climate change and air pollution strike down the rich and poor alike.

Principal Findings

- *Burning of fossil fuels and solid biomass releases hazardous chemicals into the air.*
- *Climate change caused by burning fossil fuels and other human activities poses an existential threat to Homo sapiens and contributes to mass extinction of species. In addition, air pollution caused by the same activities is a major cause of premature death globally.*

Supporting data are summarized in the attached background section. Climate change and air pollution are closely interlinked because emissions of air pollutants and climate-altering greenhouse gases and other pollutants arise largely from humanity's use of fossil fuels and biomass fuels, with additional contributions from agriculture and land use change. This interlinkage multiplies the costs arising from our current dangerous trajectory, yet it can also amplify the benefits of a rapid transition to sustainable energy and land use. An integrated plan to drastically reduce climate change and air pollution is essential.

- *Regions that have reduced air pollution have achieved marked improvements in human health as a result.*

We have already emitted enough pollutants to warm the climate to dangerous levels (warming by 1.5 °C or more). The warming as well as the droughts caused by climate change, combined with the unsustainable use of aquifers and surface water, pose grave threats to the availability of fresh water and food security. By moving rapidly to a zero-carbon energy system (replacing coal, oil, and gas with wind, solar, geothermal, and other zero-carbon energy sources), by drastically reducing emissions of all other climate-altering pollutants, and by adopting sustainable land-use practices, humanity can prevent catastrophic climate change while tackling the huge disease burden caused by air pollution and climate change.

- *We advocate a mitigation approach that factors in the low-probability–high-impact warming projections such as the one-in-twenty chance of 6 °C warming by 2100.*

Proposed Solutions

We declare that governments and other stakeholders should urgently undertake the scalable and practical solutions listed below:

1. *Health must be central to policies that stabilize climate change below dangerous levels, drive zero carbon as well as zero air pollution, and prevent ecosystem disruptions.*
2. *All nations should implement with urgency the global commitments made in Agenda 2030 (including the Sustainable Development Goals of the United Nations) and the Paris Climate Agreement.*
3. *Decarbonize the energy system as early as possible and no later than midcentury, shifting from coal, oil, and gas to wind, solar, geothermal, and other zero-carbon energy sources.*
4. *The rich should not only expeditiously shift to safe energy and land-use practices but also provide financing to the poor for the costs of adapting to climate change.*
5. *Rapidly reduce hazardous air pollutants, including the short-lived climate pollutants methane, ozone, black carbon and hydrofluorocarbons.*
6. *End deforestation and land degradation, and restore degraded lands to protect biodiversity, reduce carbon emissions, and absorb atmospheric carbon into natural sinks.*
7. *To accelerate decarbonization, there should be effective carbon pricing informed by estimates of the social costs of carbon, including the health effects of air pollution.*
8. *Promote research and development of technologies to remove carbon dioxide directly from the atmosphere for deployment if necessary.*
9. *Forge collaborations between health and climate sciences to create a powerful alliance for sustainability.*
10. *Promote behavioral changes that are beneficial for human health and protective of the environment, such as increased consumption of plant-based diets.*
11. *Educate and empower the young to become the leaders of sustainable development.*
12. *Promote an alliance with society that brings together scientists, policy makers, health care providers, faith/spiritual leaders, communities, and foundations to foster the societal transformation necessary to achieve our goals in the spirit of Pope Francis's encyclical* **Laudato Si'**.

To implement these 12 solutions, we call on health professionals to engage, educate, and advocate for climate mitigation and to undertake preventive public health actions vis-à-vis air pollution and climate change; and to inform the public of the high health risks of air pollution and climate change. The health sector should assume its obligation in shaping a healthy future. We call for a substantial improvement in energy efficiency and electrification of the global vehicle fleet and all other downstream uses of fossil fuels. Ensure that clean energy benefits also protect society's most vulnerable communities. There are numerous living laboratories—including tens of cities, many universities, Chile, California, and Sweden—that have embarked on a pathway to cut both air pollution and climate change. These thriving models have already created eight million jobs in a low-carbon economy, enhanced the well-being of their citizens, and shown that such measures can both sustain economic growth and deliver tangible health benefits for their citizens.

End of Declaration

Signatories: This declaration was signed by members of the Pontifical Academy of Sciences, including Professor Stephen Hawking (now deceased), all authors of this book, eight Nobel Laureates, faith leaders from the Church of England and the US Evangelical Association, 14 mayors from around the world, the Governor of California, and the Governor of St Luis, Argentina. For a full list of signatories, please see http://www.pas.va/content/accademia/en/events/2017/health/declaration.html

Introduction

This book provides a unique perspective on the multidimensional aspects of the health impacts of climate change and air pollution, which include the science, impacts, solutions, and advocacy of the solutions. The book is intended to add a new approach to climate change by focusing on its health impacts from a multidisciplinary understanding that can be used for academic, political, faith, community, and activism purposes. The authors of the 33 chapters have expertise in an impressive array of topics relevant to the book. We have attempted to make the text in the scientific chapters less technical and more understandable by nonscientists, yet the book has enough technical background and scientific rigor to make it relevant to higher education and professional entities and individuals for use in teaching and practice. More importantly, we have made it freely available online to reach every individual across the globe, including traditionally less involved populations and regions that are most affected by the consequences of climate change.

Even though the influence of humans on climate change has been established through extensive published evidence and reports, the connection between climate change, the health of the planet, and the impacts on human health has not received the same level of attention. Therefore, the global focus on the public health impacts of climate change is a recent area of interest triggered by two recent international initiatives, one by the World Health Organization and the other by *The Lancet*, known as the Lancet Commission on Climate Change and Health. The majority of the previous work on the health impacts of climate change was limited to epidemiological approaches and outcomes, and less focused on multidisciplinary, multifaceted collaboration between physical scientists, public health researchers, and policy makers. Further, the context of faith and ethics as the underpinning of such multidisciplinary approaches was not previously explored.

The gravity of the environmental impacts of climate change on human health, the planet's ecosystem, and the intercorrelations between them—to the degree that it is threatening our very existence and the natural systems that sustain the planet—is real. We therefore need to act immediately and in a holistic approach to be able to make a difference.

This book is not intended to recapture everything related to climate change and health that is already covered by other books. Our intent is to capture just the highlights and summarize the areas we think will resonate most with the public, faith leaders, scientists and students of science, and policy makers. All scientists, not just health scientists, need to turn their attention to the health impacts of climate for the benefit of the survival of mankind.

Part I of the book provides the big picture of the Anthropocene and nature–human interactions. We start the book with a chapter authored by a biology Nobel Laureate, who writes about the importance of protecting habitats and their living organisms to preserve our planet and our well-being (Chap. 1). This leads to a discussion on species extinction rates, which are leading to rapid loss of organisms as a result of climate change (Chap. 2). The impacts of such extinctions on human life and the planet's ecosystem are also addressed. The following chapter connects the changes in the planet to human health (Chap. 3) by providing examples of the spread of contaminating chemicals and discussing land use in addition to the direct health impacts of excessive heat, flooding, and severe weather. The economics of climate change and overuse of the Earth's resources are described insightfully and speak to the connection between the planet's sustainability and human health (Chap. 4). The author discusses the need for better and more accurate measures of wealth that take into consideration the planet's environment. The connection between air pollution and climate change is then described to explain how pollutants can cause global warming and what solutions are possible by making changes to our sources of energy (Chap. 5).

Part II is focused on air pollution, addressed from multiple angles. The evidence of the health impacts of each specific pollutant is well known and is extensively documented in the first chapter in this part, along with the associated disease burden (Chap. 6). The mechanism of these health impacts of air pollution through oxidative stress is then explained scientifically (Chap. 7). The health impacts of air pollution need to be a major component in the aim of policies and regulations related to energy, transport, technology, water, and urban planning so as to lead to a systematic shift in dealing with air pollution (Chap. 8). Furthermore, the global burden of mortality from air pollution can be substantially reduced, as demonstrated across regions and countries, by reducing residential biomass energy use and agricultural sources of ammonia, and by increasing use of new technology in transport and energy (Chap. 9).

Part III addresses fundamental aspects of sustainability in water, agriculture, and deforestation and their region-specific health impacts from climate change. This is exemplified in the Middle East, which is the area most vulnerable to climate change and its health impacts through heatwaves, water scarcity, desert storms, and their impacts on refugees in the presence of political instability (Chap. 10). A globally important topic with particular relevance to this region is water access and water quality, and how we can become sustainable in using the available global water resources (Chap. 11). Agricultural practices that are critical for food security and quality will have a detrimental health effect and require a shift toward climate-smart food systems (Chap. 12). The Amazon has been the region most central to the climate change debate because of deforestation. However, what has been overlooked is the health impacts of forest fires and air pollution on the health of the vulnerable indigenous human populations within that region (Chap. 13). A focus on these human impacts could be a more effective tool for developing environmental policy against deforestation.

Part IV provides a perspective from a number of physicians who are in clinical practice, and what they see is climate change impacts relevant to health. Important and often neglected health impacts of the consequences of climate change and extreme weather are mental illness and psychological trauma associated with such events (Chap. 14). A focus on the specific impacts of air pollution on cardiovascular diseases is missing from clinical practice and requires further attention as a cost-effective approach to limit the health costs of this causal association (Chap. 15). There is a need to explore new paradigms in health care that are holistic and incorporate well-being, the environment, and economic well-being as essential components of health care practice (Chap. 16).

Part V is focused on societal aspects of climate change. Social peace and injustice caused by environmental catastrophes are affecting the most vulnerable global populations, and there is a need to recognize and support stateless populations through specific action because we are all part of one community (Chap. 17). This is also related to global migration and the acceptance of immigrants, whether they are migrating for political reasons or reasons related to environmental and economic privations (Chap. 18). Climate change should be approached in terms of climate justice and the distribution of responsibility. Examples of actions by visionary leadership, such as those in some Latin American cities, and how they have successfully led to green urbanization despite limited resources, are described (Chap. 19).

Part VI is an important focus of this book, as it brings a perspective from Christian faith leaders who participated in the workshop that was the basis of this book. These views, like those of the physicians in Part V, are not inclusive of all communities and faiths but provide perspective and insight into why climate and health matter. For Evangelical Christians, climate change has been politically polarizing. In the first chapter in this part (Chap. 20), the president of the US National Association of Evangelicals provides action items to move beyond such polarization. He focuses on the moral and religious expectations to care for and preserve God's creation, using the term "Christian environmentalism." This view is echoed by the president of the Evangelical Environmental Network, who considers climate change the greatest moral challenge of our generation (Chap. 21). He links the prolife argument and concept to living a healthy life in a healthy environment and to tending to Earth's needs in order to maintain a relationship with God. The role of faith leaders in pointing to human responsibility in the protection of the environment as a divine mandate is also discussed in view of the fact that faith leaders have the ability to bring people together to work on significant issues (Chap. 22).

Part VII includes several chapters on solutions from the scientific perspective. Health co-benefits of mitigation are an approach that will make climate change mitigation policies more acceptable (Chap. 23). Transformative policies in energy, transportation, urban development, and food systems are required if we are to achieve the goals of slowing climate change and protecting human health (Chap. 24). It is not too late to bend the curve of global warming, and its overall impact on health is a dominant driver to raise concerns and trigger action on ten clearly described solutions, which form the topic of another book (Chap. 25). Energy poverty is closely associated with climate change and the health impacts and injustice borne by the poorest communities. However, there are viable solutions exemplified by mini-grids in indigenous and rural communities in Asia (Chap. 26). These communities can have access to renewable energy as an alternative to disruptive and unnecessary dams that submerge lands and displace people. In line with mini-grids, using sensor technology to monitor the use of clean energy cookstoves and rewarding users by payments using cell phones are practical solutions for the three billion people who rely on solid fuels (Chap. 27). This part ends with a unique perspective from the Wellcome Trust Foundation, which has embarked on a mission to fund projects on climate and human and planetary health (Chap. 28). The solutions proposed by this foundation focus on generating knowledge and impacting policy through risk-taking projects and engagement with industry, scientists, and governments.

The final section of the book, Part VIII, is focused on specific actions, with the former California Governor Brown describing how California was able to set the standards and lead the rest of the nation to address air pollution and now renewable energy (Chap. 29). US Congressman Peters describes the US politics of climate change and partisan differences, and why these differences exist (Chap. 30). He offers solutions and hope on how to counter this partisanship on climate by calling on scientists, local and state governments, the armed forces, businesses, church communities, and others to have a say and move beyond partisan politics. A Nobel Laureate in Chemistry then describes the solutions to our climate change problem—decreasing consumption, decreasing population growth, and turning to renewable solar energy—but all must be implemented at a global level rather than at a country level (Chap. 31). The 17 Sustainable Development Goals of the United Nations provide an opportunity to address climate change by making the common good and human dignity the goal so that all stakeholders come together in making it happen (Chap. 32). The final call in this part is appropriately addressed to health professionals, who have the moral authority and are trusted by society to lead the effort to address climate change and its health impacts (Chap. 33). An example from Australia is described, in which an alliance of health care professionals set up a national framework and strategy that is going to be adopted after the election of the political party that supported them.

With these eight parts in the book and its rich, wide-reaching, and diverse content and authors, we have attempted—hopefully with success—to provide something for everyone as a book that will help academics teach their courses, faith leaders preach to their congregations, policy makers communicate with their constituents, health care professionals talk to their patients, and all of the public understand what the major issues of climate change are and how they are impacting our health.

La Jolla, CA, USA   Wael K. Al-Delaimy

Contents

Part I
The Anthropocene: Human–Nature Interactions

CHAPTER 1
Complexity of Life and Its Dependence on the Environment

Werner Arber

Summary Because of the extensive scientific research carried out over the past 200 years, mainly in biology, astrophysics and geology, we have obtained increasingly better insights into the ongoing evolution of living organisms and their habitats. We have thereby acquired information on the complexity of life and its dependence on the environment, which lead us to better appreciate the need to safeguard the high diversity of both living organisms and their environments. On the one hand, all living organisms interdepend on their cohabitation with other kinds of organisms. On the other hand, the slowly changing living conditions of habitats require a living species to occasionally undergo appropriate genetic variation. Ongoing research projects continue to provide insight into evolutionary processes, which are carried out in nature by self-organization, taking care to safeguard a relatively high genetic stability and also ensuring occasional variations of living conditions and their inhabitants. The human population on our planet should avoid intervening in these naturally ongoing evolutionary processes to ensure high biodiversity and its sustainable long-term development.

Introduction

In the 1940s, scientific research revealed that the phenotypes of living organisms depend on the genetic information carried in their genomes. In subsequent decades, it became more clear that life activities also depend on environmental conditions,

---

Nobel Laureate, Medicine/Physiology 1978.

W. Arber (✉)
Molecular Microbiology, Biozentrum, University of Basel, Basel, Switzerland
e-mail: werner.arber@unibas.ch

3

W. K. Al-Delaimy, V. Ramanathan, M. Sánchez Sorondo (eds.), *Health of People, Health of Planet and Our Responsibility*, https://doi.org/10.1007/978-3-030-31125-4_1

including the living and material compositions of the habitats and various temporal impacts from the wider environment. We therefore realize that living organisms depend on both their own genetic information and various environmental conditions. The latter also exert their impact on biological evolution at the population level.

Symbiosis and Nutritional Interdependencies

The cohabitation of various kinds of living organisms can greatly facilitate the life of individual organisms. Human beings and many other eukaryotic organisms are known to strongly depend on cohabiting microorganisms in so-called microbiomes (Blaser, Bork, Fraser, Knight, & Wang, 2013). Thereby, various kinds of bacteria live in symbiosis with humans, animals or plants. Their hosts provide nutrition to the cohabiting bacteria, and the bacteria help their hosts in food digestion and other essential functions. In contrast, several pathogenic microorganisms, both bacteria and viruses, occasionally infect human beings and eukaryotic animals, making them ill. However, we have to be aware that only a minority of bacteria are pathogenic, whereas beneficial microbiomes permanently contain a large number of nonpathogenic bacteria.

Another obvious interdependency resides in the provision of nutrition to other kinds of organisms. Wild animals generally gather and hunt their food in their habitats. To improve their lives, about 10,000 years ago, human beings started to domesticate food plants and animals, which is known as agriculture. Until relatively recently, agriculture was mostly carried out manually on a small scale.

Impacts of Nonliving Environmental Factors on Terrestrial Life Conditions

Many living organisms depend, to some degree, on environmental factors such as climate, temperature, solar radiation, air pollution and weather conditions. These can vary over the course of time both intrinsically and due to the impacts of human activities. Some of these interdependencies caused by human civilization strengthen the negative impacts on the living conditions and thus also on the lives of individual beings. However, it can be quite difficult to ameliorate the situation on a global level.

Value of High Biodiversity

As a result of the long history of biological evolution, our planet benefits from a high diversity of species. As a consequence, any habitat usually contains a relatively high number of different species, and this richness can be quite beneficial for the cohabiting species. For this reason, as well as for other reasons (to be explained below), our civilization should avoid causing any negative impact on this rich biodiversity, which would also affect human life, as shown by the following two examples.

Occasionally, a segment of genetic information can be horizontally transferred from one kind of organism to another, where it can become a part of the genome and exert its activities. Horizontal gene transfer is one of the natural strategies of biological evolution (see below). It is obvious that the loss of rich biodiversity can reduce the available genetic information and thus also affect the biological evolution of other kinds of organisms.

Advancements in scientific knowledge can reveal previously unknown biological potentials of a member of the biodiversity. In order not to lose such a function upon a future extinction of the species in question, one can try to save this genetic potential by transferring it in the laboratory into another kind of organism, which can be expected to express the gene product under consideration. This can allow the investigators to isolate a product that might contribute to a technological advance for our civilization. A striking example is the horizontal transfer of the genetic potential of some spiders into appropriate bacteria to produce silk-like fibres (Rech & Arber, 2013).

Insight into the Laws of Nature for Biological Evolution

Biological evolution is a very slow, continual process; it cannot be easily perceived. As a result of the intensive research on bacterial populations, a number of molecular mechanisms are known to occasionally produce genetic variants and thus contribute to biological evolution after natural selection of mutants that demonstrate a functional improvement for living in their habitat, or sometimes for adapting to an alternative habitat. One can assign the identified individual molecular mechanisms of mutagenesis to just a few natural strategies of biological evolution (Arber, 2007). According to our knowledge to date, one of these strategies is based on local mutagenesis affecting one or a few adjacent nucleotides in the genome. The second strategy of spontaneous mutagenesis involves a genomic segment, which can become transposed, duplicated, deleted or inversed in the genome. Finally, the third strategy is based on the previously mentioned horizontal gene transfer, which allows a recipient organism to acquire a functional genome segment from a donor organism. This acquisition of foreign genetic information is a very efficient process because of the universal genetic code (i.e., the common language of different kinds of organisms).

For some of these mutagenesis events, the products of so-called evolution genes carried in the genome are involved in intragenomic DNA rearrangements. Other evolution genes limit the frequency of mutagenesis events. Some other mutagenesis processes result from the impacts of nongenetic factors, such as environmental mutagens, short-living isomeric forms of biologically active molecules or random encounters. We are aware that Mother Nature only rarely causes a spontaneous genetic variation, so the genetic potential of most individuals in a given population remains preserved. The aforementioned mutagenesis processes occur mostly randomly on the genome and randomly in time. Therefore, only a minority of novel genetic variants reveal, upon natural selection, a functional improvement.

Improvement of Life Expectancy

The knowledge acquired from intensive biological and technological research over the past 200 years has considerably improved human health and various human life facilities, including communication, travelling and the procurement of food. In only a few decades, biomedical knowledge and its applications have resulted in considerably increased human life expectancy, particularly in developed countries. It is our duty to help populations in developing countries, which often have high rates of child mortality, to also benefit from this available biomedical knowledge.

To the best of our knowledge, all living organisms have a statistical average of life expectancy. Genetic alterations or behavioural or environmental impacts can occasionally increase the life expectancy of a given species to some extent. This has recently led to discussions about the possibility of providing immortality to living organisms (including *Homo sapiens*), i.e., an eternal life (Knell & Weber, 2009; Die Frage nach der Zukunft, 2017). In my point of view, this goal is not compatible with the wonderful natural process in which biological evolution helps a species adapt to changing living conditions.

Improvement of Nutritional Diets for Better Health

It is well known that an appropriate diet provides our body with the energy required for our daily activities. The lack of a nutritional source of energy is felt as hunger. It has lately also become known that various biological activities may require so-called micronutrients, such as vitamins or some chemical elements. The lack of an essential micronutrient causes so-called hidden (unfelt) hunger and can lead to an organic mis-development. An example is a deficiency in vitamin A, which is required for the proper neuronal development of a child, both during its embryonic development and during the first few years after birth (Potrykus & Ammann, 2010). In this case, insufficient vitamin A leads to a neuronal mis-development that can cause the death of the child in its first few years or blindness.

It is important for scientific researchers to explore other adverse health effects due to a lack of sufficient micronutrients. Such knowledge may prevent mis-developments by providing the necessary specific micronutrients in due time. Ideally, this goal can be reached by enriching commonly used food crops with the required micronutrients, thus preventing hidden hunger in the people in question.

Intensive Agriculture or Protection of the Rich Biodiversity?

As mentioned previously, agriculture has been carried out manually for a long time. In many countries, a shift has occurred in recent decades to use tractors with specialized machines for cultivation, maintenance, and harvesting of food crops. These cultivations are often combined with the use of herbicides or insecticides to remove other plants and insects from large monocultures of agricultural crops. This practice can contribute to severe reductions in the rich biodiversity.

Reflection on Future Living Conditions on Our Planet Earth

As we have already discussed, a steadily, but very slowly, progressing biological evolution has resulted in the current rich biodiversity. This permanent creation originated approximately 3500 million years ago with unicellular organisms. According to astrophysical theories, the Sun is expected to continue to provide energy to its planets for approximately 4000 million years. However, we do not know how long our planet Earth will exist and contain habitats that host different kinds of living organisms. Let us hope that human civilization in its rich biodiversity will have a time horizon of about 1 million years; during this period of time, biological evolution can be expected to steadily continue slowly. As we have seen, the natural strategy of horizontal gene transfer can provide qualitative improvements to evolving organisms. For this reason, it is important to maintain the rich biodiversity on our planet Earth. Future scientific investigations are expected to allow our civilization to reach this goal by preventing activities that lead to the partial extinction of species existing in the diversity of life. In addition, we should also take care not to entirely use up essential raw materials by our life activities.

From Quantitative to Qualitative Growth

Our planet Earth has a constant size and therefore a limited capacity to host living organisms. In view of the conceptual relevance of a rich biodiversity, a given species of living organisms should not overgrow other kinds of organisms. The aforementioned interdependence of different kinds of organisms, both for their life activities and for their biological evolution, is of high conceptual relevance to the living world. A logical conclusion from this insight is that no species should quantitatively overgrow the others. However, a qualitative growth would always be welcome. Applying this concept to the human population is of primary importance for the healthy long-term persistency of human civilization on our planet.

Conclusion

As members of the human species, we have the privilege of a high intellectual capacity. Thus, we realize that our life interdepends on the presence and activities of many other species of living organisms. Based on these long-term insights, the Christian religions attribute to *Homo sapiens* the role of a good shepherd of creation, who should take care of other kinds of organisms and their habitats. Following this conceptual task, we are expected to protect the presence of a rich diversity of other kinds of living organisms. This essential conclusion is based both on a religious tradition and on more recently acquired scientific knowledge.

As shepherds, we have the responsibility to protect the long-term living conditions of the planet, including the natural process of biological evolution. In addition, it is our task to bring the world view described here to the attention of all human beings on our planet. This requires a well-qualified education applied to human societies worldwide.

References

Arber, W. (2007). Genetic variation and molecular evolution. In R. A. Meyers (Ed.), *Genomics and genetics* (Vol. 1, pp. 385–406). Weinheim: Wiley-VCH.

Blaser, M., Bork, P., Fraser, C., Knight, R., & Wang, J. (2013). The microbiome explored: Recent insights and future challenges. *Nature Reviews Microbiology, 11*, 213–217.

Die Frage nach der Zukunft. In: zur Debatte 4/2017. Themen und Tagungen der Katholischen Akademie, Bayern.

Knell, S., & Weber, M. (2009). *Menschliches Leben*. Berlin, Germany: Walter de Gruyter.

Potrykus, I., & Ammann, K. (Eds.). (2010). Transgenic plants for food security in the context of development. *Special Issue of New Biotechnology, 27*, 445–717.

Rech, E. L., & Arber, W. (2013). Biodiversity as a source for synthetic domestication of useful specific traits. *Annals of Applied Biology, 162*, 141–144.

CHAPTER 2
Biological Extinction and Climate Change

Peter H. Raven

Summary What are the current dimensions of biological diversity? Taxonomists have described approximately two million species of eukaryotic organisms. Many more remain unknown, and the global total may approximate 12 million species or more. Generally, for the past 65 million years, the rate of extinction of these species appears to have been ~0.1 extinctions per million species per year. Now, however, as a result of human activities, it has increased by about 1000 times, to ~100 species per million per year. We are losing species at about 1000 times the rate at which new ones are evolving. Many species are local and particularly liable to extinction, with climate change and increasing human-related pressures of all kinds pushing very strongly on life as it exists. Most of the species cannot be saved by forming parks and protected areas or away from their natural habitats unless human pressures are lessened by general action among nations, a prospect that is not being well realized at present. The strong call for the preservation of biodiversity in the encyclical *Laudato Si'* represents the kind of ethical responsibility that must be adopted if there is to be any hope for the survival of our civilization.

Background

This chapter presents the most up-to-date information on biological extinction. Its rate and intensity have made it increasingly urgent for us to create a sustainable planet. Our individual and collective greed and seeming inability or unwillingness to face reality are preventing our coming to grips with the problem, but there is no other solution than common action for the common good.

We depend on the communities of living organisms that have formed over hundreds of millions of years for our survival, and on the properties of individual kinds

P. H. Raven (✉)
Missouri Botanical Garden, St. Louis, MO, USA
e-mail: peter.raven@mobot.org

© The Author(s) 2020   11
W. K. Al-Delaimy, V. Ramanathan, M. Sánchez Sorondo (eds.), *Health of People, Health of Planet and Our Responsibility*, https://doi.org/10.1007/978-3-030-31125-4_2

of organisms for almost every aspect of our lives. There is no reason to think that the dimensions of the extinction crisis we are facing now will leave communities functioning properly to hold steady the qualities of the atmosphere, soils, or water, and no reason to think that we have already recognized all of the kinds of organisms that would be useful to us if we understood them and their properties better. That is why the following words expressed by Harvard biologist E.O. Wilson (1984, p. 121) are so very important for us to ponder:

> The worst thing that will *probably* happen—in fact is already well underway—is not energy depletion, economic collapse, conventional war, or the expansion of totalitarian governments. As terrible as these catastrophes would be for us, they can be repaired in a few generations. The one process now going on that will take millions of years to correct is loss of genetic and species diversity by the destruction of natural habitats. This is the folly our descendants are least likely to forgive us.

Climate change is one of the major forces that will drive biological extinction in our time and by doing so will have an extraordinary and lasting detrimental effect on our collective health. In turn, any loss of species caused by other factors, including those discussed in this chapter, will have further detrimental effects not only now but permanently (Isbell et al., 2017). We depend on other species and the ecosystems they comprise for the air we breathe, the soil in which we grow our crops, the ability to absorb pollution, all of our food, many of our medicines, a major proportion of our building materials, and the beauty that surrounds us when we appreciate and protect it. In short, there is no factor that will harm human health and our ability to live sustainably on this planet more than the loss of biological diversity. While climate change and pollution are proving to be hugely damaging to human health directly, as amply illustrated by the rich diversity of chapters presented in this book, the loss of biological diversity and the natural systems it drives will prove much worse over the longer run. Biological extinction will be the worst legacy of climate change and pollution, as well as the most damaging to the continuation of our sustainable lives on earth and to our ability to maintain our health and our civilization in the near term.

It is difficult to estimate the current rate of extinction because in fact we know so little about the living world, even though our lives depend completely on its healthy functioning. As we shall discuss in this chapter, we have recognized and given names to a relatively small proportion of the total number of organisms on earth. We cannot even begin to understand the harm we are doing to ourselves by ignoring the loss of as many as half of them during the remainder of this century. Overall, the account presented here is based on the reviews of extinction presented in our February 2017 Study Week on Biological Extinction (Dasgupta & Ehrlich, 2019; Pimm & Raven, 2019) and in the outstanding paper in the same session by Lenton (2019). In the earth's 4.54 billion year history, life originated early and invaded the land about 400 million years ago. Our species, *Homo sapiens*, was first discovered in African fossils that were 300,000 years old, migrated to Eurasia at least 60,000 years ago, and reached all habitable continents by at least 15,000 years ago. Our spread and ultimate increase, discussed below, has had and is continuing to have a massive negative effect on all other forms of life on earth except for those we encourage for our own use.

How Many Species Exist Now?

Our summary paper (Pimm & Raven, 2019), with others presented at the PAS/PASS meeting on biological extinction in February 2017, attempted to estimate the number of species that exist as well as to predict their rate of extinction. Most of the references for conclusions offered by us in that symposium will not be repeated here.

Science has discovered and named approximately two million species or organisms to date, and most of them remain poorly known. Judging from well-known groups of eukaryotes and their overall patterns of distribution, we might assume that at least 12 million species may actually exist. For marine eukaryotes, estimates vary greatly (Appeltans et al., 2012; Mora, Tottensor, Adl, Simpson, & Worm, 2011), with further investigations clearly necessary to arrive at realistic estimates. When it comes to prokaryotes, we cannot even guess the numbers of species reasonably; Locey and Lennon (2016) recently speculated that the total number of species of Eubacteria and Archaea might even amount to an astonishing one trillion. It would be well worth putting more effort into estimating their diversity and geographical specificity because they, as much or even perhaps more than eukaryotes, are clearly of fundamental importance for us both ecologically and economically (Montgomery & Bicklé, 2015). Whatever we learn about them is very likely to serve us well.

Most species that exist now are likely to become extinct before they are discovered, particularly in the tropics; thus, it will be important to choose our research and conservation directions very carefully during the coming decades. The record we shall be able to assemble is likely to be a limited representation of the richness of life as we know it now.

Factors Driving Extinction

The key factors driving massive biological extinction in our time have resulted from the continuing, rapid growth of the human population in the roughly 11,700 years of the Holocene Epoch, a relatively warm time following the most recent glacial maximum. Early in the Holocene, our ancestors mastered agriculture. Over the subsequent 10,000 years or so, the total human population has grown from only about one million people to more than 7.6 billion; our numbers are estimated to keep growing to about 9.9 billion by mid-century (www.prb.org). In contrast, when people began cultivating crops, there were fewer than 100,000 living in all of Europe! With the development of agriculture, however, early farmers began to be able to live together permanently in villages, towns, and ultimately cities. The inhabitants of such settlements could store enough food to carry them through unfavorable times. Living for the long term together in groups, individuals could for the first time specialize in various professions—the activities that together form the basis of our modern civilization. Everywhere in the world, forests began to be cut rapidly, with disastrous effects on biological diversity that have been greatly accelerated recently (e.g., Betts et al., 2017).

Particularly since the acceleration of the Industrial Revolution some 250 years ago, with its subsequent intensification of all our activities, the pollutants that we have produced have poisoned soils, waters, and the air we breathe, damaging our health and the natural systems that support us. Ultimately, the gasses our activities generate have begun to alter the world's atmosphere markedly and to drive major climate change that became evident about 75 years ago. Agriculture, urban sprawl, deforestation, and all of the activities associated with human population growth; the spread of invasive plants and animals, pests and pathogens; and the hunting and gathering of plants and animals from nature, have all damaged natural populations and sometimes driven them to extinction.

As our numbers have grown, our desire for enhanced consumption has grown even more rapidly. The Global Footprint Network (*www.footprintnetwork.org*) estimates that we are using approximately 175% of the world's sustainable productivity now, up from an estimated 70% in 1970—a situation that clearly cannot be sustained, and one in which the stronger nations will be led ever more forcefully to drain the potentially renewable resources from the poorer and weaker ones. Biological communities provide the basis of our lives, and the levels of species loss from them that we are experiencing therefore pose dangers for the sustainable functioning of life on earth (Duffy et al., 2017). Today, nearly 40% of Earth's land surface is devoted to farming and grazing, most of the latter unsustainable. One can only imagine how many species must have disappeared while agriculture was spreading over wide regions like the intensively cultivated, broadly defined Mediterranean region of western Eurasia and North Africa. At the same time, invasive species of animals and plants, pests and pathogens, are spreading around the world in ever-increasing numbers, out-competing others in the new regions that they have reached and often driving them to extinction.

Another significant driver of biological extinction, recognized from the 1950s onward, is anthropogenic climate change; methane, enormous quantities of carbon dioxide and other greenhouse gasses are being emitted into the atmosphere continuously. Reviewed elsewhere, the changes that we are experiencing are extremely dangerous, and our collective will to deal with them remains dangerously limited, with greed always a more attractive alternative for many. There is a great deal of urgency in coping strongly with this enormous threat, and Pope Francis' powerful encyclical *Laudato Si'* warns all of humanity to attend to the problem for our common benefit.

Extinction Rates

With this review of the current factors driving extinction, we are enabled to return to a calculation of extinction rates now and in the future. The calculations that we presented in our earlier review were explicitly based on current records, the fossil record, and comparisons with rates of the origin of new species as determined by molecular phylogenies (Pimm & Raven, 2019). We express our estimates in terms of extinctions per million species-years (E/MSY). With a historical extinction rate of approximately 0.1 extinctions per million species per year, we have now reached at least 1000 times that level, with the rates increasing rapidly.

Comparing International Union for Conservation of Nature (IUCN) Red List of Threatened Species (www.iucnredlist.org/) estimates of the likelihood of extinction for the species they include with other extinction probabilities leads to an estimate that perhaps 20% of all species in these groups may be in danger of extinction over the next several decades. At present observed extinction rates and a total of 12 million species of eukaryotes, about 60,000 of them might become extinct over the next 40 years. In sharp contrast, if 20% of all species were actually to become extinct during this period, 40 times that many, amounting to some 2.4 million, would have to be lost. Such a rate would amount to losing 60,000 species per year (higher in the future, lower now, of course). Considering that unknown species or poorly known ones are more likely to be rare and therefore on the average more likely to be in danger of extinction than those that have been rated, these huge numbers might actually be attained, but that will depend to a great extent on the nature and scale of human activities in the future. At any rate, the estimate is staggering!

Most species have relatively small ranges, and those that do are on the average in greater danger of extinction than common, widespread ones. Even the widespread ones are virtually never distributed uniformly and densely within the boundaries of their ranges, although that is often an implicit assumption in calculations. In addition, local species are usually also scarce even where they occur (Pimm & Raven, 2019). In his definition of biological "hot spots"—areas of high concentrations of endemic unusual species—Myers (1988) by definition included only those areas that had already lost 70% or more of their original vegetation to human activities and which had at least 2000 species of plants found nowhere else. Certainly such areas are likely to become important centers for extinction as human activities increase in intensity and in volume and conservations select and pursue the best goals for conservation they can define—intrinsically limited. Hotspots, therefore, are especially critical areas for future extinction, and ones where adequate conservation plans would be especially helpful in lowering future rates of extinction (Pimm & Raven, 2000). In comparing 25 hotspots using the data on original area, endemic bird or mammal species, and the current remaining natural vegetation (Crooks et al., 2017; Pimm & Raven, 2000), dismal futures are predicted for a high proportion of the species now living in these hotspots.

Considering the increasing number of factors that are driving species to extinction, it indeed seems plausible that 20% of them may disappear during the next several decades. If that is the case, and given the rates of extinction that would have been attained during the next few decades, it seems possible that a proportion of species amounting to roughly half of the total may vanish from nature by the end of the century (Pimm & Raven, 2017). Detailed analyses are provided by Ceballos et al. (2015, 2017) and Dasgupta and Ehrlich (2019), which support the idea that the rate of population is proceeding much more rapidly than we have realized. In addition, the well-documented shrinking of many populations of birds (Gross, 2015; Vorisek, Gregory, Van Strien, & Meyling, 2008; http://www.stateofthebirds.

org/2017/wp-content/uploads/2016/04/2017-state-of-the-birds-farm-bill.pdf;
http://www.stateofthebirds.org/2016/wp-content/uploads/2016/05/SotB_16-04-26-
ENGLISH-BEST.pdf) and insects (e.g., Hallmann et al., 2017; Vogel, 2017) appear
to speak to the same point—human beings are simply taking up so much of the
space that the numbers of other kinds of organisms must decrease and keep decreasing unless we find and deploy the means we need to maintain a sustainable world.

Regardless of what I consider to be a largely semantic problem, which is whether
we have entered a major extinction event comparable with the earlier five the world
has experienced (e.g., Baranosky et al., 2011), we are clearly living in times that
have the potential to be catastrophic for our own future. In such a time, we must not
lose hope, but rather strengthen our efforts to the extent to deal with scientifically
demonstrated trends as effectively as we possibly can!

The least well understood factor that is driving future extinction is global climate
change, a problem with which we are just starting to come to grips intellectually.
Such rapid and catastrophic events would, of course, often be outside of the possibility of adaptation for the species concerned. Although the mechanisms underlying
the relationship between climate change and extinction are complex (Pereira et al.,
2010), some detailed, regional modelling exercises have been carried out in Australia
(Williams, Jackson, & Kutzbach, 2007) and South Africa (Erasmus et al., 2002).
These have led to predictions of the extinction of many species with restricted
ranges during the coming decades in those areas. We do not understand yet how
much climate change will add to biological extinction, and that of course depends
in part on what actions we manage to take to slow down the change itself. Under any
scenario, however, the current habitats of those species restricted to the southern
edges of the southern continents will simply disappear, as will the habitats of those
species restricted to the higher elevations in mountains worldwide, although they
would have some opportunity for dispersal. Certainly, the rich arrays of endemic
plants and other organisms that occur along the southern borders of Australia and
Africa are in extreme danger of extinction from climate change.

Extinction rates during periods of rapid climate shift in the Pleistocene were not
especially high, except in areas where humans were actively hunting populations of
specific animals. As Lenton (2019) has pointed out, this observation would tend to
suggest climate-caused extinction rates when the changes have been relatively
extreme. A detailed review of estimates of extinction rates likely to occur in relation
to climate change (Bellard et al., 2012) has shown a large degree of uncertainty with
a number of factors still to be considered in detail. Thus, Lenton (2019) inquired to
what extent the projected high levels of species loss would compromise the function
of ecosystems and the biosphere, and found that massive changes are likely to occur.
Changes of this magnitude could well result in the loss of entire biomes by the tipping of critical balance points. The massive changes in precipitation levels that are
projected as part of global climate change, as from the Green Sahara of 6000 years
ago to the sere desert of today, or from dense forests to deserts and savannas, could
result from the overall pattern of change.

For the seas, projections of extinction rates are being developed slowly, with a
relatively small proportion of marine organisms rated by the IUCN for degree of

threat. It is clear, however, that climate change with its accompanying acidification have become key factors in marine extinction. Overfishing is the most important of the obvious factors affecting extinction in the seas. Pollution from the land, histori- cally chemical but now including the massive deposits of plastic, remains important and must certainly also be controlled if we want to maintain the productive seas that exist today. Our relatively poor knowledge of the extent of marine eukaryotic biodi- versity makes estimates of extinction difficult, too, as Webb and Mindel (2015) have stressed.

Fortunately, efforts to measure and to ameliorate the environmental damage we are causing to the seas are growing (e.g., Sullivan et al., 2019). Certainly specific communities, such as coral reefs, attuned to narrow temperature ranges, appear to be in extreme danger of collapse (Hoegh-Guldberg et al., 2007; Hughes et al., 2017; Lucht et al., 2006). Marine anoxia resulting from the runoff of nitrogen and phos- phorus from agricultural systems on land is leading to increased ocean warming and thus further reducing the capacity of the water to absorb oxygen. Summarizing, overfishing, habitat destruction, climate change, pollution, and ocean acidification (e.g., DeWeerdt, 2017) are all contributing to extinction in the seas.

The possibility of even higher rates of climate warming than are generally assumed to be likely would of course have a much more damaging effect on biologi- cal diversity than anything discussed thus far in this chapter. A continuing disregard for our common welfare could lead to global temperate increases of 6 °C; if we reach such levels, or even higher ones, the consequences would be staggering. For exam- ple, Malin, Pinsky, and Jon Payne (personal communication) have pointed out that 10 °C warming at the time of the end Permian extinction drove more than 75% of the genera then in existence to extinction. This amounted to the most extensive extinc- tion event in the history of our planet, permanently changing the character of life on earth. Regardless of whether we are wise enough to stop the process somewhere short of those levels, we are clearly facing major losses of biological diversity. That loss in turn will have an immediate and long-lasting effect on our civilization and our collective health both directly and indirectly. We must find effective ways to address the loss before it is too late (e.g., Lovejoy, 2017; Tilman et al., 2017).

Conclusions

Nothing we are doing to damage the functioning of the living world and the stability of its ecosystems will do more harm than biological extinction. Curbing global climate change will be indispensable, but it is only one of the steps that we will need to take collectively if we wish to curb extinction and maintain a stable planet for our succes- sors. Increasing human population and especially increased human greed for ever- higher levels of consumption will greatly increase future rates of extinction through the creation of a warmer world and in many other ways as well. We have estimated that as many as 2.4 million species of organisms, 20% of the estimated total, could become extinct during the next several decades as human pressures of all kinds intensify and presently sustainable systems collapse, and that perhaps half of the estimated total number of species could be gone by the end of the twenty-first century.

As Dasgupta and Ehrlich (2019) have concluded, seeking to limit the proximate causes of extinction will not ultimately prevail unless the basic drivers are addressed—"continued population growth, policies seeking economic growth at any cost, overconsumption by the rich, and racial, gender, political, and economic inequality (including failure to redistribute). Collectively addressing these are possibly the greatest challenges civilization has ever faced."

Achievement of the goals just outlined clearly must be taken as our common responsibility; without doing so, we simply have not got a chance. Learning about how to achieve these goals and then working to achieve them ultimately will be responsible for our survival. We must learn, teach, act, and vote in such a way as to advance our common cause, the search for a sustainable world. Doing so will require a degree of humility, compassion, and love that we have yet to exhibit, but which is indicated very clearly in our topical encyclical *Laudato Si'*. We must follow the example of Pope Francis in contemplating the beauty and importance of life as it is and the urgency of working to preserve it for the future.

Acknowledgments I have received helpful comments and discussion of this chapter from V. Ramanathan and Partha Dasgupta; Douglas H. Erwin, Smithsonian Institution; John F. Fitzpatrick, Cornell Laboratory of Ornithology; Jane Lubchenco and Jenna Sullivan, Oregon State University; Guy Midgley, Stellenbosch University; Camille Parmesan, University of Texas; Stuart Pimm, Duke University; Melin Pinsky, Rutgers University; Hugh Possingham, Nature Conservancy.

References

Appeltans, W., et al. (2012). The magnitude of global marine species diversity. *Current Biology, 22*, 2189–2202.

Bellard, C., Bertelsmeier, C., Leadley, P., Thuiller, W., & Courchamp, F. (2012). Impacts of climate change on the future of biodiversity. *Ecology Letters, 15*, 365–377.

Betts, M. G., Wold, C., Ripple, W. J., Phalan, B., Millers, K. A., Duarte, A., et al. (2017). Global forest loss disproportionately erodes biodiversity intact landscapes. *Nature, 547*, 441–447.

Ceballos, G., Ehrlich, P. R., Baranosky, A. D., García, A., Pringle, M., & Palmer, T. M. (2015). Accelerated human-induced species loss: Entering the sixth mass extinction. *Science Advances, 1*, e1400253.

Ceballos, G., Ehrlich, P. R., & Dirzo, R. (2017). Biological annihilation via the ongoing sixth mass extinction signaled by vertebrate population losses and declines. *Proceedings of the National Academy of Sciences of the United States of America, 114*, 6089–6096.

Crooks, K. R., Burdett, C. L., Theobald, C. M., King, S. R. B., Di Marco, M., Rondinini, C., et al. (2017). Quantification of habitat fragmentation reveals extinction risk in terrestrial mammals. *Proceedings of the National Academy of Sciences of the United States of America, 114*, 7635–7640.

Dasgupta, P., & Ehrlich, P. R. (2019). Why we are in the Sixth Extinction and what it means to humanity. In Dasgupta, P., P.H. Raven., P.H., & Anna McIvor (eds.), *Biological extinction: New perspectives* (pp. 262–284). Cambridge, UK: Cambridge University Press.

DeWeerdt, S. (2017). Sea change. *Nature, 550*, S56–S58.

Duffy, J. E., Goodwin, C. M., & Cardinale, B. J. (2017). Biodiversity effects in the wild are common and as strong as key drivers of productivity. *Nature, 549*, 261–264.

Erasmus, B., Erasmus, F. N., Van Jaarsveld, A. S., Chown, S. L., Kshatriya, M., & Wessels, K. J. (2002). Vulnerability of South African animal taxa to climate change. *Global Change Biology, 8*, 679–693.

Gross, M. (2015). Europe's bird populations in decline. *Current Biology, 25*, R483–R489.

Gustafson, P., Raven, P. H., & Ehrlich, P. R. (2020) (in press). *Population, agriculture, and biodiversity*. Columbia, MI: University of Missouri Press.

Hallmann, C. A., Sorg, M., Jongejans, E., Siepel, H., Hofland, N., Schwan, H., et al. (2017). More than 75 percent decline over 27 years in total flying insect biomass in protected areas. *PLoS One, 12*, e0185809. https://doi.org/10.1371/journal.pone

Hoegh-Guldberg, O., et al. (2007). Coral reefs under rapid climate change and ocean acidification. *Science, 318*, 1737–1742.

Hughes, T. P., Barnes, M. L., Bellwood, D. R., Cinner, J. E., Cumming, G. S., Jackson, J. B. C., et al. (2017). Coral reefs in the Anthropocene. *Nature, 546*, 82–90.

Isbell, F., Gonzalez, A., Loreau, M., Cowles, J., Díaz, S., Hector, A., et al. (2017). Linking the influence and dependence of people on biodiversity across scales. *Nature, 546*, 65–72.

Lenton, T. M. (2019). Biodiversity and climate change: From creator to victim. In Dasgupta, P., P. H. Raven., P. H., & Anna McIvor (eds.), *Biological extinction: New perspectives* (pp. 34–79). Cambridge, UK: Cambridge University Press.

Locey, K. J., & Lennon, J. T. (2016). Scaling laws predict global microbial diversity. *Proceedings of the National Academy of Sciences of the United States of America, 113*, 5070–5975.

Lovejoy, T. E. (2017). Extinction tsunami can be avoided. *Proceedings of the National Academy of Sciences of the United States of America, 114*, 8440–8441.

Lucht, W., Schaphoff, S., Erbrecht, T., Heyder, U., & Cramer, W. (2006). Terrestrial vegetation redistribution and carbon balance under climate change. *Carbon Balance and Mangement 1*, 6.

Montgomery, D. R., & Bicklé, A. (2015). *The hidden half of nature: The microbial roots of life and health*. New York: W.W. Norton & Company.

Mora, C., Tottensor, D. P., Adl, S., Simpson, A. G. B., & Worm, B. (2011). How many species are there on earth and in the ocean? *PLoS Biology*. https://doi.org/10.1371/journal.pbio.1001127

Myers, N. (1988). Threatened biotas: "Hot spots" in tropical forests. *Environmentalist, 8*, 187–208.

Pereira, H. M., Leadley, P. W., Proença, V., Alkemade, R., Scharlemann, J. P., Fernandez-Manjarrés, J. F., et al. (2010). Scenarios for global biodiversity in the 21st century. *Science, 330*, 1496–1501.

Pimm, S. L. (2009). Climate disruption and biodiversity. *Current Biology, 19*, 595–601.

Pimm, S. L., & Raven, P. H. (2000). Biodiversity: Extinction by numbers. *Nature, 403*, 843–845.

Pimm, S. L., & Raven, P. H. (2017). Fate of the world's plants. *Trends in Ecology and Evolution, 32*, 317–320.

Pimm, S. L. & Raven, P. H. (2019). The state of the world's biodiversity. In Dasgupta, P., P. H. Raven., P.H., & Anna McIvor (eds.), *Biological extinction: New perspectives* (pp. 80–112). Cambridge, UK: Cambridge University Press.

Sullivan, J. M., Constant, V., & Lubchenco, J. (2019). Extinction threats to life in the ocean and opportunities for their amelioration. In P. Dasgupta, P. H. Raven, & Anna McIvor (Eds.), *Biological extinction: New perspectives* (pp. 113–137). Cambridge, UK: Cambridge University Press.

Tilman, D., Clark, M., Williams, D. R., Kimmel, K., Polasky, S., & Packer, C. (2017). Future threats to biodiversity and pathways to their prevention. *Nature, 546*, 73–81.

Vogel, G. (2017). Where have all the insects gone? *Science, 356*, 576–579.

Vorisek, P., Gregory, R. D., Van Strien, A. J., & Meyling, A. G. (2008). Population trends of 48 common terrestrial bird species in Europe: Results from the Pan-European Common Bird Monitoring Scheme. *Revista Catalana d'Ornitologia, 24*, 4–14.

Webb, T. J., & Mindel, B. L. (2015). Global patterns of extinction risk in marine and non-marine systems. *Current Biology, 25*, 506–511.

Williams, J. W., Jackson, S. T., & Kutzbach, J. E. (2007). Projected distributions of novel and disappearing climates by 2100 AD. *Proceedings of the National Academy of Sciences of the United States of America, 104*, 5738–5742.

Wilson, E. O. (1984). *Biophilia*. Cambridge, MA: Harvard University Press.

CHAPTER 3
Sustaining Life: Human Health–Planetary Health Linkages

Howard Frumkin

Summary Our planet is infinitely beautiful, complex, and precious. As each of us is well aware, our species has begun to alter the planet, sometimes inadvertently, sometimes negligently or even recklessly. Ecologists often ask, "How do our actions undermine nature?" We need also to invert the question, asking "How do the planetary changes to which we contribute undermine our own health and well-being?"

This query lies at the heart of the emerging field of Planetary Health. The premise of Planetary Health is this: the earth system changes we have wrought are now so far-reaching that they drive a substantial, and increasing, proportion of the global burden of disease. Because we depend on stable earth systems to survive and thrive, environmental destruction amounts to self-destruction. We need to understand these consequences. Based on that understanding, we need to reexamine many of our core assumptions about our place on the earth. Finally, we need to deploy both scientific insight and conceptual reorientation to inspire, and to guide, corrective action.

This chapter begins with several examples of the ways in which planetary changes threaten human health and well-being—the human dimensions of planetary boundaries, as described by Will Steffen and colleagues at the Stockholm Resilience Centre (Steffen, Richardson, Rockstrom et al., 2015). The first is climate change. While the consequences of climate change range broadly, from air pollution to mental health to infectious disease, this chapter focuses on just one aspect: the effects of heat—an apparently straightforward hazard that when explored reveals great complexity. But even if there were no climate change, we would still need the field of Planetary Health, because other global-scale changes are also affecting health and well-being. Three examples discussed here are chemical contamination, land use changes, and biodiversity loss. For each of these four examples, I submit,

H. Frumkin (✉)
University of Washington, Seattle, WA, USA
e-mail: frumkin@uw.edu

© The Author(s) 2020
W. K. Al-Delaimy, V. Ramanathan, M. Sánchez Sorondo (eds.), *Health of People, Health of Planet and Our Responsibility*, https://doi.org/10.1007/978-3-030-31125-4_3

the science paints a clear and sobering, if not yet complete, picture: global environmental change threatens human health and well-being.

Reviewing the relevant science behind these four examples raises far-reaching questions. What are the implications of Planetary Health for the conceptual paradigms we use to apprehend the natural world and our place in it? To what extent does conventional biomedical ethics provide a sufficient ethical framework for considering these phenomena? The chapter ends with a consideration of these questions.

Science

Of the planetary changes that affect human health, *climate change* is perhaps the best recognized. Climate change threatens human health through many pathways, as graphically depicted by the Lancet Commission on Health and Climate Change (Watts, Adger, Agnolucci et al., 2015). Some are primary and direct, such as those due to heat and severe weather events. Others are secondary, such as infectious disease risks that are mediated by earth system changes. Still others are tertiary, such as armed conflict and population displacement, mediated by more diffuse and complex social disruptions (McMichael, 2013). Of course, these categories overlap considerably. Every important category of human health outcome is affected: infectious diseases, noncommunicable diseases, traumatic injuries, mental distress and illness. Consider the effects of heat. Excessive heat causes a well-known cascade of medical consequences, from the relatively mild and self-limited heat rash and heat cramps, to more severe heat exhaustion and potentially fatal heat stroke. People die during heat waves, sometimes in frightful numbers. Those who are most vulnerable are the very young and very old, the poor and socially marginalized, those who live alone, and people with certain medical conditions (Kovats & Hajat, 2008). Most of the deaths that occur with extreme heat are not recognized or certified as heat-related; they occur from underlying illnesses such as cardiovascular disease. Heat waves can have substantial population impacts—70,000 deaths during the 2003 European heat wave (Robine, Cheung, Le Roy et al., 2008), 54,000 in Russia during the 2010 heat wave (Revich, 2011), and thousands per year in India—not well quantified—during the summers of 2010, 2013, 2015, and 2016 (Mazdiyasni, AghaKouchak, Davis et al., 2017). People can adapt to heat, but only up to a point (Hanna & Tait, 2015)—and with more frequent, more intense, and longer-lasting heat events coming, that point will be increasingly exceeded (Im, Pal, & Eltahir, 2017; Jones et al., 2015; Lelieveld et al., 2016; Mueller, Zhang, & Zwiers, 2016; Pal & Eltahir, 2015).

But heat threatens health and well-being in less obvious ways. Heat may predispose people to violence and crime; crime rates rise on hotter days (Anderson, 2001; Gamble & Hess, 2012; Schinasi & Hamra, 2017). Heat may predispose people to self-harm as well; some data suggest an association between heat and suicide (Dixon & Kalkstein, 2016; Dixon, Sinyor, Schaffer et al., 2014; Kim, Kim, Honda et al., 2016; Williams, Hill, & Spicer, 2015). Heat seems to increase the risk of kidney disease (Glaser, Lemery, Rajagopalan et al., 2016)—a problem that especially targets workers in low-income countries who toil in high temperatures without adequate rehydration (Tawatsupa, Lim, Kjellstrom, Seubsman, & Sleigh, 2012; Wesseling, Aragon, Gonzalez et al., 2016). In hot conditions, it is more difficult to protect food and water from bacterial contamination; the incidence of diarrheal diseases such as Shigella and Campylobacter rises during hot weather (Kovats et al., 2004; Levy, Woster, Goldstein, & Carlton, 2016). Heat compromises the quality of sleep (Obradovich, Migliorini, Mednick, & Fowler, 2017); sleep deprivation, in turn, is a risk factor for inflammatory conditions, metabolic and neuroendocrine abnormalities, and cardiovascular disease (Cappuccio, Cooper, D'Elia, Strazzullo, & Miller, 2011; Irwin, Olmstead, & Carroll, 2016; Knutson, Spiegel, Penev, & Van Cauter, 2007). People reduce their physical activity during hot weather (Obradovich & Fowler, 2017); being more sedentary increases the risk for cardiovascular disease (Biswas, Oh, Faulkner et al., 2015; Lee et al., 2012). Similarly, heat reduces work capacity (Dunne, Stouffer, & John, 2013; Kjellstrom et al., 2016), to the point that economic output may decline substantially in very hot places—compounding poverty in such places. Workers are affected in another way: workplace injuries rise during hot weather (McInnes et al., 2017; Otte Im Kampe, Kovats, & Hajat, 2016). In these and other ways, both direct and indirect, through mechanisms that range from sleep loss to food poisoning to deepening of poverty, heat threatens health and well-being—and heat is just one of the pathways through which climate change undermines public health.

A second planetary change is the *widespread dissemination of synthetic chemicals* through the world's ecosystems. Plastics, a family of polymers of ethylene, propylene, styrene, vinyl chloride, and other simple molecules, are emblematic. Large-scale production of plastics began in the years after World War II. Global production of plastics has now reached 300 million metric tons per year, or about 40 kg for each man, woman, and child on earth, and accounts for about four percent of the global petroleum supply (Thompson, Moore, vom Saal, & Swan, 2009). That may not sound like much. But given the current average life expectancy of 71 years, this means that, on average, each of us will account for over 2800 kg of plastic production during the course of our lives, or 30 or 40 times our body weight.

Plastics are persistent. If Michelangelo had sipped water from plastic bottles while painting the ceiling of the Sistine Chapel 500 years ago, those plastic bottles would still exist. As one commentator put it, with just a bit of hyperbole, "every piece of plastic ever made still exists today" (Every piece of plastic ever made still exists today, 2015). That is not completely true; after four or five centuries, some plastics do degrade, and of the roughly 8.3 billion metric tons of plastic that have been produced to date, a bit over a tenth has been incinerated (Geyer, Jambeck, & Law, 2017). But that still means that for every human being now alive, there is, somewhere in the world, about one metric ton of plastic. Some of it is still in use, but well over half of it is waste. Some of this is in large pieces, some is pellet-sized, and some has broken down to microplastics.

Where is it? Some of the waste is in landfills. Some ends up in waterways—a common and heartbreaking sight in many of the world's major cities, especially in low- and middle-income countries. Plastic generally floats, so it travels with currents and accumulates in eddies, forming patches that can be as large and imposing as islands. From streams, rivers, and harbors, the waste winds its way to oceans, where it continues to move under the influence of wind and currents. According to one recent estimate, between 4.8 and 12.7 million metric tons of plastic waste entered the oceans in 2010 (Jambeck, Geyer, Wilcox et al., 2015).

The plastic seems to go everywhere, making this a global problem (Barnes, Galgani, Thompson, & Barlaz, 2009). It has been found in the deepest ocean trenches, in arctic ice, on the shores of remote islands. According to the United Nations Environment Programme, "Microplastics have been detected in environments as remote as a Mongolian mountain lake and deep sea sediments deposited five kilometres below sea level" (UNEP, 2016, p. 32). The world's ocean systems tend to concentrate waste in the great ocean gyres of the North and South Pacific, the North and South Atlantic, and the Indian Ocean. The highest concentration of plastic debris ever measured on a beach was on remote Henderson Island, in the South Pacific, 5000 km from the nearest significant human habitation—but on the western edge of the South Pacific gyre (Lavers & Bond, 2017).

The plastics permeate ecosystems. We now know that some birds confuse plastic bits with food, and ingest the plastic—but cannot digest it. An estimated 90% of seabirds now have plastic in their bodies (Wilcox, Van Sebille, & Hardesty, 2015). The albatross is one such bird. Traditionally a sign of good luck for sailors, later a symbol of guilt in Samuel Taylor Coleridge's *Rime of the Ancient Mariner*, it takes on new metaphorical power in the Anthropocene, as a graphic warning of our environmental recklessness—the seabird over the ocean as the Planetary Health equivalent of the canary in the coal mine (Millow, Mackintosh, Lewison, Dodder, & Hoh, 2015).

Nor are birds the only affected species. In one recent study, investigators purchased fish and shellfish from markets in Indonesia and California, and assessed them for plastic debris. They found it in 28% of the individual fish and 55% of species sampled in Indonesia, and in 25% of individual fish and 67% of species sampled in the USA (Rochman, Tahir, Williams et al., 2015). When we eat fish, we are eating plastic.

When ecosystems are infused with plastics, human health can suffer (Efferth & Paul, 2017). Importantly, about 7% of plastic, on average, consists not of the polymers themselves, but of additives such as plasticizers, fillers, and flame retardants (Geyer et al., 2017). Moreover, plastics and microplastics, especially polypropylenes, polyethylenes, and polystyrenes, are efficient carriers of synthetic organic chemicals such as polychlorinated biphenyls (PCBs) and other PAHs, which adsorb to their surfaces (Lee, Shim, & Kwon, 2014; Rochman, Hoh, Hentschel, & Kaye, 2013; Rochman, Manzano, Hentschel, Simonich, & Hoh, 2013; Teuten, Saquing, Knappe et al., 2009). Many of these chemicals are highly persistent in the environment, hence the term Persistent Organic Pollutants, or POPs. Indeed, environmental loading with such chemicals is a defining feature of the Anthropocene (Diamond, 2017; Diamond, de Wit, Molander et al., 2015).

POPs are fat-soluble. They bioconcentrate within the fat of organisms such as fish, cattle, and polar bears, and they are biomagnified as they move from lower to higher trophic levels of food webs (Corsolini & Sara, 2017; Mackay & Fraser, 2000). POPs have become widely distributed across the globe, even in ecosystems such as the Arctic, far from where they were ever produced or used (Hung, Katsoyiannis, Brorstrom-Lunden et al., 2016; Rigét, Bignert, Braune, Stow, & Wilson, 2010).

When POPs are widely distributed through ecosystems, human exposure is unavoidable. Measurement of population tissue levels of POPs has revealed nearly ubiquitous body burdens in human populations (Fisher, Arbuckle, Liang et al., 2016; Fourth National Report on Human Exposure to Environmental Chemicals. Updated Tables, 2017; Koch & Calafat, 2009; Porta, Puigdomenech, Ballester et al., 2008; Pumarega, Gasull, Lee, López, & Porta, 2016). Many of these chemicals are biologically active, affecting neurodevelopmental, endocrine, metabolic, and other delicately balanced systems in humans and in animals. Consider the many chemicals collectively known as endocrine disrupters, which either block or activate receptors in sex hormone, thyroid, and other pathways. Synthetic organic chemicals that act in this manner include PCBs, bisphenols (e.g., bisphenol A, or BPA), organochlorine pesticides, brominated flame retardants, and perfluorinated substances (perfluorooctanoic acid, or PFOA, and perfluorooctane sulfonate, or PFOS).

Evidence suggests that these exposures play a role in several noncommunicable diseases, through both epigenetic and non-epigenetic mechanisms (Barouki, Gluckman, Grandjean, Hanson, & Heindel, 2012; Hou, Zhang, Wang, & Baccarelli, 2012; Meeker, Sathyanarayana, & Swan, 2009; Vandenberg, Colborn, Hayes et al., 2012). POPs exposure has been associated with metabolic conditions such as adiposity, insulin resistance, and dyslipidemias (Cano-Sancho, Salmon, & La Merrill, 2017; Grandjean, Henriksen, Choi et al., 2011; Lee et al., 2011; Lee et al., 2011; Lee, Lee, Song et al., 2006). One recent study of First Nation Inuit natives in the Canadian Arctic found that those with the highest levels of blood PCBs were 1.9–3.5 times more likely to have diabetes than those with the lowest levels (Singh & Chan, 2017). The concept of chemical obesogens is well established (Cano-Sancho et al., 2017; Darbre, 2017; Wassenaar & Legler, 2017). POPs exposure has also been associated with increased risk of some cancers, especially non-Hodgkin's lymphoma (Freeman & Kohles, 2012) and hormone-responsive cancers such as those of the breast, ovaries, and prostate; to date, the animal evidence is more extensive than the human epidemiologic evidence (Gore, Chappell, Fenton et al., 2015; Soto & Sonnenschein, 2010). POPs may increase the risk of thyroid disease, neurobehavioral disorders, and reproductive dysfunction (Gore et al., 2015).

There are many features of global chemical contamination we do not fully understand: its geographic extent, its pathways through ecosystems, its full impacts on human health, to what extent these exposures account for rising trends in some diseases, where are the boundaries we should not cross. But as the Lancet Commission on Pollution and Health recently made clear (Landrigan, Fuller, Acosta et al., 2017), this planetary change does no favors to ourselves or to ecosystems.

A third example of global environmental change that affects human health is *land use change*. The pressures of a growing human population and its resource demands have led to dramatic expropriation of land and water globally (Foley, DeFries, Asner et al., 2005; Lambin, Turner, Geist et al., 2001; Turner, Lambin, & Reenberg, 2007). As much as half the world's ice-free land surface has been transformed, much of it converted from natural grassland or forest to cropland and pasture for grazing (Turner et al., 2007). These changes have far-reaching effects, on such domains as biodiversity (Newbold, Hudson, Arnell et al., 2016; Newbold, Hudson, Hill et al., 2015). They also have far-reaching effects on human health. Consider two examples.

The effects of land use change and the resulting ecosystem alterations on infectious disease risk have long been appreciated and described for diseases as diverse as malaria, schistosomiasis, hantavirus pulmonary syndrome, Lyme disease, and West Nile virus (Bauch, Birkenbach, Pattanayak, & Sills, 2015; Pongsiri, Roman, Ezenwa et al., 2009). Many mechanisms contribute, from changes in microbial ecology to increased contact between humans and vectors.

A recently published example is instructive. A team of investigators based at the University of Vermont was interested in the effect of degraded watersheds on children's health. They studied nearly 300,000 children under 5 years of age in 35 countries (Herrera, Ellis, Fisher et al., 2017). Their primary exposure measures were of watershed quality—the hydrologic influence of upstream livestock, people, and tree cover on downstream water quality. Their primary outcome of interest was diarrhea—an important condition, as it is the second leading cause of death among children in the under-5 age group (Pruss-Ustun, Bartram, Clasen et al., 2014), responsible for about 448,000 deaths in 2016 (Global Burden of Disease, 2017). The Demographic and Health Surveys program provided health and socioeconomic data, and geolocated climate and watershed data came from the WaterWorld model. After controlling for socioeconomic factors, the presence of improved water and sanitation, and other potential confounders, the investigators found that more intact tree cover in the upstream watershed reduced rural children's probability of diarrhea. Thirty percent more tree cover offered roughly the same protection as improved sanitation infrastructure (but not as much as wealth, education, or an improved water supply). Healthy tree cover in a watershed is good for human health.

A second example of the impact of land use change on health comes from palm oil production in South Asia. Production is increasing as a result of demand for biofuels in Europe and food in India, Indonesia, and China. Oil palm plantations affect local ecology; they are consistently less biodiverse than primary forests, with only about 50% of the vertebrate species found in primary forests (Savilaakso, Garcia, Garcia-Ulloa et al., 2014). To clear tropical forests in Indonesia for palm oil production (as well as for other purposes, such as timber plantations), fire is commonly used; the resulting smoke blows in defined ways, affecting populations in Indonesia and the Malay Peninsula. This smoke, containing fine particulate matter, is an established risk factor for cardiovascular mortality; in fact, the smoke from such fires accounts for over 250,000 global deaths annually (Johnston, Henderson,

Chen et al., 2012; Marlier, DeFries, Kim et al., 2015). A team of investigators based at Harvard has studied this phenomenon over recent years, innovatively combining data on land types, land use, fire occurrence, wind patterns, smoke composition, and health outcomes (Kim, Jacob, Mickley et al., 2015; Koplitz, Mickley, Marlier et al., 2016; Marlier et al., 2015; Marlier, DeFries, Voulgarakis et al., 2013; Spracklen, Reddington, & Gaveau, 2015). The Indonesian fires were found to cause, on average, approximately 11,000 excess regional deaths each year, but in a pattern that varies considerably with such factors as El Niño (Marlier et al., 2013). In an especially bad year, 2015, the toll was an order of magnitude higher, at just over 100,000 excess deaths (Koplitz et al., 2016). In addition, the smoke is not the only health impact of palm oil production. The loss of forests and the combustion of peat contribute to climate change. Furthermore, dietary palm oil contains highly saturated fatty acids, which have been proposed as a risk factor for heart disease and other noncommunicable diseases (Basu et al., 2013; Ismail, Maarof, Ali et al., 2018).

The final example of global environmental change I will mention is *biodiversity loss*. Biodiversity loss has accelerated dramatically during the Anthropocene (Cardinale, Duffy, Gonzalez et al., 2012; Newbold et al., 2016). While this has numerous potential impacts on human health (Bernstein, 2014), two examples are especially illustrative: pollinator loss and fisheries depletion.

Pollination by insects is an important form of reproduction for more than 35% of the annual global food production by volume. At least 87 major food crops, and up to 40% of the world's supply of some micronutrients, such as vitamin A, depend on pollination by insects (Klein, Vaissière, Cane et al., 2007). Pollinators are declining in many parts of the world for a combination of reasons, including habitat loss, pesticide use, and parasitic infestation. Pollinator loss can reduce the amount of fruits, vegetables, nuts, and seeds in the diet, contributing to vitamin A and folate deficiencies. A recent analysis projected that a 50% loss of pollination would cause about 700,000 additional deaths worldwide, mostly as a result of increased ischemic heart disease and stroke due to reduced fruit and vegetable consumption (Smith, Singh, Mozaffarian, & Myers, 2015).

Fisheries depletion has emerged as a global problem, with about 90% of fisheries now at or beyond maximum sustainable levels of exploitation (FAO, 2016). Climate change will intensify this problem in the coming decades (Comte & Olden, 2017). For many populations, fish are a leading dietary source of protein, micronutrients (often in highly bioavailable form), and omega-3 fatty acids (mainly from oily fish). Dietary omega-3 fatty acids may reduce ischaemic heart disease risk, although evidence remains inconsistent (Balk & Lichtenstein, 2017; Rangel-Huerta & Gil, 2017). In addition, through gene regulation, anti-inflammatory effects, or other mechanisms, omega-3 fatty acids may play a role in preventing and/or treating other conditions such as cancer (Lee, Sim, Lee, & Na, 2017) and arthritis (Senftleber, Nielsen, Andersen et al., 2017). However, reductions in fish stocks could limit these potential benefits. For example, the UK is unable to meet healthy diet guidelines for its population from its domestic catch, and fish intake fell to only 19% of the recommended level in 2012 (Thurstan & Roberts, 2014). One study projected that more than 10% of the global population could face micronutrient and fatty-acid deficiencies due to fish declines over coming decades, especially in low-and middle-income nations near the equator (Golden, Allison, Cheung et al., 2016). Fortunately, there is some indication that with aggressive management, collapsed fisheries can recover, as may be happening with the North Atlantic cod fishery, which abruptly collapsed in 1992. Aquaculture is to some extent replacing wild fish catches, but most aquaculture (70% in 2012) depends on external feedstocks that may be unsustainable (FAO, 2016). Challenges such as disease in farmed fish (Stentiford, Neil, Peeler et al., 2012; Stentiford, Sritunyalucksana, Flegel et al., 2017), chemical contamination of farmed fish (Hamilton et al., 2005; Jacobs, Covaci, & Schepens, 2002), genetic contamination of wild fish stocks (Cognetti, Maltagliati, & Saroglia, 2006), and pollution of waters near fish farms (Cao, Wang, Yang et al., 2007) must be addressed if sustainable aquaculture is to be achieved.

These four sets of examples—climate change, chemical contamination, land use changes, and biodiversity loss—together paint a compelling picture. Human disruptions of planetary systems threaten human health and well-being. Those who are most at risk from these disruptions are those who live each day with high levels of risk, due to poverty and deprivation.

Ways of Thinking and Knowing

This has been a grim litany. Indeed, planetary changes are threatening human health and well-being—even civilization as we know it—on a frightening scale. But the news is not all bad. Some planetary changes may offer benefits, such as higher agricultural yields, or more physical activity, in cold areas that are becoming warmer. More importantly, the steps we need to take to address planetary challenges such as climate change yield a range of co-benefits, such as stronger communities, cleaner air and more wholesome food. In the meantime, the threats I have described call on us to change our thinking, as scientists and as global citizens, in some fundamental ways—addressing what the Rockefeller Foundation–Lancet Commission on Planetary Health called *imagination challenges* (conceptual and empathy failures) and *research and information challenges* (knowledge failures) (Whitmee, Haines, Beyrer et al., 2015).

First, we need to learn to *acknowledge, and live within, limits* (Butler, 2017). We have all lived our lives during the last half of the twentieth century and the first part of the twenty-first century—a time of plenty unprecedented in human history. What enabled this, of course, was our profligate consumption of energy and materials, propelled by ever more powerful technologies. The years during and after World War II were a time of widespread limits, even in wealthy countries, and of course, those who are poor face limits every day. But for the most part, we are no more conscious of our extravagance than fish are conscious of water. And as we have learned from recent behavioral economics research, we humans are wired *not to* confront limits (Lorenzoni, Nicholson-Cole, & Whitmarsh, 2007; Shu & Bazerman, 2011; Weber, 2017). We are biased toward short-term thinking and we discount future consequences. We have a status-quo bias; we cling to what we have and are averse to making what we see as risky change. We are lousy at probabilistic thinking. We need to work to replace profligacy with restraint, extravagance with modesty, wastefulness with thrift.

Second, we need *systems thinking*. Every serious thinker about sustainability, or climate change, or human affairs in general, knows that reductionist, linear thinking misses the mark. Nowhere is this truer than in Planetary Health. But still, university departments, government agencies, and foundations organize themselves in silos, without enough cross-links. Too many donors remain infatuated with silver bullet technical solutions, rather than aiming for patient, long-term adaptive management of complex systems. In my own field, medicine, why do we still require prospective students to have studied chemistry, physics and calculus, but not ecology and evolutionary biology? We need thinking habits and institutional arrangements to be based on systems.

Third, we need a sense of *urgency*. Our house is on fire, and we need to act accordingly. Our scientific discourse is typically sober and dispassionate, but passion in this setting is no sin; rather, it is a necessity. Fifty years ago, discussing a very different emergency, the Vietnam War, Martin Luther King wrote words that are deeply appropriate today: "We are now faced with the fact, my friends, that tomorrow is today. We are confronted with the fierce urgency of now. In this unfolding conundrum of life and history, there is such a thing as being too late. Procrastination is still the thief of time. Life often leaves us standing bare, naked, and dejected with a lost opportunity. The tide in the affairs of men does not remain at flood—it ebbs. We may cry out desperately for time to pause in her passage, but time is adamant to every plea and rushes on. Over the bleached bones and jumbled residues of numerous civilizations are written the pathetic words, 'Too late'. There is an invisible book of life that faithfully records our vigilance or our neglect. Omar Khayyam is right: 'The moving finger writes, and having writ moves on'" (Beyond Vietnam, 1967). As it was then for Reverend King, so it is now for us. We must confront the "fierce urgency of now."

Fourth, we need *intellectual humility*. Modern science and technology, animated by positivism for the last few centuries, have delivered untold benefits to humankind. But they have also brought wicked problems and profound dilemmas. Science and technology do not offer all the answers. We scientists need to be open to other intellectual traditions, to faith traditions (Bingham, 2016; Francis, 2015; Hayhoe & Farley, 2009) (and I am so appreciative of Pope Francis and of the Pontifical Academy of Sciences for engaging the challenges of climate change and health, and for convening us) and to the traditional ecological knowledge of indigenous peoples (Finn, Herne, & Castille, 2017; Gomez-Baggethun, Corbera, & Reyes-Garcia, 2013). This will take us beyond our comfort zones—the right place to be when we need to be stretching toward solutions.

Ethics

Let me close with a word about ethics. As a medical student and young physician, I learned and took seriously the principles of biomedical ethics: respect for autonomy, non-maleficence, beneficence, and justice. These principles guided some very sound decisions: Do not treat a patient without his or her consent. Do not withhold information from a patient. These are all well and good, but they are not nearly enough. Biomedical ethics has been far too confined to clinics and hospitals (Macpherson, 2013).

We need an ethics that recognizes everybody's right to have access to the clinic in the first place—and not only to the clinic, but to all of the precursors needed for a fully expressed and healthy life: education, housing, political voice. This is an ethics of social justice. It is essential in Planetary Health for both practical and moral reasons; those who are most vulnerable to all the threats I discussed are the disenfranchised, those without voice, those who must be at the center of any system of ethical principles (Shue, 2014). We need an ethics that recognizes intergenerational responsibility. Our obligations extend well beyond our own lifetimes (Weston, 2008).

Finally, we need an ethic that extends from humans to the more-than-human world (Leopold, 1949; Rolston, 1989). As much as we bear responsibility for each other, we bear responsibility for other species, and for ecosystems. At the very least, this responsibility is derivative from conventional ethics; Planetary Health science makes clear that care for people intrinsically includes care for the earth. But it may go further, suggesting an ethical system grounded in our place as part of the larger world we share.

Conclusion

We are living in a Golden Age. Prosperity has increased around the globe, poverty has been reduced, and human health has never been better. But, in the words of the Rockefeller Foundation–Lancet Commission on Planetary Health, "we have been mortgaging the health of future generations to realize economic and development gains in the present" (Whitmee et al., 2015). As scientists, we have increasingly uncovered and quantified the human health impacts of disrupting earth systems and crossing planetary boundaries. We need to continue this task, with ever-greater sophistication and accuracy, creating and using scientific frameworks appropriate to the complexity of the challenges, and with intellectual humility. Just as important, we need to act on the science, with passion for sustaining life, with urgency, with a drive for social and environmental justice, ever mindful of the great privilege we have: to choose the legacy we leave.

References

Anderson, C. A. (2001). Heat and violence. *Current Directions in Psychological Science, 10,* 33–38.

Balk, E. M., & Lichtenstein, A. H. (2017). Omega-3 fatty acids and cardiovascular disease: Summary of the 2016 agency of healthcare research and quality evidence review. *Nutrients, 9,* pii: E865. https://doi.org/10.3390/nu9080865

Barnes, D. K. A., Galgani, F., Thompson, R. C., & Barlaz, M. (2009). Accumulation and fragmentation of plastic debris in global environments. *Philosophical Transactions of the Royal Society B: Biological Sciences, 364,* 1985–1998.

Barouki, R., Gluckman, P. D., Grandjean, P., Hanson, M., & Heindel, J. J. (2012). Developmental origins of non-communicable disease: Implications for research and public health. *Environmental Health: A Global Access Science Source, 11,* 42.

Basu, S., Babiarz, K. S., Ebrahim, S., Vellakkal, S., Stuckler, D., & Goldhaber-Fiebert, J. D. (2013). Palm oil taxes and cardiovascular disease mortality in India: Economic-epidemiologic model. *British Medical Journal (Clinical Research Ed), 347,* f6048.

Bauch, S. C., Birkenbach, A. M., Pattanayak, S. K., & Sills, E. O. (2015). Public health impacts of ecosystem change in the Brazilian Amazon. *Proceedings of the National Academy of Sciences of the United States of America, 112,* 7414–7419.

Bernstein, A. S. (2014). Biological diversity and public health. *Annual Review of Public Health, 35,* 153–167.

Beyond Vietnam. (1967, April 4). Riverside Church, New York City. Retrieved from https://ratical. org/ratville/JFK/MLKapr67.html

Bingham, S. G. (2016). Faith and science working together on climate change. *Eos, Transactions American Geophysical Union, 97,* https://doi.org/10.1029/2016EO050243

Biswas, A., Oh, P. I., Faulkner, G. E., et al. (2015). Sedentary time and its association with risk for disease incidence, mortality, and hospitalization in adults: A systematic review and meta-analysis. *Annals of Internal Medicine, 162,* 123–132.

Butler, C. D. (2017). Limits to growth, planetary boundaries, and planetary health. *Current Opinion in Environmental Sustainability, 25,* 59–65.

Cano-Sancho, G., Salmon, A. G., & La Merrill, M. A. (2017). Association between Exposure to p,p'-DDT and Its metabolite p,p'-DDE with obesity: Integrated systematic review and meta-analysis. *Environmental Health Perspectives, 125,* 096002.

Cao, L., Wang, W., Yang, Y., et al. (2007). Environmental impact of aquaculture and counter-measures to aquaculture pollution in China. *Environmental Science and Pollution Research International, 14,* 452–462.

Cappuccio, F. P., Cooper, D., D'Elia, L., Strazzullo, P., & Miller, M. A. (2011). Sleep duration predicts cardiovascular outcomes: A systematic review and meta-analysis of prospective studies. *European Heart Journal, 32,* 1484–1492.

Cardinale, B. J., Duffy, J. E., Gonzalez, A., et al. (2012). Biodiversity loss and its impact on humanity. *Nature, 486,* 59–67.

Cognetti, G., Maltagliati, F., & Saroglia, M. (2006). The risk of 'genetic pollution' in Mediterranean fish population related to aquaculture activities. *Marine Pollution Bulletin, 52,* 1321–1323.

Comte, L., & Olden, J. D. (2017). Climatic vulnerability of the world's freshwater and marine fishes. *Nature Climate Change, 7,* 718–722.

Corsolini, S., & Sara, G. (2017). The trophic transfer of persistent pollutants (HCB, DDTs, PCBs) within polar marine food webs. *Chemosphere, 177,* 189–199.

Darbre, P. D. (2017). Endocrine disruptors and obesity. *Current Obesity Reports, 6,* 18–27.

Diamond, M. L. (2017). Toxic chemicals as enablers and poisoners of the technosphere. *The Anthropocene Review, 4,* 72–80.

Diamond, M. L., de Wit, C. A., Molander, S., et al. (2015). Exploring the planetary boundary for chemical pollution. *Environment International, 78,* 8–15.

Dixon, P. G., & Kalkstein, A. J. (2016). Where are weather-suicide associations valid? An examination of nine US counties with varying seasonality. *International Journal of Biometeorology, 62*, 685–697.

Dixon, P. G., Sinyor, M., Schaffer, A., et al. (2014). Association of weekly suicide rates with temperature anomalies in two different climate types. *International Journal of Environmental Research and Public Health, 11*, 11627–11644.

Dunne, J. P., Stouffer, R. J., & John, J. G. (2013). Reductions in labour capacity from heat stress under climate warming. *Nature Climate Change, 3*, 563–566.

Efferth, T., & Paul, N. W. (2017). Threats to human health by great ocean garbage patches. *The Lancet Planetary Health, 1*, e301–e303.

FAO. (2016). *The state of world fisheries and aquaculture*. Rome: Food and Agriculture Organization.

Finn, S., Herne, M., & Castille, D. (2017). The value of traditional ecological knowledge for the environmental health sciences and biomedical research. *Environmental Health Perspectives, 125*, 085006.

Fisher, M., Arbuckle, T. E., Liang, C. L., et al. (2016). Concentrations of persistent organic pollutants in maternal and cord blood from the Maternal-Infant Research on Environmental Chemicals (MIREC) Cohort Study. *Environmental Health, 15*, 59.

Foley, J. A., DeFries, R., Asner, G. P., et al. (2005). Global consequences of land use. *Science, 309*, 570–574.

Freeman, M. D., & Kohles, S. S. (2012). Plasma levels of polychlorinated biphenyls, non-Hodgkin lymphoma, and causation. *Journal of Environmental and Public Health, 2012*, 258981.

Gamble, J. L., & Hess, J. J. (2012). Temperature and violent crime in Dallas, Texas: Relationships and implications of climate change. *Western Journal of Emergency Medicine, 13*, 239–246.

Geyer, R., Jambeck, J. R., & Law, K. L. (2017). Production, use, and fate of all plastics ever made. *Science Advances, 3*.

Glaser, J., Lemery, J., Rajagopalan, B., et al. (2016). Climate change and the emergent epidemic of CKD from heat stress in rural communities: The case for heat stress nephropathy. *Clinical Journal of the American Society of Nephrology, 11*, 1472–1483.

Global Burden of Disease. (2017). Retrieved from http://ghdx.healthdata.org/gbd-results-tool

Golden, C. D., Allison, E. H., Cheung, W. W., et al. (2016). Nutrition: Fall in fish catch threatens human health. *Nature, 534*, 317–320.

Gomez-Baggethun, E., Corbera, E., & Reyes-Garcia, V. (2013). Traditional ecological knowledge and global environmental change: Research findings and policy implications. *Ecology and Society: A Journal of Integrative Science for Resilience and Sustainability, 18*, 72, https://doi.org/10.5751/ES-06288-180472

Gore, A. C., Chappell, V. A., Fenton, S. E., et al. (2015). EDC-2: The endocrine society's second scientific statement on endocrine-disrupting chemicals. *Endocrine Reviews, 36*, E1–e150.

Grandjean, P., Henriksen, J. E., Choi, A. L., et al. (2011). Marine food pollutants as a risk factor for hypoinsulinemia and type 2 diabetes. *Epidemiology (Cambridge, Mass), 22*, 410–417.

Hamilton, M. C., Hites, R. A., Schwager, S. J., Foran, J. A., Knuth, B. A., & Carpenter, D. O. (2005). Lipid composition and contaminants in farmed and wild salmon. *Environmental Science & Technology, 39*, 8622–8629.

Hanna, E. G., & Tait, P. W. (2015). Limitations to thermoregulation and acclimatization challenge human adaptation to global warming. *International Journal of Environmental Research and Public Health, 12*, 8034–8074.

Hayhoe, K., & Farley, A. (2009). *A climate for change: Global warming facts for faith-based decisions*. New York: FaithWords.

Herrera, D., Ellis, A., Fisher, B., et al. (2017). Upstream watershed condition predicts rural children's health across 35 developing countries. *Nature Communications, 8*, 811.

Hou, L., Zhang, X., Wang, D., & Baccarelli, A. (2012). Environmental chemical exposures and human epigenetics. *International Journal of Epidemiology, 41*, 79–105.

Hung, H., Katsoyiannis, A. A., Brorstrom-Lunden, E., et al. (2016). Temporal trends of Persistent Organic Pollutants (POPs) in Arctic air: 20 years of monitoring under the Arctic Monitoring and Assessment Programme (AMAP). *Environmental Pollution (Barking, Essex: 1987), 217,* 52–61.

Im, E.-S., Pal, J. S., & Eltahir, E. A. B. (2017). Deadly heat waves projected in the densely populated agricultural regions of South Asia. *Science Advances, 3,* e1603322.

Irwin, M. R., Olmstead, R., & Carroll, J. E. (2016). Sleep disturbance, sleep duration, and inflammation: A systematic review and meta-analysis of cohort studies and experimental sleep deprivation. *Biology Psychiatry, 80,* 40–52.

Ismail, S. R., Maarof, S. K., Ali, S. S., Ali, A., Atkin, S. L., (2018) Systematic review of palm oil consumption and the risk of cardiovascular disease. *PLoS One* 13(2):e0193533

Jacobs, M. N., Covaci, A., & Schepens, P. (2002). Investigation of selected persistent organic pollutants in farmed Atlantic salmon (*Salmo salar*), salmon aquaculture feed, and fish oil components of the feed. *Environmental Science & Technology, 36,* 2797–2805.

Jambeck, J. R., Geyer, R., Wilcox, C., et al. (2015). Plastic waste inputs from land into the ocean. *Science, 347,* 768–771.

Johnston, F. H., Henderson, S. B., Chen, Y., et al. (2012). Estimated global mortality attributable to smoke from landscape fires. *Environmental Health Perspectives, 120,* 695–701.

Jones, B., Oneill, B. C., McDaniel, L., McGinnis, S., Mearns, L. O., & Tebaldi, C. (2015). Future population exposure to US heat extremes. *Nature Climate Change, 5,* 652–655.

Kim, P. S., Jacob, D. J., Mickley, L. J., et al. (2015). Sensitivity of population smoke exposure to fire locations in Equatorial Asia. *Atmospheric Environment, 102,* 11–17.

Kim, Y., Kim, H., Honda, Y., et al. (2016). Suicide and ambient temperature in East Asian Countries: A time-stratified case-crossover analysis. *Environmental Health Perspectives, 124,* 75–80.

Kjellstrom, T., Briggs, D., Freyberg, C., Lemke, B., Otto, M., & Hyatt, O. (2016). Heat, human performance, and occupational health: A key issue for the assessment of global climate change impacts. *Annual Review of Public Health, 37,* 97–112.

Klein, A.-M., Vaissière, B. E., Cane, J. H., et al. (2007). Importance of pollinators in changing landscapes for world crops. *Proceedings of the Royal Society A: Mathematical, Physical and Engineering Sciences, 274,* 303–313.

Knutson, K. L., Spiegel, K., Penev, P., & Van Cauter, E. (2007). The metabolic consequences of sleep deprivation. *Sleep Medicine Reviews, 11,* 163–178.

Koch, H. M., & Calafat, A. M. (2009). Human body burdens of chemicals used in plastic manufacture. *Philosophical Transactions of the Royal Society B: Biological Sciences, 364,* 2063–2078.

Koplitz, S. N., Mickley, L. J., Marlier, M. E., et al. (2016). Public health impacts of the severe haze in equatorial Asia in September–October 2015: Demonstration of a new framework for informing fire management strategies to reduce downwind smoke exposure. *Environmental Research Letters, 11,* 094023.

Kovats, R. S., Edwards, S. J., Hajat, S., Armstrong, B. G., Ebi, K. L., & Menne, B. (2004). The effect of temperature on food poisoning: A time-series analysis of salmonellosis in ten European countries. *Epidemiology and Infection, 132,* 443–453.

Kovats, R. S., & Hajat, S. (2008). Heat stress and public health: A critical review. *Annual Review of Public Health, 29,* 41–55.

Lambin, E. F., Turner, B. L., Geist, H. J., et al. (2001). The causes of land-use and land-cover change: Moving beyond the myths. *Global Environmental Change, 11,* 261–269.

Landrigan, P. J., Fuller, R., Acosta, N. J. R., et al. (2017). The Lancet Commission on pollution and health. *The Lancet, 391,* 462–512.

Lavers, J. L., & Bond, A. L. (2017). Exceptional and rapid accumulation of anthropogenic debris on one of the world's most remote and pristine islands. *Proceedings of the National Academy of Sciences of the United States of America, 114,* 6052–6055.

Lee, D. H., Lee, I. K., Song, K., et al. (2006). A strong dose-response relation between serum concentrations of persistent organic pollutants and diabetes: Results from the National Health and Examination Survey 1999–2002. *Diabetes Care, 29*, 1638–1644.

Lee, D. H., Lind, P. M., Jacobs Jr., D. R., Salihovic, S., van Bavel, B., & Lind, L. (2011). Polychlorinated biphenyls and organochlorine pesticides in plasma predict development of type 2 diabetes in the elderly: The prospective investigation of the vasculature in Uppsala Seniors (PIVUS) study. *Diabetes Care, 34*, 1778–1784.

Lee, D. H., Steffes, M. W., Sjodin, A., Jones, R. S., Needham, L. L., & Jacobs Jr., D. R. (2011). Low dose organochlorine pesticides and polychlorinated biphenyls predict obesity, dyslipidemia, and insulin resistance among people free of diabetes. *PLoS One, 6*, e15977.

Lee, H., Shim, W. J., & Kwon, J. H. (2014). Sorption capacity of plastic debris for hydrophobic organic chemicals. *The Science of the Total Environment, 470-471*, 1545–1552.

Lee, I. M., Shiroma, E. J., Lobelo, F., Puska, P., Blair, S. N., & Katzmarzyk, P. T. (2012). Effect of physical inactivity on major non-communicable diseases worldwide: An analysis of burden of disease and life expectancy. *The Lancet, 380*, 219–229.

Lee, J. Y., Sim, T. B., Lee, J. E., & Na, H. K. (2017). Chemopreventive and chemotherapeutic effects of fish oil derived omega-3 polyunsaturated fatty acids on colon carcinogenesis. *Clinical Nutrition Research, 6*, 147–160.

Lelieveld, J., Proestos, Y., Hadjinicolaou, P., Tanarhte, M., Tyrlis, E., & Zittis, G. (2016). Strongly increasing heat extremes in the Middle East and North Africa (MENA) in the 21st century. *Climatic Change, 137*, 245–260.

Leopold, A. (1949). *A sand county almanac, and sketches here and there.* New York: Oxford University Press.

Levy, K., Woster, A. P., Goldstein, R. S., & Carlton, E. J. (2016). Untangling the impacts of climate change on waterborne diseases: A systematic review of relationships between diarrheal diseases and temperature, rainfall, flooding, and drought. *Environmental Science & Technology, 50*, 4905–4922.

Lorenzoni, I., Nicholson-Cole, S., & Whitmarsh, L. (2007). Barriers perceived to engaging with climate change among the UK public and their policy implications. *Global Environmental Change, 17*, 445–459.

Mackay, D., & Fraser, A. (2000). Bioaccumulation of persistent organic chemicals: Mechanisms and models. *Environmental Pollution, 110*, 375–391.

Macpherson, C. C. (2013). Climate change is a bioethics problem. *Bioethics, 27*, 305–308.

Marlier, M. E., DeFries, R. S., Kim, P. S., et al. (2015). Fire emissions and regional air quality impacts from fires in oil palm, timber, and logging concessions in Indonesia. *Environmental Research Letters, 10*, 085005.

Marlier, M. E., DeFries, R. S., Voulgarakis, A., et al. (2013). El Nino and health risks from landscape fire emissions in southeast Asia. *Nature Climate Change, 3*, 131–136.

Mazdiyasni, O., AghaKouchak, A., Davis, S. J., et al. (2017). Increasing probability of mortality during Indian heat waves. *Science Advances, 3*, e1700066.

McInnes, J. A., Akram, M., MacFarlane, E. M., Keegel, T., Sim, M. R., & Smith, P. (2017). Association between high ambient temperature and acute work-related injury: A case-crossover analysis using workers' compensation claims data. *The Scandinavian Journal of Work, Environment & Health, 43*, 86–94.

McMichael, A. J. (2013). Globalization, climate change, and human health. *The New England Journal of Medicine, 368*, 1335–1343.

Meeker, J. D., Sathyanarayana, S., & Swan, S. H. (2009). Phthalates and other additives in plastics: Human exposure and associated health outcomes. *Philosophical Transactions of the Royal Society B: Biological Sciences, 364*, 2097–2113.

Millow, C. J., Mackintosh, S. A., Lewison, R. L., Dodder, N. G., & Hoh, E. (2015). Identifying bioaccumulative halogenated organic compounds using a nontargeted analytical approach: Seabirds as sentinels. *PLoS One, 10*, e0127205.

Mueller, B., Zhang, X., & Zwiers, F. W. (2016). Historically hottest summers projected to be the norm for more than half of the world's population within 20 years. *Environmental Research Letters, 11*, 044011.

Newbold, T., Hudson, L. N., Arnell, A. P., et al. (2016). Has land use pushed terrestrial biodiversity beyond the planetary boundary? A global assessment. *Science, 353*, 288–291.

Newbold, T., Hudson, L. N., Hill, S. L., et al. (2015). Global effects of land use on local terrestrial biodiversity. *Nature, 520*, 45–50.

Obradovich, N., & Fowler, J. H. (2017). Climate change may alter human physical activity patterns. *Nature Human Behavioriour, 1*, 0097.

Obradovich, N., Migliorini, R., Mednick, S. C., & Fowler, J. H. (2017). Nighttime temperature and human sleep loss in a changing climate. *Science Advances, 3*, e1601555.

Otte Im Kampe, E., Kovats, S., & Hajat, S. (2016). Impact of high ambient temperature on unintentional injuries in high-income countries: A narrative systematic literature review. *BMJ Open, 6*, e010399.

Planting Peace (2015). *Every piece of plastic ever made still exists today*. Topeka, KS: Planting Peace. Retrieved October 11, 2019 from https://www.plantingpeace.org/2015/05/plastic-footprint/

Pope Francis. (2015).Laudato Si. Vatican City.

Pal, J. S., & Eltahir, E. A. B. (2015). Future temperature in southwest Asia projected to exceed a threshold for human adaptability. *Nature Climate Change, 6*, 197–200.

Pongsiri, M. J., Roman, J., Ezenwa, V. O., et al. (2009). Biodiversity loss affects global disease ecology. *BioScience, 59*, 945–954.

Porta, M., Puigdomenech, E., Ballester, F., et al. (2008). Monitoring concentrations of persistent organic pollutants in the general population: The international experience. *Environment International, 34*, 546–561.

Pruss-Ustun, A., Bartram, J., Clasen, T., et al. (2014). Burden of disease from inadequate water, sanitation and hygiene in low- and middle-income settings: A retrospective analysis of data from 145 countries. *Tropical Medicine & International Health: TM & IH, 19*, 894–905.

Pumarega, J., Gasull, M., Lee, D.-H., López, T., & Porta, M. (2016). Number of persistent organic pollutants detected at high concentrations in blood samples of the United States population. *PLoS One, 11*, e0160432.

Rangel-Huerta, O. D., & Gil, A. (2017). Omega 3 fatty acids in cardiovascular disease risk factors: An updated systematic review of randomised clinical trials. *Clinical Nutrition (Edinburgh, Scotland), 37*, 72–77.

Revich, B. A. (2011). Heat-wave, air quality and mortality in European Russia in summer 2010: Preliminary assessment. *Ekologiya Cheloveka (Journal of Human Ecology), 7*, 3–9.

Rigét, F., Bignert, A., Braune, B., Stow, J., & Wilson, S. (2010). Temporal trends of legacy POPs in Arctic biota, an update. *Science of the Total Environment, 408*, 2874–2884.

Robine, J. M., Cheung, S. L., Le Roy, S., et al. (2008). Death toll exceeded 70,000 in Europe during the summer of 2003. *Comptes Rendus Biologies, 331*, 171–178.

Rochman, C. M., Hoh, E., Hentschel, B. T., & Kaye, S. (2013). Long-term field measurement of sorption of organic contaminants to five types of plastic pellets: Implications for plastic marine debris. *Environmental Science & Technology, 47*, 1646–1654.

Rochman, C. M., Manzano, C., Hentschel, B. T., Simonich, S. L., & Hoh, E. (2013). Polystyrene plastic: A source and sink for polycyclic aromatic hydrocarbons in the marine environment. *Environmental Science & Technology, 47*, 13976–13984.

Rochman, C. M., Tahir, A., Williams, S. L., et al. (2015). Anthropogenic debris in seafood: Plastic debris and fibers from textiles in fish and bivalves sold for human consumption. *Scientific Reports, 5*, 14340.

Rolston, H. (1989). *Environmental ethics: Duties to and values in the natural world*. Philadelphia, PA: Temple University Press.

Savilaakso, S., Garcia, C., Garcia-Ulloa, J., et al. (2014). Systematic review of effects on biodiversity from oil palm production. *Environmental Evidence, 3*, 4.

Schinasi, L. H., & Hamra, G. B. (2017). A time series analysis of associations between daily temperature and crime events in Philadelphia, Pennsylvania. *Journal of Urban Health, 94*, 892–900.

Senftleber, N. K., Nielsen, S. M., Andersen, J. R., et al. (2017). Marine oil supplements for arthritis pain: A systematic review and meta-analysis of randomized trials. *Nutrients, 9*, pii: E42. https://doi.org/10.3390/nu9010042

Shu, L. L., & Bazerman, M. H. (2011). Cognitive barriers to environmental action: Problems and solutions. In P. Bansal & A. J. Hoffman (Eds.), *The Oxford handbook of business and the natural environment* (pp. 161–175). Oxford, UK: Oxford University Press.

Shue, H. (2014). *Climate justice: Vulnerability and protection*. Oxford: Oxford University Press.

Singh, K., & Chan, H. M. (2017). Persistent organic pollutants and diabetes among Inuit in the Canadian Arctic. *Environment International, 101*, 183–189.

Smith, M. R., Singh, G. M., Mozaffarian, D., & Myers, S. S. (2015). Effects of decreases of animal pollinators on human nutrition and global health: A modelling analysis. *The Lancet, 386*, 1964–1972.

Soto, A. M., & Sonnenschein, C. (2010). Environmental causes of cancer: Endocrine disruptors as carcinogens. *Nature Reviews Endocrinology, 6*, 363–370.

Spracklen, D. V., Reddington, C. L., & Gaveau, D. L. A. (2015). Industrial concessions, fires and air pollution in Equatorial Asia. *Environmental Research Letters, 091001*, 10.

Steffen, W., Richardson, K., Rockstrom, J., et al. (2015). Planetary boundaries: Guiding human development on a changing planet. *Science, 347*, 1259855.

Stentiford, G. D., Neil, D. M., Peeler, E. J., et al. (2012). Disease will limit future food supply from the global crustacean fishery and aquaculture sectors. *Journal of Invertebrate Pathology, 110*, 141–157.

Stentiford, G. D., Sritunyalucksana, K., Flegel, T. W., et al. (2017). New paradigms to help solve the global aquaculture disease crisis. *PLoS Pathogens, 13*, e1006160.

Tawatsupa, B., Lim, L. L., Kjellstrom, T., Seubsman, S. A., & Sleigh, A. (2012). Association between occupational heat stress and kidney disease among 37,816 workers in the Thai Cohort Study (TCS). *Journal of Epidemiology/Japan Epidemiological Association, 22*, 251–260.

Teuten, E. L., Saquing, J. M., Knappe, D. R., et al. (2009). Transport and release of chemicals from plastics to the environment and to wildlife. *Philosophical Transactions of the Royal Society B: Biological Sciences, 364*, 2027–2045.

Thompson, R. C., Moore, C. J., vom Saal, F. S., & Swan, S. H. (2009). Plastics, the environment and human health: Current consensus and future trends. *Philosophical Transactions of the Royal Society B: Biological Sciences, 364*, 2153–2166.

Thurstan, R. H., & Roberts, C. M. (2014). The past and future of fish consumption: Can supplies meet healthy eating recommendations? *Marine Pollution Bulletin, 89*, 5–11.

Turner, B. L., Lambin, E. F., & Reenberg, A. (2007). The emergence of land change science for global environmental change and sustainability. *Proceedings of the National Academy of Sciences of the United States of America, 104*, 20666–20671.

U.S. Centers for Disease Control and Prevention. (2017). *Fourth national report on human exposure to environmental chemicals. Updated Tables*. Retrieved from https://www.cdc.gov/exposurereport/index.html

UNEP. (2016). *UNEP Frontiers Report: Emerging issues of environmental concern*. Nairobi: United Nations Environment Programme.

Vandenberg, L. N., Colborn, T., Hayes, T. B., et al. (2012). Hormones and endocrine-disrupting chemicals: Low-dose effects and nonmonotonic dose responses. *Endocrine Reviews, 33*, 378–455.

Wassenaar, P. N. H., & Legler, J. (2017). Systematic review and meta-analysis of early life exposure to di(2-ethylhexyl) phthalate and obesity related outcomes in rodents. *Chemosphere, 188*, 174–181.

Watts, N., Adger, W. N., Agnolucci, P., et al. (2015). Health and climate change: Policy responses to protect public health. *The Lancet, 386*, 1861–1914.

Weber, E. U. (2017). Breaking cognitive barriers to a sustainable future. *Nature Human Behaviour,* *1*, 0013.

Wesseling, C., Aragon, A., Gonzalez, M., et al. (2016). Kidney function in sugarcane cutters in Nicaragua—A longitudinal study of workers at risk of Mesoamerican nephropathy. *Environmental Research, 147,* 125–132.

Weston, B. H. (2008). Climate change and intergenerational justice: Foundational reflections. *Vermont Journal of Environmental Law, 9,* 375–430.

Whitmee, S., Haines, A., Beyrer, C., et al. (2015). Safeguarding human health in the Anthropocene epoch: Report of The Rockefeller Foundation—Lancet Commission on planetary health. *The Lancet, 386,* 1973–2028.

Wilcox, C., Van Sebille, E., & Hardesty, B. D. (2015). Threat of plastic pollution to seabirds is global, pervasive, and increasing. *Proceedings of the National Academy of Sciences of the United States of America, 112,* 11899–11904.

Williams, M. N., Hill, S. R., & Spicer, J. (2015). Do hotter temperatures increase the incidence of self-harm hospitalisations? *Psychology, Health & Medicine, 2,* 1–10.

CHAPTER 4
How Do Our Actions Undermine Nature?

Partha Dasgupta

Summary The global growth experience since the end of the Second World War has given us two conflicting messages. On the one hand, if we look at the state of the biosphere (fresh water, ocean fisheries, the atmosphere as a carbon sink—more generally, ecosystems), there is strong evidence that the rates at which we are utilizing them are unsustainable. For example, the rate of biological extinctions globally today is 10–1000 times the average rate over the past several million years (the "background rate"). The mid-twentieth century years are acknowledged to have been the beginnings of an era that environmental scientists now call the Anthropocene (Vosen, 2016), during which the processes that define the biosphere are being altered enormously (see Waters et al. 2016).

On the other hand, it is argued by many that just as previous generations in the West and (and more recently in the Far East) invested in science and technology, education, and machines and equipment so as to bequeath to the present generation the ability to achieve high living standards, we in turn can make investments that would assure still higher living standards in the future. In 1950, global income per capita was approximately 3500 international dollars (at 2011 prices) and world population was about 2.5 billion. In 2015, the corresponding figures were 15,000 international dollars and 7.5 billion. A somewhat-greater-than 12-fold increase in global income over a 65-year period is unprecedented, that too starting at a 3500 international dollars base. The years immediately following the Second World War are routinely praised by commentators for being the start of the Golden Age of Capitalism (Micklethwait and Wooldridge (2000), Ridley (2010), Norberg (2016), and Pinker (2017) are a sample of books with that message).

P. Dasgupta (✉)
Faculty of Economics and the Centre for the Study of Existential Risk, University of Cambridge, Cambridge, UK
e-mail: pd10000@cam.ac.uk

© The Author(s) 2020
W. K. Al-Delaimy, V. Ramanathan, M. Sánchez Sorondo (eds.), *Health of People, Health of Planet and Our Responsibility*, https://doi.org/10.1007/978-3-030-31125-4_4

We should not be surprised that the Anthropocene and the Golden Age of Capitalism began at about the same time. We should also not be surprised that the conflicting signals of the 65 years following 1950, particularly the potentially irreversible changes to the biosphere, do not receive much airing by economic commentators. That is because contemporary models of economic growth and development (e.g. Helpman, 2004) largely ignore their damaging impacts on the workings of the biosphere.

Part I

The Idea of Sustainable Development

Recently, a group of economists have studied the tension inherent in the conflicting intuitions by appealing to the idea of "sustainable development", a term coined in the famous Brundtland Report (World Commission on Environment and Development, 1987). By sustainable development, the Commission meant "development that meets the needs of the present without compromising the ability of future generations to meet their own needs". In this reading, sustainable development requires that, relative to their respective demographic bases, each generation should bequeath to its successor at least as large a *productive base* as it had inherited from its predecessor. If a generation follows this prescription, then the economic possibilities facing its successor would be no worse than those it faced when inheriting productive assets from its predecessor.

The problem is that the rule leaves open the question of how the productive base is to be measured. We are thus in need of an index whose movements over time track the sustainability of development programmes. Prominent attempts at constructing ways to assess sustainability have been less than satisfactory because they did not arrive at their favoured indices from a well-articulated notion of sustainable development (Fleurbaey & Blanchet, 2013; Stiglitz, Sen, & Fitoussie, 2009).

Wealth and Well-Being

In this chapter intergenerational well-being refers to not only the well-being of those who are present today, but also the well-being of people in the future. The term indicates an aggregate measure of the flow of personal well-beings across time and the generations. Now, if sustainable development is meant to indicate that intergenerational well-being should not decline over time, then the index that signals whether an economy is on a sustainable development path is an inclusive measure of wealth (see Arrow et al., 2004; Arrow et al., & Oleson, 2012, 2013; Dasgupta & Mäler, 2000). Formally, *if intergenerational well-being increases over time, inclusive wealth also increases over time; and if intergenerational well-being*

decreases over time, inclusive wealth also decreases over time. We call this relationship the *wealth/well-being equivalence theorem.* Inclusive wealth and intergenerational well-being are not the same thing, but they move in the same way. The
wealth/well-being equivalence theorem tells us that neither GDP nor the United
Nations' Human Development Index should be used as an index for determining
whether an economy is on a path of sustainable development.

Inclusive wealth is the dynamic version of income. It is the social worth of an
economy's stock of produced capital (roads, buildings, machines), human capital
(health, education), and natural capital (ecosystems, sub-soil resources). Wealth is a
stock, whereas income (e.g., GDP) is a flow. In a stationary economy, the two
amount to the same thing; however, they can point in different directions when an
economy is not in a stationary state.

To better appreciate the notion of inclusive wealth, imagine someone is asked to
estimate their personal wealth. The individual would most likely turn first to financial assets (savings in the bank, stocks and bonds) and the properties he owns (house
and belongings, e.g.). He would use their market value to compute wealth. If
pressed, he would acknowledge that his future earnings at work should be included,
and he would estimate that part of his wealth by making a forecast of the flow of his
(post-tax) earned incomes and adding them over the working life that is ahead of
him, using perhaps a market interest rate to discount future earnings. If he were
pressed no further, he would probably stop there and agree that his earned incomes
represent returns on the human capital he has accumulated (sociality, education,
skills, health). He would also agree that wealth is important to him because it determines the opportunities he has to shape his life—the activities he can engage in, the
commodities he can purchase for pleasure, and so on. But he would probably over-
look that his taxes go to pay for the public infrastructure he uses, and he would
almost certainly not mention the natural environment he makes use of daily, free
of charge.

The notion of wealth in which a society should be interested is far wider. Inclusive
wealth is the social worth of the economy's entire stock of assets. Assets are often
called by a more generic name, "capital goods", so we may use the terms inter-
changeably. Assets offer potential streams of goods and services over time; the more
durable an asset, the more lasting is the potential stream. Time is built into an asset.
That explains why an economy's inclusive wealth at a point in time is able to reflect
the flow of well-being across time and the generations.

Accounting Prices and the Social Environment

An asset's *accounting price* is the social worth of the stream of goods and services
a society is able to obtain from a marginal unit of it. A mangrove forest is a habitat
for fish populations. It is also a recurrent source of timber for inhabitants, and it
protects people from storms and tsunamis. The forest's accounting price measures
its worth to society. An economy's institutions and politics are factors determining

the social value of its assets because they influence what people are able to enjoy from them. The value of the house a family inhabits is not independent of whether society is at peace.

Accounting prices can be very different from market prices. The difference between an asset's accounting price and its market price reflects a distortion in the economy and should be eliminated if possible. To give an example, as the market price of fish in the open seas is zero, fishermen harvesting them ought to be charged for doing so. The charge, or tax, in this case is the accounting price of fish in their natural habitat. It may be even be judicious to impose a quota on fishing, but quotas are only an extreme form of taxation (zero tax per unit up to the quota, a prohibitive tax beyond it).

The assets whose accounting value adds up to inclusive wealth (i.e. produced capital, human capital, and natural capital) should be distinguished from an economy's social environment and practices. The latter pair comprises the intangible medium in which goods and services are produced and allocated across persons, time, and the generations. The social environment consists of the laws and social norms that provide people with incentives to choose one course of actions rather than another; it includes the workings of social and economic institutions such as families, firms, communities, charities, and government; it includes the play of politics; and it includes personal motivations and norms of conduct. The social environment is the seat of mutual trust. A strengthening of trust facilitates enterprise and exchange, thus enhancing personal well-being and thereby inclusive wealth.

Sustainable Development Goals

The practical significance of the equivalence between intergenerational well-being and inclusive wealth was lost on the framers of the Sustainable Development Goals (SDGs), which were adopted by the United Nations General Assembly in September 2015. The UN has made a commitment to attain the goals by 2030. The SDGs are noble goals; they are to be lauded. Seventeen in number, the goals range from poverty eradication and improvements in education and health, to the protection of global assets that include the oceans and a stable climate. Each is of compelling importance. But neither the SDGs nor their background documents mention the need to move to a system of national accounts that contains estimates of inclusive wealth. Without that move, however, there would be no way for governments to check that the economic measures they take to meet the international agreement would not jeopardize the sustainability of those goals. If inclusive wealth (adjusted for population and the distribution of wealth) increases as governments try to meet the SDGs, the SDGs will be sustainable; if it declines, the SDGs will be unsustainable. It could be that the goals are reached in the stipulated time period but are not sustainable because the development paths nations follow erode productive capacities beyond repair. The supporting documents of the United Nations' Sustainable Development Goals do not tell us how to check that the goals are being met in a sustainable way. Managi and Kumar (2018), in their *Inclusive Wealth Report 2018*, have found that 44 out of the 140 countries in their sample experienced a decline in (inclusive) wealth per capita since 1998, even though GDP (read, "income") per capita increased in all but a handful of them. The tension I alluded to is expressed quantitatively in that study.

Part II

What Is Investment?

Today, the most commonly used measure of people's health is longevity. We routinely talk about investment in a country's health and commend it when longevity rises. But does an increase in longevity amount to investment?

The word "investment", as customarily used, embodies a sense of robust activism. But that is only because national income statisticians have traditionally limited the term's use to the accumulation of produced capital. When a government invests in roads, the picture that is drawn is one of bulldozers levelling the ground and tarmac being laid by men in hard hats. That is investment. In Part I, we found it necessary to extend the notion of capital beyond produced capital to include human capital and natural capital. So we were obliged to stretch the notion of "net investment" to mean any decision that raises future well-being. To leave a forest alone so that it can grow is in our extended sense to invest in the forest. Net investment is positive when trees grow. To allow a fishery to restock under natural conditions is to invest in the fishery. Net investment in the fishery is positive when the stock increases. And so on.

That could suggest that investment amounts to deferred consumption; however, the matter is subtler. Providing additional food to undernourished people via, for example, food guarantee schemes not only increases their current well-being, it enables them also to be more productive in the future and to live longer. Because their human capital increases, the additional food intake should count also as net investment. Note, though, that food intake by the well-nourished does not alter their nutritional status, which means the intake is consumption, not investment. By "net investment" in an asset we mean the value of the change in the stock of that asset over the past period.

To illustrate, consider a closed, egalitarian economy with constant population. Suppose in a given year it invests 40 billion dollars in produced capital, spends 20 billion dollars on primary education and health care, and depletes and degrades its natural capital by 70 billion dollars. The economy's System of National Accounts (SNA) would record the 40 billion dollars as investment ("gross capital formation"), the 20 billion dollars as consumption, and would typically remain silent on the 70 billion dollars of loss in stocks of natural capital (e.g. wetlands, the atmosphere). Proper accounting methods in contrast would reclassify the 20 billion dollars as expenditure in the formation of human capital ("investing in the young", as the saying goes) and the 70 billion dollars as depreciation of natural capital. Aggregating them and assuming that expenditure on education is a reasonable approximation of gross human-capital formation, we would conclude that, owing to the depreciation of natural capital, the economy's wealth will have declined over the year by 10 billion dollars—and that is before taking note of the depreciation of produced capital and human capital. The moral we should draw is that development was unsustainable that year.

Wealth in India: Estimates

Arrow et al. (2012) put the wealth/well-being equivalence theorem to work by estimating the change in wealth per capita over the period of 1995–2000 in Brazil, China, India, United States, and Venezuela.[1] The choice of countries was in part designed to reflect different stages of economic development and in part to focus on particular resource bases. Because of an absence of data, the authors did not study wealth inequality within countries. Their publications are like reconnaissance exercises. They explore the land mostly in the dark; you know they have got it wrong, but you have reasons to believe they are in the right territory.

Table 4.1 reproduces their findings for India. The value of produced capital in 1995, amounting to $1530 per head, was calculated from government publications on past capital investments. The implicit assumption was that prices used by the government to record expenditures are reasonable approximations of shadow (true price) prices. The value of education per person ($6420) was estimated on the basis of a functional relationship between wage differences and differences in levels of education.

No data were available for calculating the contribution of health to labour productivity and current well-being. For that reason, the authors studied longevity only. Its accounting price was estimated from the value of a statistical life (VSL), which is commonly obtained from the willingness-to-pay for a marginal reduction in the risk of death. Recent work suggests VSL in India is approximately $500,000. Arrow et al. (2012) showed that under a set of simplifying assumptions VSL equals the value of health per person [row (3), column (1)]. They then estimated the value of a statistical life-year and used that to value the increase in life expectancy between 1995 and 2000 [row (3), column (2)].

Four categories of natural capital were included in the study: forests (valued for their timber), oil and minerals, land, and carbon concentration in the atmosphere. Like institutions and knowledge, atmospheric carbon was interpreted to be an "enabling" asset, which is why it is excluded from columns (1) and (2) but included in the estimate of the change in wealth over the 5-year period.

---

[1] IHDP-UNU/UNEP (2012, 2014) used the same framework to measure wealth in 120 countries.

Table 4.1 Per capita wealth and its growth in India, 1995–2000 (2000 US$)

	(1)	(2)	(3)	(4)
	1995 stock	2000 stock	Change (1995–2000)	Growth rate (% per year)
(1) Produced capital	1530	2180	650	7.30
(2) Human capital, 1 (education)	6420	7440	1020	3.00
(3) Human capital, 2 (health	500,000	503,750	3750	0.14
(4) Natural capital	2300	2280	−20	−0.15
(5) Oil (net capital gains)			−140	
(6) Carbon damage			−90	
(7) Total	510,250	515,650	5400	0.20
(8) *TFP*				1.84
(9) Wealth per capita				2.04

Source: Arrow et al. (2012), Table 5 (modified)

The value of land was taken from World Bank publications. Using market prices for timber and oil and minerals, the shadow value of natural capital in 1995 was estimated to be $2300 per person [row (4), column (1)]. Because of the lack of relevant data, the figure did not include the value of all the many ecological services that forests provide. Moreover, ecosystems such as fisheries, wetlands, mangroves, and water bodies are missing from Table 4.1. That means $2300 is an underestimate, in all probability seriously so. Adding the figures, wealth per capita in 1995 was found to be $510,250 [row (5), column (1)].

The population in India grew at an average annual rate of 1.74%. Column (3) records changes in per capita capital stocks over the period in question; and column (4) presents the corresponding annual rates of change. The former is embellished by two factors. First, India is a net importer of oil, whose real price rose during the period. The capital losses owing to that increase amounted to a wealth reduction in India, which was calculated to be $140 per person [row (5), column (3)]. Secondly, during 1995–2000, global carbon emissions into the atmosphere exceeded 38 billion tons. At the levels of concentration prevailing in 1995 (380 parts per million), carbon was a global "public bad". The theory of public goods says that the loss to India over the period would have been global emissions times the shadow price of carbon specific to India. In their base case, Arrow et al. (2012) took the global shadow price to be negative $50 per ton. The loss to India per ton of carbon emissions was taken to be 5% of the global shadow price, which is negative $2.50. This amounted to a loss per person of $90 [row (6), column (3)].

Row (7) records the change in wealth per capita in India over the period of 1995–2000. It translates to 0.20% a year, a figure so near to zero as to be alarming. However, the estimate does not include improvements in knowledge and institutions. Arrow et al. (2012) modelled the latter as "enabling assets" and interpreted improvements in them as growth in total factor productivity (*TFR*), which in India has been estimated to be 1.84% a year (row (8)). Based on a formula the authors derived for including the residual in wealth calculations, row (9) records the annual rate of growth of wealth per head in India during 1995–2000 as having been 2.04%.

The composition of wealth in Table 4.1 does not have direct implications for policy. A mere study of the relative magnitudes of the different forms of wealth would not tell us their relative importance. Suppose, for example, that the value of asset i swamps all other forms of capital, by a factor of 1000. That does not mean investment ought to be directed at further increases in i, for we do not know the costs involved in doing so. Only social cost–benefit analysis, using the same shadow prices as are estimated for sustainability analysis, would tell the evaluator which investment projects are socially desirable.

Taken at face value, Table 4.1 reveals a number of interesting characteristics of India's economic development during the final years of the twentieth century. It is helpful to highlight the most striking:

1. Of the four types of capital comprising measured wealth, produced capital is the smallest. Even though the value of natural capital in both years is in all likelihood a serious underestimate, it was considerably greater in 1995 than reproducible capital.
2. The rapid growth of produced capital (7.30% a year), as against a 0.15% annual rate of decline of natural capital, meant that by 2000 their stocks were nearly identical.
3. In 1995, human capital in the form of education was more than four times that of produced capital. However, the ratio declined over the 5-year period owing to a slower growth in education.
4. Health swamps all other forms of wealth. It was some two orders of magnitude larger than all other forms of wealth combined in 1995, in what was then a low-income country; this is unquestionably the most striking result of the exercise. That the finding is a cause for surprise is, however, no reason for dismissing it. Health has been much discussed in the development literature but has not been valued within the same normative theory as produced capital. There was no basis for a prior expectation of what the finding would be once health was placed in the same normative footing as other forms of wealth. Health dominates because of the high figure for V reported in the empirical literature. Longevity matters to people everywhere and matters greatly. In democratic societies, that should count.
5. Growth in wealth per capita in India has been largely a consequence of *TFP* growth (the "residual"). However, contemporary estimates of the residual should be treated with the utmost suspicion because they are based on models that do not include natural capital as factors of production. If the rate at which natural capital is degraded were to increase over a period of time, *TFP* growth obtained from regressions based on those models would be overestimated. The implication is more than just ironic. The regressions would misinterpret degradation of the environment as increases in knowledge and improvements in institutions. Worse still, the greater the under-coverage of natural capital, the greater the bias in the estimate of *TFP* growth. By plundering Earth, *TFP* could be raised by as much as the authorities like.

References

Arrow, K. J., Dasgupta, P., Goulder, L., Daily, G., Ehrlich, P. R., Heal, G. M., et al. (2004). Are we consuming too much? *Journal of Economic Perspectives, 18*, 147–172.

Arrow, K. J., Dasgupta, P., Goulder, L. H., Mumford, K. J., & Oleson, K. (2012). Sustainability and the measurement of wealth. *Environment and Development Economics, 17*, 317–355.

Arrow, K. J., Dasgupta, P., Goulder, L. H., Mumford, K. J., & Oleson, K. (2013). Sustainability and the measurement of wealth: Further reflections. *Environment and Development Economics, 18*, 504–516.

Becker, G. S., Philipson, T. J., & Soares, R. S. (2005). The quantity and quality of life and the evolution of world inequality. *American Economic Review, 95*, 227–291.

Dasgupta, P., & Mäler, K.-G. (2000). Net national product, wealth, and social well-being. *Environment and Development Economics, 5*, 69–93.

Fleurbaey, M., & Blanchet, D. (2013). *Beyond GDP: Measuring welfare and assessing sustainability*. New York: Oxford University Press.

Helpman, E. (2004). *The mystery of economic growth*. Cambridge, MA: Belknap Press.

IHDP-UNU/UNEP. (2012/2014). *Inclusive wealth report*. Cambridge: Cambridge University Press.

Klenow, P. J., & Rodriguez-Clare, A. (2005). Externalities and growth. In P. Aghion & S. Durlauf (Eds.), *Handbook of economic growth*. Amsterdam: North Holland.

Landes, D. S. (1998). *The wealth and poverty of nations*. New York: W.W. Norton.

Managi, S., & Kumar, P. (2018). *Inclusive wealth report 2018*. London: Routledge.

Micklethwait, J., & Wooldridge, A. (2000). *A future perfect: The challenge and promise of globalization*. New York: Random House.

Mokyr, J. (2002). *The gifts of Athena: Historical origins of the knowledge economy*. Princeton, NJ: Princeton University Press.

Mokyr, J. (2016). *The culture of growth: The origins of the modern economy*. Princeton, NJ: Princeton University Press.

Norberg, J. (2016). *Progress: Ten reasons to look forward to the future*. London: One World.

Pinker, S. (2017). *Enlightenment now: The case for reason, science, humanism, and progress* New York: Allen Lane.

Putnam, R. D., Leonardi, R., & Nanetti, R. Y. (1993). *Making democracy work: Civic traditions in modern Italy*. Princeton, NJ: Princeton University Press.

Ridley, M. (2010). *The rational optimist: How prosperity evolves*. London: 4th Estate.

Stiglitz, J. E., Sen, A., Fitoussi, J.-P., & Commission on the Measurement of Economic Performance and Social Progress. (2009). *Mismeasuring our lives: Why GDP doesn't add up*. New York: The New Press.

Vosen, P. (2016). Anthropocene pinned down to post war period. *Science, 353*, 852–853.

Waters, C. N., Zalasiewicz, J., Summerhayes, C., Barnosky, A. D., Poirier, C., Galuszka, A., et al. (2016). The Anthropocene is functionally and stratigraphically distinct from the Holocene. *Science, 351*, aad2622-(1-10).

World Bank. (2011). *World development indicators*. Washington, DC: World Bank.

World Commission on Environment and Development—The Brundtland Report. (1987). *Our common future*. New York: Oxford University Press.

CHAPTER 5
Climate Change, Air Pollution, and Health: Common Sources, Similar Impacts, and Common Solutions

Veerabhadran Ramanathan

Summary We are living in the Anthropocene. Human beings have become a major force by massively polluting the air we breathe as well as the entire atmosphere, which maintains the climate in a habitable zone. Close to one trillion tons of air and climate pollutants are blanketing the earth, and trillions of additional tons will be added this century. Millions of people are dying prematurely every year due to air pollution. If climate pollutant emissions are allowed to continue well into the twenty-first century, global warming and climate change can pose existential threats to *Homo sapiens* and many other species. The dominant sources of air pollution and climate change are the same: (1) combustion of fossil fuels and biomass for energy; and (2) agriculture, including livestock. Both air pollution and climate change have catastrophic impacts on human health, exposing billions of people to toxic pollution, deadly heat waves, floods, droughts, and fires. Basically, fossil fuel has become an outdated fuel. There is still time to mitigate and avoid the worst consequences. An integral strategy to mitigate air pollution and climate pollution is required because drastic emission cuts in climate pollutants, such as by switching from fossil fuels to abundantly available renewable fuels, will also reduce air pollution. Less wasteful use of fertilizers and greater consumption of a plant-based diet are also required. Technical solutions have to be bolstered by societal transformation solutions to solve the problem in time and transition to a safer Anthropocene.

Science of Air Pollution and Climate Change

Air pollution and climate change have been the subjects of rigorous scientific enquiry for more than a century. The focus of air pollution studies has been primarily on the impacts on human and ecosystem health. Climate change studies have

V. Ramanathan (✉)
Scripps Institution of Oceanography, University of California at San Diego,
La Jolla, CA, USA
e-mail: vramanathan@ucsd.edu

© The Author(s) 2020   49
W. K. Al-Delaimy, V. Ramanathan, M. Sánchez Sorondo (eds.), *Health of People,
Health of Planet and Our Responsibility*, https://doi.org/10.1007/978-3-030-31125-4_5

largely focused on impacts on the physical climate system, including temperature, weather, glaciers, and sea level rise, among others. Only during the last two decades has the integrated nature of the air pollution and climate change problems been recognized; that is, both air pollution and climate change affect public health as well as the health of the ecosystem. The dominant sources of air pollution and climate change are the same: burning of fossil fuels; burning of solid biomass, including forests; and agriculture emissions. Air pollutants also have a strong influence on climate change. Likewise, weather extremes caused by climate change affect health in major ways. These linkages between air pollution, climate change, and health of the people and the ecosystem are explored in this chapter, as well as many other chapters in this book.

The primary climate warming pollutant gases are carbon dioxide, methane, ozone, nitrous oxide, and halocarbons. The primary air pollutants are particles and ozone. The primary air pollution particles are sulfates, nitrates, black carbon (soot), and organics. All particles intercept sunlight and cause a dimming of sunlight at the surface. Black carbon absorbs sunlight and is one of the largest climate warming pollutants. All other particles act like mirrors and reflect sunlight back to space, which causes cooling.

Air Pollutants: Air pollution particles are very small in size, ranging from nanometers (billionth of a meter) to 10 μm (micrometers). A micrometer is a millionth of a meter. For reference, a human hair ranges from 20 μm to larger sizes. Air pollution particles sampled from the air typically consist of sulfates and nitrates from fossil fuel burning; black carbon from diesel combustion and biomass burning; organic carbon from fossil fuel combustion and biomass burning; and mineral dust and sea salt particles from natural sources. These particles are too tiny to see with our eyes. The particles scatter sunlight, which gives a hazy appearance to the sky; it is this hazy sky that we perceive as air pollution. Some particles, such as black carbon, also trap sunlight in blue and UV wavelengths, which gives a brownish appearance to the hazy sky and is called brown clouds.

Ozone is also an air pollutant, but it is a gas. Human activities do not emit ozone, but they emit pollutant gases such as carbon monoxide, nitrogen oxides (NOx), nitrous oxide, and methane, which produce ozone in the air. Ozone is also a major climate warming pollutant.

Air pollution is transported far and wide by atmospheric winds. For example, air pollution from the east coast of China travels within 5–7 days to the west coast of North America; air pollution from the east coast of the United States travels to Europe in less than five days. The lifetime of an air pollution particle in the air ranges from a few days to a few weeks depending on the location, the season, and the altitude. The particles are removed from the air either by rain or by directly depositing on the surface. These are important facts because solutions to the air pollution problem often require local, regional, national, and even global actions.

Climate Pollutants: The major climate pollutants are gases that absorb and emit infrared (heat) radiation. Because their lifetimes range from a decade to centuries, greenhouse gases cover the planet like a blanket. A blanket on a cold winter night keeps us warm by trapping body heat. Likewise, the greenhouse blanket in the air traps the infrared heat emitted by the surface and the atmosphere and warms the planet. The magnitude of the infrared heat trapping can be estimated from the quantum mechanics of the gas molecules and verified by laboratory data. These infrared heat-trapping gases are called greenhouse gases.

Greenhouse gases have multiple sources. The sources of carbon dioxide, which is a major climate pollutant, are fossil fuel combustion and biomass burning (mainly deforestation and cooking with firewood). For methane, the sources include fossil fuel (natural gas) production and transport, agriculture (rice cultivation), cattle breeding, farm manure, landfills, and sewage. Nitrous oxide is primarily from fertilizers, whereas the main sources of halocarbons are refrigerants. Greenhouse gases stay in the air for months (ozone) to decades (methane and some halocarbons) to a century (nitrous oxide and some halocarbons). Carbon dioxide is the most insidious climate pollutant because it has multiple time scales, ranging from decades to centuries to thousands of years. For example, we are likely still breathing carbon dioxide emitted in the nineteenth century by the improved steam engine invented by the famous British engineer James Watt. In that sense, carbon dioxide is immortal when judged by human time scales. Because of their long lifetimes (more than a decade), greenhouse gases (except ozone) emitted on any corner of the planet can travel to the rest of the planet, trap heat, and impact the climate globally. That is why the phenomenon is called *global warming*.

Air Pollutants and Climate Pollutants: Similar Impacts

Both air pollution and climate change have large negative impacts on the health of people and of the ecosystem.

Air Pollution Impacts

Health Impacts of Air Pollution: When inhaled, tiny nano- to micrometer sized particles enter the lungs, the bloodstream, and even the brain (GBD 2013 Risk Factors Collaborators 2015). Air pollution is the fourth largest risk factor to global health, after high blood pressure, dietary risks, and smoking. It also is the biggest environmental health risk factor.

Epidemiologists have classified air pollution particles under two categories: ambient (outdoor) air pollution (AAP) and household air pollution (HAP). In principle because of wind-driven mixing, AAP and HAP should be similar. However, in rural households in Asia, Africa, and South America, cooking is mostly done by burning solid fuels such as firewood, coal, dung, and crop residues in rudimentary stoves. The incomplete combustion of solid fuels in such stoves gives rise to smoke consisting of black carbon and organic carbon particles. Due to poor ventilation, the particles are concentrated inside homes, and this has a major impact on the health of the poorest three billion who rely on such cooking. Every year, at least three million succumb to indoor inhalation of HAP. Ultimately, however, the smoke particles escape outdoors, become part of ambient air pollution, and kill even more.

The inhalation of outdoor air pollution causes millions (7–9.5 million) of premature deaths in addition to exposing more than six billion people to adverse health effects. The health effects of air pollution are documented extensively in many chapters of this book.

Ecosystem Impacts of Air Pollution: The impacts of air pollution on ecosystem health include the following:

- Soil acidification due to the acidity of precipitation (acid rain) caused by sulfuric and nitric acids, mostly from fossil fuel combustion;
- Eutrophication of lakes and rivers by NOx (from fertilizers and fossil fuels);
- More than one hundred million tons of damaged crops each year due to ozone exposure (mainly fossil fuels).

Climate Impacts of Air Pollutants: Air pollution has major impacts on the climate, particularly on precipitation.

1. *Global Dimming, Monsoon Weakening, and Drying:* Air pollutant particles intercept sunlight and reduce the amount of sunlight at the surface, which is commonly known as *Global Dimming*. The dimming leads to a decrease in evaporation from the oceans, which results in a decrease in precipitation, thus leading to droughts. The deadly multi-decadal Sahelian drought during the 1960s to the 1980s was attributed in part to the dimming of the North Atlantic Ocean by sulfate and nitrate aerosols from Europe. This asymmetric dimming of the North Atlantic shifted the equatorial rain belt southwards, away from the Sahelian latitude belt (Rotstayn & Lohmann, 2002). Numerous studies have attributed the decrease in the Indian summer monsoon rains to dimming by widespread atmospheric brown clouds (Ramanathan et al., 2005), which consist of black carbon (soot), sulfates, nitrates, and organics particles that preferentially form over the northern Indian Ocean compared with the southern Indian Ocean.
2. *Warming Particles:* In addition, black carbon (soot) aerosols from diesel and wood burning are the strongest absorbers (per mass) of solar radiation and contribute to warming (Ramanathan & Carmichael, 2008). For example, a ton of black carbon from diesel vehicles has the same warming effect as that of 2000 tons of carbon dioxide. The deposition of black carbon has been shown to significantly enhance the solar heating of snow-capped surfaces over Tibetan-Himalaya glaciers, arctic sea ice, and glaciers (Xu et al., 2016).
3. *Cooling Effects:* Aerosols also have a large cooling effect on the climate. Sulfates, nitrates, and some organics particles primarily scatter sunlight, some of which is reflected back to space instead of being absorbed by the surface, thus exerting a global cooling effect. This aerosol cooling effect has masked (offset) as much as 40% of the warming effect of greenhouse gases.

4. *Disruption of Global Weather Patterns:* Because of their short lifetimes, the cooling as well as the warming aerosols are more concentrated over land than over oceans, as well as more concentrated over the northern hemisphere (because it is more populated) than over the southern hemisphere. This asymmetric effect disrupts the weather patterns because they are driven by temperature gradients between the land and ocean and between the north and south. One such disruption is associated with a tropics-wide circulation called the Hadley circulation, which originates in the tropical region and culminates in the sub-tropical regions (20° latitude to 35° latitude) by bringing dry upper atmosphere air downwards. This descending dry air is responsible for many of the major widespread desert regions, such as the Sahara and other major arid regions in Middle East.

Climate Pollution Impacts

Since 1750, we have emitted 2.2 trillion tons of carbon dioxide, 45% of which is still in the atmosphere. In addition, methane has been emitted from fossil fuels and a number of other sources, halocarbons (CFCs, HFCs, etc.) from refrigerators and air conditioners, ozone-producing pollutant gases from vehicle exhaust, and nitrous oxide from fertilizers. In addition, the black carbon emitted by diesel vehicles and biomass burning warms the planet. Among these non-CO_2 climate warming pollutants, methane, halocarbons, nitrous oxide, and black carbon are popularly known as super pollutants; they are, per ton of emissions, 25 (methane) to 2000 times (halocarbons and black carbon from diesel) more potent than CO_2. Among these super pollutants, methane, HFCs (one form of halocarbon), and black carbon are known as short-lived climate pollutants because their lifetimes are much shorter than that of CO_2, which ranges from a century to thousands of years. Methane's lifetime is approximately 11 years, the lifetimes of HFCs range from 10 to 20 years, whereas the lifetime of black carbon ranges from days to weeks.

To date, the climate has warmed by approximately 1 °C. The magnitudes of warming are always referred to in terms of the temperature of the planet before 1900. The contribution of CO_2 to this warming is about 50%, with non-CO_2 pollutants accounting for the other 50%. However, future warming is expected to be dominated by CO_2 increases.

Present and Future Global Warming: Climate models are able to simulate the observed surface warming of 1 °C. This by itself is not sufficient to validate the simulations or verify the science of climate change. However, climate models are also able to simulate the following observed climate changes: the penetration of this warming to ocean depths of up to a kilometer, the warming of the atmosphere up to about 15 km, increased humidity in the atmosphere, the retreat of the arctic sea ice, and increases in sea level, among other changes. In addition, climate models predicted these observed changes decades ago, back in the 1970s and 1980s (e.g.,

Madden & Ramanathan, 1980). The warming has been predicted (Ramanathan & Xu, 2010) to pass 1.5 °C by 2030 and exceed 2 °C by 2050.

Global Warming to Climate Change: As the atmosphere warms, humidity increases by about 7% near the surface and 10–20% aloft. Atmospheric thermodynamic energy increases in proportion to the increase in humidity. The increase in energy leads to stronger storms and more intense hurricanes, and these extreme events alter the global wind patterns. When the storms get stronger in some locations, it leads to drying and less precipitation in other locations. This redistribution happens in such a manner that wet regions get wetter and dry regions get drier. The dry regions are typically in the sub-tropical regions between 25N and 40N. Because of the intensification of the water cycle of the atmosphere, global warming leads to a substantial increase in the severity of storms, floods, droughts, fires, and heat waves, among other extreme events.

Predictions for the future show a range of probabilities, as illustrated in Fig. 5.1. The red and brown curves are the projections for the baseline scenarios with unchecked emissions. The green curve is the projected warming for 2100 with the

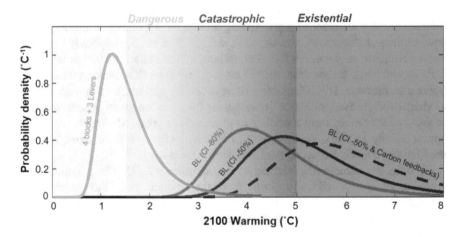

Fig. 5.1 Projected warming for four different scenarios from preindustrial to 2100 as adapted from Xu and Ramanathan (2017). The warming is given in terms of the probability distribution instead of a single value because of uncertainties in climate feedbacks, which could make the warming larger or smaller than the central value shown by the peak probability density value. The three curves on the right side indicated by BL (for baseline) denote the projected warming in the absence of climate policies. BL (CI-80%) represents the scenario in which the energy intensity (the ratio of energy use to economic output) of the economy decreases by 80% compared with its value for 2010. In BL (CI-50%), the energy intensity decreases by only 50%. These scenarios bound the energy growth scenarios proposed by IPCC-WGIII (2014). The right extreme curve is the BL (CI-50%) case. The carbon cycle feedback uses IPCC recommended values for the reduction in CO_2 uptake by the oceans as a result of the warming, the release of CO_2 by melting permafrost, and the release of methane by wetlands

mitigation measures described in a companion chapter in this book (Chap. 25). With unchecked emissions, the end-of-century warming can be in the range of 2.5–8 °C.

Risk Categories of Climate Change: The warming has been categorized (following Xu and Ramanathan 2017 and the climate risk assessment IPCC reports) under three risk categories: *dangerous risk (>1.5 °C), catastrophic risk (>3 °C), and existential risk (>5 °C).*

Dangerous Risk (1.5–3 °C): Based on data presented in Xu and Ramanathan (2017), the 1.5 °C threshold will be crossed by 2030 with a 50% probability.

- The 1 °C warming witnessed as of now has already led to the following impacts: The likelihood of extreme events (defined as 3-sigma events), including heat waves, has increased tenfold in recent decades (WMO, 2016). Human beings are extremely sensitive to heat stress. For example, the 2003 European heat wave led to about 70,000 premature mortalities (Mitchell et al., 2016). Hurricanes have become about 25% more intense (Holland & Bruyère, 2013). Dry areas worldwide have increased from about 15% during mid-twentieth century to about 30% by the first decade of the twenty-first century. The number of disastrous floods have increased from less than 50 per year during the mid-twentieth century to more than 150 during the first decade of the twenty-first century.
- With an increase of another 0.5 °C above the 1 °C warming we have already experienced, the underlying dynamic and thermodynamic forces that linked the 1 °C warming with weather extremes described previously will intensify by another 50%. Warming in the range of 1.5 °C will subject half of the world's population every other year to regional summer mean temperatures that exceed those of the historically hottest summers (Mueller et al., 2016).
- During the last interglacial period about 130,000 years ago, called the Eemian, the planet was warmer than the preindustrial climate by less than 1 °C. However, this small warming was sufficient enough to raise the sea level by about 6 m.
- During the first decade of the twenty-first century, tropical forests have become net sources of CO_2 (Baccini, 2017). If this persists and expands to other forests, forests can become a large source of carbon pollution.
- The last, but very important, issue concerns climate justice: Numerous island nations, including the Maldives and low-lying nations like Bangladesh, will be facing existential threats with a sea level rise of just 1–2 m.

Catastrophic Risks (3–5 °C): With unchecked emissions, there is a 10% probability of the warming reaching catastrophic levels by 2050 and a greater than 50% probability of reaching catastrophic levels before 2100. Catastrophic risks have been defined by some economists as follows: The magnitude of the change is too large and the velocity of that change is too rapid for adaptation. The first major concern with the 3–5 °C warming is the issue of tipping points. Several studies have concluded that 3–5 °C global warming is likely to be the threshold for tipping points such as the collapse of the western Antarctic ice sheet, disappearance of the sea ice, disruption of the deep-water circulation in the North Atlantic, dieback of Amazon rainforests and boreal forests, collapse of the West African monsoon, and world forests turning into net sources (instead of sinks) of carbon dioxide, among others. The category of "Catastrophic Risk" is also justified by the *burning embers* diagram from IPCC, in which it categorized climate risks under the following five reasons for concern (RFCs): risks to natural systems, risks of extreme weather events, distribution of impacts between regions of the world, aggregate impacts, and risks of large-scale discontinuities. For warming in the range of 3–5 °C, all RFCs were ranked as a high-risk category, with two of them ranked as very high. The IPCC burning embers diagram does not extend beyond 5 °C. Warming in the range of 3–5 °C can have the following health impacts on people and ecosystems:

- Extreme heat waves with effective (combination of temperature and humidity) temperatures greater than 55 °C (131 °F) are projected (Russo, Sillmann, & Sterl, 2017) to be routine phenomena when global warming exceeds 4 °C.
- Exposing more than 7 billion people to deadly heat stress (Mora et al., 2017).
- Exposing 2.5 billion people to vectorborne diseases (Proestos et al., 2015).
- Subjecting approximately 20% (Urban, 2015) to 37% (Thomas et al., 2004) of species to extinction. Climate change interacting with other established drivers of extinction, such as habitat destruction and commercial exploitation of valuable species, can lead to another mass extinction involving 75% of species (Barnosky, 2015).
- About 30–44% of the planet's land areas would be exposed to drying (Cook, Smerdon, Seager, et al., 2014) with warming of 3–4 °C, accompanied by severe drought conditions over Europe, the eastern United States, southeast Asia, and most of the Amazon region. Food security will be severely threatened.

Existential Risks (>5 °C): With unchecked emissions of climate pollution, there is at least a 10% probability for climate change to pose existential threats to *Homo sapiens* and all other species.

IPCC risk discussions, as shown in the burning embers diagram, do not extend beyond 5 °C. When all of the risks described under the dangerous and catastrophic risk categories are summed, it is clear that the entire planet and all of the global population will be severely impacted. The last time the planet was warmer by 5–8 °C (as shown in Fig. 5.1) was 35 million years ago, when the CO_2 concentration was around 1000 parts per million (Pagani et al., 2011). The planet was ice free, including over the Antarctic, with an implied sea level rise of more than 50 m. It is the combination of multiple catastrophic changes that compelled us (Ramanathan et al., 2017; Xu & Ramanathan, 2017) to conclude that warming in excess of 5 °C will pose existential threats.

Ethics and Justice Issues: The catastrophic and existential risks raise major ethical issues. More than 50% of climate pollution is caused by the wealthiest one billion people, while the poorest three billion people emit only 5% of the climate pollution (Dasgupta & Ramanathan, 2014; Ramanathan, PAS-PASS workshop, 2014). As described in numerous IPCC reports and other reports, the poor will suffer the worst consequences of climate change. Already today, more than 20 million people are displaced each year from weather extreme events. In addition to this intra-generational equity issue, there is huge inter-generational equity because climate change impacts, such as the melting of glaciers and sea level rise, are irreversible for centuries or longer, affecting generations unborn.

Concluding Remarks: Common Solutions

Both air pollution and anthropogenic climate change problems have been studied for more than a century. However, air pollution science has enjoyed dramatic policy successes. Air pollution emissions have been reduced dramatically, by as much as 70% for some pollutants (Samet, Burke, & Goldstein, 2017) in the USA and in other industrialized nations, by deliberate policy actions. The policy actions in the USA were mainly motivated by human health impacts. Climate change, however, is still awaiting enforceable and mandatory laws to mitigate emissions of greenhouse gases and other climate pollutants.

Miraculously, there is still time to avoid the sort of catastrophic outcomes described in this chapter. What has been overlooked is the fact that integration of the solutions to mitigate climate pollutant emissions will also mitigate air pollution emissions. I will give two major examples:

1. Switching from fossil fuels to renewables, such as solar and wind, and banning biomass burning will not only eliminate carbon dioxide emissions but will also eliminate emissions of sulfates and black carbon, as well as decrease NOx emissions by more than 80%.
2. Likewise, giving renewable energy access to the poorest three billion people can phase out the burning of solid biomass fuels and pave the way for a carbon-neutral pathway for the poorest three billion people (or more in the future).

In a companion chapter in this book (Chap. 25), I describe the solutions recommended by a group of 50 researchers from the 10 campuses of the University of California (Ramanathan et al., 2016) and a group of 33 international experts (Ramanathan et al., 2017).

References

Baccini, A., Walker, W., Carvalho, L., Farina, M., Sulla-Menashe, D., & Houghton, R. A. (2017) Tropical forests are a net carbon source based on aboveground measurements of gain and loss. *Science, 358*, 230–234.

Barnosky, A. (2015). Transforming the global energy system is required to avoid the sixth mass extinction. *MRS Energy & Sustainability, 2*, E10. https://doi.org/10.1557/mre.2015.11

Cook, B. I., Smerdon, J. E., Seager, R., & Coats, S. (2014). Global warming and 21st century drying. *Climate Dynamics, 43*, 2607–2627. https://doi.org/10.1007/s00382-014-2075-y

Dasgupta, P., & Ramanathan, V. (2014). Pursuit of the common good, *Science, 345*, 1457–1458. https://doi.org/10.1126/science.1259406

Fu, R. (2015). Global warming-accelerated drying in the tropics. *Proceedings of the National Academy of Sciences of the United States of America, 112*, 3593–3594. https://doi.org/10.1073/pnas.1503231112

GBD 2013 Risk Factors Collaborators (2015). Global, regional, and national comparative risk assessment of 79 behavioural, environmental and occupational, and metabolic risks or clusters of risks in 188 countries, 1990–2013: A systematic analysis for the Global Burden of Disease Study 2013. *Lancet, 386*, 2287–2323.

Holland, G., & Bruyère, C. (2013). Recent intense hurricane response to global climate change. *Climate Dynamics, 42*. https://doi.org/10.1007/s00382-013-1713-0

Madden, R. A., & Ramanathan, V. (1980). Detecting climate change due to increasing CO_2 in the atmosphere. *Science, 209*, 763–768.

Mitchell, D., Heaviside, C., Vardoulakis, S., Huntingford, C., Masato, G., Guillod, B. P., et al. (2016). Attributing human mortality during extreme heat waves to anthropogenic climate change. *Environmental Research Letters, 11*, 074006.

Mora, C., Dousset, B., Caldwell, I. R., Powell, F. E., Geronimo, R. C., Bielecki, C. R., et al. (2017). Global risk of deadly heat. *Nature Climate Change, 7*, 501–506. https://doi.org/10.1038/nclimate332

Mueller, B., Zhang, X., & Zwiers, F. W. et al. (2016). Historically hottest summers projected to be the norm for more than half of the world's population within 20 years. *Environmental Research Letters, 11*, 044011.

Pagani, M., Huber, M., Liu, Z., Bohaty, S.M., Henderiks, J., Sijp, W., et al. (2011). The role of carbon dioxide during the onset of Antarctic glaciation. *Science, 334*, 1261–1264

Proestos, Y., Christophides, G. K., Ergüler, K., Tanarhte, M., Walkdock, J., & Lelieveld, J. (2015). Present and future projections of habitat suitability of the Asian tiger mosquito, a vector of viral pathogens, from global climate simulation. *Philosophical Transactions of the Royal Society, B: Biological Sciences, 370*, 1–16.

Ramanathan, V., Chung, C., Kim, D., Bettge,T., Buja, L., Kiehl, J. T. et al. (2005). Atmospheric brown clouds: Impacts on South Asian climate and hydrological cycle. *Proceeding of the National Academy of Science of the United States of America, 102*, 5326–5333.

Ramanathan, V., & Carmichael, G. (2008). Global and regional climate changes due to black carbon. *Nature Geoscience, 1*, 221–227.

Ramanathan, V., & Xu, Y. (2010). The Copenhagen Accord for limiting global warming: Criteria, constraints, and available avenues. *Proceedings of the National Academy of Sciences of the United States of America, 107*, 8055–8062.

Ramanathan, V., Allison, J., Auffhammer, M., Auston, D., Barnosky, A. D., Chiang, L., et al. (2016). *Chapter 1. Bending the curve: Ten scalable solutions for carbon neutrality and climate stability. Collabra, 2* (p. 15). Oakland: University of California Press. Retrieved February 9, 2020 from https://doi.org/10.1525/collabra.55

Ramanathan, V., Molina, M. J., Zaelke, D., Borgford-Parnell, N., Xu, Y., Alex, K., et al. (2017). *Full: Well under 2 degrees celsius: Fast action policies to protect people and the planet from extreme climate change*. Washington, DC: Institute of Governance and Sustainable Development. http://www.igsd.org/wp-content/uploads/2018/03/HLS-Well-Under-2C-VPAS.pdf

Rotstayn, L.D., & Lohman, U. (2002). Tropical rainfall trends and the aerosol indirect effect. *Journal of Climate, 15*, 2103–2115. https://doi.org/10.1175/1520-0442(2002)015<3C2103:TRTATI>3E2.0.CO2

Russo, S., Sillmann, J., & Sterl, A. (2017). Humid heat waves at different warming levels. *Scientific Reports, 7*. https://doi.org/10.1038/s41598-017-07536-7

Samet, J. M., Burke, T., & Goldstein, B. (2017). The Trump administration and the environment—Heed the science. *New England Journal of Medicine, 376*, 1182–1188. https://doi.org/10.1056/NEJMms1615242

Thomas, C., Alison, C., Green Rhys, E., Michel, B., Beaumont Linda, J., & Collingham Yvonne, C. (2004). Extinction risk from climate change. *Nature, 427*, 145–148. https://doi.org/10.1038/nature02121

Urban, M. C. (2015). Climate change. Accelerating extinction risk from climate change. *Science (New York, NY), 348*, 571–573. https://doi.org/10.1126/science.aaa4984

WMO (2016). Global Climate in 2011–2015. Available at https://public.wmo.int/en/resources/library/global-climate-2011%E2%80%932015

Xu, Y., Ramanathan, V., & Washington, W. M. (2016). Observed high-altitude warming and snow cover retreat over Tibet and the Himalayas enhanced by black carbon aerosols. *Atmospheric Chemistry and Physics, 16*, 1303–1315.

Xu, Y., & Ramanathan, V. (2017). Well below 2°C: Mitigation strategies for avoiding dangerous to catastrophic climate changes. *Proceedings of the National Academy of Sciences of the United States of America, 114*, 10315–10323.

Part II
Air Pollution, Climate Change, and Health: The Underlying Science and Impacts

CHAPTER 6
Air Pollution: Adverse Effects and Disease Burden

Jonathan M. Samet

Introduction

Summary In 2017, the scientific evidence was certain: ambient air pollution (i.e., contamination of outdoor air consequent to man's activities) is a major cause of morbidity (ill health) and premature mortality (early death). While the rise of ambient air pollution is relatively recent, air pollution has probably had adverse effects on human health throughout history. In fact, the respiratory tract, which includes the nose, throat and lungs, has a remarkable system of defense mechanisms to protect against inhaled particles and gases. The use of fire for heating and cooking came with exposure to smoke, an exposure that persists today for the billions who use biomass fuels for cooking and heating. The rise of cities concentrated the emissions of pollutants from dwellings and industry and led to air pollution, which was likely affecting health centuries ago. Continued industrialization and also electric power generation brought new point sources of pollution into areas adjacent to where people lived and worked. During the twentieth century, cars, trucks, and other fossil fuel–powered vehicles became a ubiquitous pollution source in higher-income countries and created a new type of pollution—photochemical pollution, or "smog"—which was first recognized in the Los Angeles air basin in the 1940s. The unprecedented growth of some urban areas to form "megacities," such as Mexico City, São Paulo, London, and Shanghai, has led to unrelenting air pollution from massive vehicle fleets and snarled traffic and from polluting industries and coal-burning power plants. With population growth and urbanization, ever more megacities are anticipated; the current total of cities with a population over 10 million has now reached 31.

This chapter provides an overview of the current state of knowledge concerning the adverse health effects of ambient air pollution and the associated burden of premature death. It draws on various authoritative reviews, particularly those car-

J. M. Samet (✉)
Colorado School of Public Health, University of Colorado, Aurora, CO, USA
e-mail: Jon.Samet@ucdenver.edu

© The Author(s) 2020   63
W. K. Al-Delaimy, V. Ramanathan, M. Sánchez Sorondo (eds.), *Health of People, Health of Planet and Our Responsibility*, https://doi.org/10.1007/978-3-030-31125-4_6

ried out by the US Environmental Protection Agency in support of revisions to the National Ambient Air Quality Standards (NAAQS), which apply to the major outdoor pollutants. Emphasis is given to airborne particulate matter, a ubiquitous type of ambient pollution that derives from multiple sources; the most widely used indicators are PM_{10} and $PM_{2.5}$ (particulate matter less than 10 and 2.5 μm in aerodynamic diameter, respectively) and to ozone. The chapter also addresses the burden of disease attributable to air pollution, documenting the target for air quality improvement and the global gains in public health that can be made.

While the emphasis here is on ambient air pollution that has adverse effects at local and regional levels, climate change also comes from air pollution with greenhouse gases. The connections are direct; the same combustion processes that release particles and toxic gases also generate carbon dioxide.

The Health Effects of Air Pollution

Overview Ambient air pollution comprises a complex and dynamic mixture of gaseous and particulate air pollutants. The array of health effects linked to ambient air pollution is broad and includes increased risk for respiratory infections, exacerbation of asthma, and chronic obstructive pulmonary disease (COPD—a disease involving destruction of the lung structure) and cardiac (heart) events, contributions to development of major chronic diseases (coronary heart disease, COPD, and cancer), and impaired lung growth and respiratory symptoms during childhood. Additional adverse health outcomes are under investigation: autism and other neurodevelopmental disorders, adverse reproductive outcomes, and more rapid "brain aging," for example. There are several general mechanisms underlying these health effects, particularly oxidant stress, an excess of reactive molecules, and a heightened inflammatory state, given the oxidative nature of ambient air pollution. The increased risk for cancer causally linked to air pollution likely comes primarily from the presence of specific carcinogens (i.e., cancer-causing agents) in ambient air pollution (e.g., polycyclic aromatic hydrocarbons, and inflammation); particles collected in outdoor air are mutagenic, which means that they can damage DNA (IARC, 2015). Table 6.1 provides a listing of major pollutants and associated health effects, as well as some of the current standards and guidelines for controlling their concentrations.

While the problem of air pollution was noted centuries ago, the contemporary era of research on air pollution and health and evidence-driven air quality regulation and management began in the mid-twentieth century following a series of episodes of very high pollution with disastrous health consequences (Brimblecombe, 1987). The most dramatic was the London Fog of 1952, which caused thousands of excess deaths and prompted some of the first epidemiological studies of the health effects of air pollution (Fig. 6.1) (Bell & Davis, 2001). In the United States, recognition of the public health dimensions of air pollution also began in the mid-twentieth century, driven by the rising problem of smog in southern California, episodes of visibly high pollution in major cities, and the 1948 air pollution episode in Donora, Pennsylvania, which caused 20 excess deaths and thousands of illnesses in one small town.

Table 6.1 Major ambient air pollutions: sources, health effects, and regulations[a]

	Source types and major sources	Health effects	Regulations and guidelines
Lead	Primary	Accumulates in organs and tissues. Learning disabilities, cancer, damage to the nervous system	*U.S. NAAQS:*
	Anthropogenic: leaded fuel (phased out in some locations such as the United States), lead batteries, metal processing		Quarterly average: 1.5 μg/m^3
			WHO Guidelines:
			Annual: 0.50 μg/m^3
Sulfur dioxide	Primary	Lung impairment, respiratory symptoms. Precursor to PM. Contributes to acid precipitation	*U.S. NAAQS:*
	Anthropogenic: combustion of fossil fuel (power plants), industrial boilers, household coal use, oil refineries		Annual arithmetic mean: 0.03 ppm (80 μg/m^3)
	Biogenic: decomposition of organic matter, sea spray, volcanic eruptions		24-h average: 0.14 ppm (365 μg/m^3)
			WHO Guidelines:
			10-min average: 500 μg/m^3
			Annual: 20 μg/m^3
Carbon monoxide	Primary	Interferes with delivery of oxygen. Fatigue, headache, neurological damage, dizziness	*U.S. NAAQS:*
	Anthropogenic: combustion of fossil fuels (motor vehicles, boilers, furnaces)		1-h average: 35 ppm (40 mg/m^3)
			8-h average: 9 ppm (10 mg/m^3)
			WHO Guidelines:
			15-min average: 100 mg/m^3
			30-min average: 60 mg/m^3
			1-h average: 30 mg/m^3

(continued)

Table 6.1 (continued)

	Source types and major sources	Health effects	Regulations and guidelines
Particulate matter[b]	Primary and secondary	Respiratory symptoms, decline in lung function, exacerbation of respiratory and cardiovascular disease (e.g., asthma), mortality	*U.S. NAAQS*:
	Anthropogenic: burning of fossil fuel, wood burning, natural sources (e.g., pollen), conversion of precursors (NO_x, SO_x, VOCs)		PM_{10}
			24-h average 150 $\mu g/m^3$
	Biogenic: dust storms, forest fires, dirt roads		$PM_{2.5}$
			Annual arithmetic mean: 15 $\mu g/m^3$
			24-h average: 35 $\mu g/m^3$
			WHO Guidelines:
			PM_{10}
			Annual: 20 $\mu g/m^3$
			24-h average: 50 $\mu g/m^3$
			$PM_{2.5}$:
			Annual: 10 $\mu g/m^3$
			24-h average: 25 $\mu g/m^3$
Nitrogen oxides	Primary and secondary	Decreased lung function, increased respiratory infection	*U.S. NAAQS* for NO_2:
	Anthropogenic: fossil fuel combustion (vehicles, electric utilities, industry), kerosene heaters	Precursor to ozone. Contributes to PM and acid precipitation	Annual arithmetic mean: 0.053 ppm (100 $\mu g/m^3$)
	Biogenic: biological processes in soil, lightning		Related to compliance with NAAQS for ozone.
			WHO guidelines for NO_2:
			1-h average: 200 $\mu g/m^3$
			Annual: 40 $\mu g/m^3$

Pollutant	Type	Sources	Health effects	U.S. NAAQS / Regulations
Tropospheric ozone	Secondary	Formed through chemical reactions of anthropogenic and biogenic precursors (VOCs and NO_x) in the presence of sunlight	Decreased lung function, increased respiratory symptoms, eye irritation, bronchoconstriction	*U.S. NAAQS:* 1-h average: 0.12 ppm (235 µg/m³). Applies in limited areas. 8-h average: 0.075 ppm (147 µg/m³) *WHO guidelines:* 8-h average: 100 µg/m³
"Toxic" pollutants ("Hazardous") (e.g., asbestos, mercury, dioxin, some VOCs)	Primary and secondary	Anthropogenic: industrial processes, solvents, paint thinners, fuel	Cancer, reproductive effects, neurological damage, respiratory effects	EPA rules on emissions for more than 80 industrial source categories (e.g., dry cleaners, oil refineries, chemical plants). EPA and state rules on vehicle emissions
Volatile organic compounds (e.g., benzene, terpenes, toluene)	Primary and secondary	Anthropogenic: solvents, glues, smoking, fuel combustion. Biogenic: vegetation, forest fires	Range of effects, depending on the compound. Irritation of respiratory tract, nausea, cancer. Precursor to ozone. Contributes to PM.	EPA limits on emissions. EPA toxic air pollutant rules. Related to compliance with NAAQS for ozone.
Biological pollutants (e.g., pollen, mold, mildew)	Primary	Biogenic: trees, grasses, ragweed, animals, debris. Anthropogenic systems, such as central air conditioning, can create conditions that encourage production of biological pollutants.	Allergic reactions, respiratory symptoms, fatigue, asthma	

Source: Bell ML, Samet JM. Air Pollution. In Frumkin H. (ed) Environmental health: From Global to Local. 2010:387–415. Figure 12.1

[a]This table lists only a sample of the sources and health effects associated with each pollutant. Additionally, health effects may be the result of characteristics of the pollutant mixture rather than the independent effects of a pollutant. Additional legal requirements often apply, such as state regulations

[b]Sources and effects of PM can differ by size

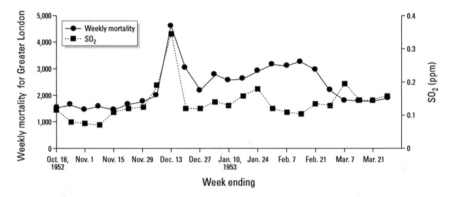

Fig. 6.1 Approximate weekly mortality and SO$_2$ concentrations for Greater London, 1952–1953

Multiple investigational approaches have been used to characterize the health effects of ambient air pollution. Initially, the dramatic pollution episodes made clear that high levels of air pollution caused excess mortality, particularly in elderly people with chronic diseases and in infants and young children. As air pollution levels declined with regulation, increasing emphasis was placed on understanding the quantitative risks of air pollution so that air quality standards could be set that would be protective of public health. In other words, researchers did studies to understand by how much risk changes as air pollution increases or decreases. Epidemiological studies (i.e., research based in populations) were critical for that purpose. Cohort or longitudinal studies were most informative. Such studies involved following participants over time, estimating pollution exposures, and tracking health events; analyses focused on quantifying the risks associated with air pollution exposure during follow-up. For example, the Children's Health Study in Southern California tracked lung growth and respiratory health in school children from communities having a range of pollution concentrations (Gauderman et al., 2015). Now in progress for two decades, the study has shown that higher levels of air pollution slow lung growth and that reduction of pollution enhances it.

These epidemiological approaches are complemented by toxicological studies that provide insights into the mechanisms by which air pollution causes adverse health effects. Such evidence is critical in reaching causal conclusions on the adverse health effects of air pollution. In the past, toxicological studies often involved exposure of animals to a single pollutant, such as ozone, to isolate the pollutant's effect from those of other pollutants present in the air pollution mixture. For studying some pollutants, human volunteers have inhaled the pollutants in an exposure chamber over a short interval and their responses closely monitored. Additionally, pollutants are also studied in cell systems; these systems are likely to gain increasing prominence as new, sophisticated systems probe gene expression of different kinds of cells following exposure.

There are a number of national groups that periodically assess the evidence on adverse effects of air pollution and provide guidance to the setting of standards and guidelines. In the United States, the Environmental Protection Agency carries out reviews of the evidence as the basis for renewal of major air quality standards (the National Ambient Air Quality Standards or NAAQS) on a 5-year cycle. Reviews are conducted by the United Kingdom, the European Commission and other nations. The World Health Organization releases air quality guidelines, which are currently being updated. In reviewing the evidence, a judgment that the findings are strong enough to infer a causal relationship has great weight for regulation. The health risks associated with major air pollutants are reviewed below.

While air pollution research and regulation generally focuses on specific pollutants, air pollution outdoors is a complex mixture. Effects attributed to a single pollutant, particularly when studied in the "real world" context, may reflect the toxicity of the mixture as indexed by a particular pollutant. Ambient particulate matter (PM), for example, comes from myriad sources and is emitted as a primary pollutant from combustion and other sources; it is also formed through chemical transformations of gaseous pollutants, such as the formation of particulate nitrates from gaseous nitrogen oxides. The mixture of pollutants formed from vehicle emissions, generally referred to as traffic-related air pollution, may have specific toxicity beyond that of well-studied individual components.

Particulate Matter The literature on the health effects of particles is enormous, comprising many epidemiological and toxicological studies (U.S. Environmental Protection Agency, 2009). With regard to ambient air pollution, the risks of particulate matter have assumed great prominence because particles are widely monitored and used as the principal indicator for estimating the burden of morbidity and premature mortality attributable to air pollution. Particles are a robust indicator of ambient air pollution because of their myriad sources and the contributions of sulfur and nitrogen oxides and organic compounds to secondary particle formation. Particles in outdoor air have numerous natural and man-made sources. The man-made sources are diverse and include power plants, industry, and motor vehicles, including diesel-powered vehicles that emit particles in the size range that penetrates into the lung. In areas where biomass fuels are used, the contributions of indoor combustion to outdoor air pollution may be substantial.

Particles in outdoor air span a wide range of sizes (Fig. 6.2) and are highly diverse in composition and physical characteristics, including size as indicated by aerodynamic diameter. Thus, $PM_{2.5}$ includes those particles less than 2.5 μm in aerodynamic diameter, a size band that contains most man-made particles in outdoor air and also the particles of a size that can reach the smaller airways and air sacs of the lungs. The very small ultrafine particles, which include freshly generated combustion particles, are another set of particles of concern. Much research has been done on characteristics of particles that determine toxicity. Hypotheses on determinants of particle toxicity have focused on acidity, transition metals such as iron that can cause damaging injury, organic compounds, bioaerosols, and size; as a further complication, different characteristics could be relevant to different health outcomes. However, in spite of extensive toxicological and epidemiological research, the evidence is not yet sufficiently definitive to link particular characteristics to toxicity. Making such linkages would be helpful for targeting control approaches.

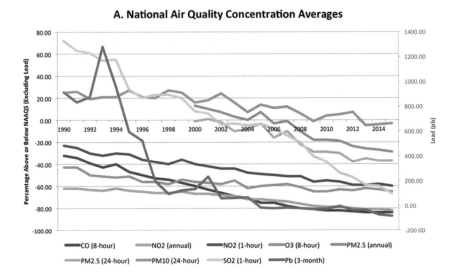

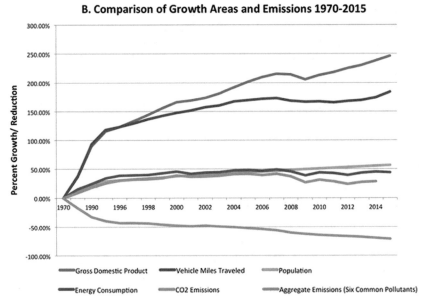

Fig. 6.2 Changes in economic indicators, pollutant emissions, and concentrations of criteria pollutants over time. *Note*: Shown are changes in criteria pollutant concentrations during 1990-2015 (*Panel A*) and changes in economic growth indicators and emissions during 1970-2015 (*Panel B*). (*Source*: Drawn from EPA data at https://gispub.epa.gov/air/trendsreport/2016/)

There has been extensive epidemiological and toxicological investigation of the effects of particles on health since the air pollution disasters of mid-century. The toxicological studies have used approaches including exposing volunteers to generated particles or concentrated air particles, animal exposures, and diverse in vitro assays. This extensive body of evidence shows that particles are injurious and indicates mechanisms by which particles could cause adverse effects on the respiratory and cardiovascular systems. The epidemiological studies have grown in size and in the sophistication of their methodology. Recent studies involve large, national-level populations, such as all people in the United States who are 65 years and older, and estimation of exposures at all household addresses using models that incorporate available monitoring data, satellite information, and land-use data, such as on roadways and manufacturing. Table 6.2 lists the findings of the most recent comprehensive review of the evidence on airborne particles (the Integrated Science Assessment) by the US Environmental Protection Agency. Both short-term and long-term adverse effects were found to be caused by particulate matter. The most recent epidemiological studies continue to find associations of $PM_{2.5}$ with increased risk for mortality, even at contemporary concentrations in the United States. In a recent study utilizing the Medicare data for persons 65 years of age and older, over 60 million people, Di et al. showed increased mortality at annual averages below 12 μg/m, the current annual standard in the United States[3] (Di et al., 2017).

One analysis has suggested that reductions in particulate air pollution during the 1980s and 1990s have led to measurable improvements in life expectancy in the United States. The researchers estimated that for each 10 μg/m³ reduction in air pollution over this period, the average gain in life expectancy was 0.61 years (about 7.3 months). The authors concluded that as much as 15% of the total life expectancy increase seen during this time in the United States was attributable to the air pollution reductions.

This strong evidence on the adverse health effects of particulate matter has led to ever tighter ambient air quality standards (Table 6.1). Nonetheless, adverse health effects are still observed and much of the world's population is exposed to high concentrations at which adverse effects are certain.

Ozone Ozone is a specific gas that has been studied for its toxicity using toxicological approaches. It is also used as an indicator of photochemical pollution, or "smog," which is the complex oxidant mixture produced by the action of sunlight on hydrocarbons and nitrogen oxides. Smog has become a worldwide problem as vehicle fleets have grown. The problem of tropospheric (ground-level) ozone pollution is distinct from the problem of depletion of the strato-pheric (high-level) ozone layer.

Ozone, a highly reactive molecule, has been extensively investigated using toxicological approaches that have included exposures of human volunteers and short- and long-term exposures of animals (U.S. EPA, 2013). The human studies have involved exposures of volunteers, generally young and healthy, to concentrations of ozone found in urban areas in the United States and elsewhere. Collectively, the studies show that exposures of up to 6–8 h with intermittent exercise result in temporary drops in lung function and that some individuals have greater susceptibility

Table 6.2 Summary of causal determinations for airborne particles by exposure duration and health outcome

Size fraction	Exposure	Outcome	Causality determination
PM$_{2.5}$	Short-term	Cardiovascular effects	Causal
		Respiratory effects	Likely to be causal
		Central nervous system	Inadequate
	Long-term	Mortality	Causal
		Cardiovascular effects	Causal
		Respiratory effects	Likely to be causal
		Mortality	Causal
		Reproductive and developmental	Suggestive
		Cancer, mutagenicity, genotoxicity	Suggestive
PM$_{10-2.5}$	Short-term	Cardiovascular effects	Suggestive
		Respiratory effects	Suggestive
		Central nervous system	Inadequate
		Mortality	Suggestive
	Long-term	Cardiovascular effects	Inadequate
		Respiratory effects	Inadequate
		Mortality	Inadequate
		Reproductive and developmental	Inadequate
		Cancer, mutagenicity, genotoxicity	Inadequate
Ultrafine particles	Short-term	Cardiovascular effects	Suggestive
		Respiratory effects	Suggestive
		Central nervous system	Inadequate
		Mortality	Inadequate
	Long-term	Cardiovascular effects	Inadequate
		Respiratory effects	Inadequate
		Mortality	Inadequate
		Reproductive and developmental	Inadequate
		Cancer, mutagenicity, genotoxicity	Inadequate

Source: U.S. EPA (2009). Table 2-6

to ozone. While the effects are transient, they are of sufficient magnitude in some people (loss of around 10% of function) to be considered adverse. In some of the studies, the lungs have been sampled and evidence of inflammation was found by measuring concentrations of molecules that reflect the tissue's response. In experimental animals, sustained low-level exposure damages the small airways and leads to early changes of COPD; thus, there is concern about permanent structural alteration in ozone-exposed populations. In human studies, asthmatics have not been shown to have increased susceptibility to ozone compared with non-asthmatics.

Epidemiological studies provide coherent evidence on the short-term effects of ozone on respiratory health. There is also evidence from daily time-series studies (studies examining day-to-day variations in death counts in relationship to variations in pollution levels) that ozone increases the risk for mortality. There is inconsistent evidence for cardiovascular effects and a just-completed exposure study of older persons with cardiovascular disease did not find adverse effects.

Reflecting the evidence on short-term effects on lung function, standards for ozone concentrations are directed at brief time spans (Table 6.1). Given the range of susceptibility of the population, it is likely that feasibly achieved standards will not protect the full population from adverse respiratory effects.

Nitrogen Oxides Gaseous nitrogen oxides are produced by combustion processes and also contribute to the formation of aerosols. Nitrogen dioxide (NO_2), an oxidant gas, is the indicator that is generally monitored. The principal source of NO_2 in outdoor air is motor vehicle emissions, and NO_2 is considered to be a useful indicator of traffic-related air pollution in urban environments. Power plants and industrial sources may also contribute. The health effects of NO_2 emitted into outdoor air probably come mainly from the formation of secondary pollutants, including ozone and particles. NO_2, along with hydrocarbons, is an essential precursor of ozone and the nitrogen oxides also form acidic nitrate particles.

Nitrogen dioxide itself has been studied in animal models and in clinical studies. It can reach the small airways and air sacs of the lung because of its low solubility. The toxicological evidence at high exposures has raised concern that NO_2 exposure can impair lung defenses against infectious agents such as viruses and cause airway inflammation, thereby increasing the risk for respiratory infections. Supporting epidemiological research is lacking and population studies directed at NO_x are complicated by its role in the formation of ozone and its presence in the complex mixture of traffic-related pollutants. Human exposure studies have been performed to investigate the immediate effects of NO_2 on persons with asthma. Nitrogen dioxide could plausibly increase airway responsiveness (the extent to which the airways constrict when irritated) by causing airway inflammation. The findings of the exposure studies have been inconsistent, but suggest that some people with asthma may be susceptible.

Carbon Monoxide Carbon monoxide (CO) is an invisible gas formed by incomplete combustion. It is a prominent indoor pollutant with sources including biomass fuel combustion and space heating with fossil fuels. At high levels indoors, fatal CO poisoning may result. Outdoors, vehicle exhaust is the major source and concentrations are highly variable, reflecting vehicle density and traffic patterns. Urban locations with high traffic density ("hot spots") tend to have the highest concentrations. The toxicity of CO comes from its tight binding to hemoglobin, which carries oxygen in the blood, and the resulting reduction of oxygen delivery to tissues. Exposures to CO can be assessed by using the level of carboxyhemoglobin as a marker of exposure or by measuring the concentration of CO in the breath.

Because of the reduction of oxygen delivery, persons with cardiovascular disease are considered to be at greatest risk from CO exposure; research has focused on CO and adverse effects in this susceptible group. The research has used an approach referred to as a "clinical study"; volunteers with cardiovascular disease are exposed to levels of CO of interest and clinical measurements made, such as by taking an electrocardiogram. The evidence from such studies indicates that CO exposure leads to earlier evidence of myocardial ischemia (inadequate oxygenation of the heart) following exposure compared with unexposed controls. There are other potential susceptible groups: fetuses and persons with COPD may also be harmed by CO, and normal persons may have reduced oxygen uptake during exercise at low levels of CO exposure.

The exposure studies have provided robust evidence for standards for CO, which are based on brief time windows, reflective of the handling of CO in the body. In higher-income countries, outdoor levels of CO have fallen greatly over recent decades as controls have greatly reduced emissions (see Fig. 6.3). Nonetheless, CO may be a concern in some high-traffic locations. Less is known about CO exposure in middle- and low-income countries, where ambient CO may be added to indoor exposure from biomass fuel combustion.

Sulfur Oxides Sulfur oxides are generated by combustion of fuels containing sulfur, such as coal, crude petroleum, and diesel, and by smelting operations. The water-soluble gas, sulfur dioxide or SO_2, is the indicator that is generally monitored. However, other sulfur oxides are emitted and the sulfur oxides undergo transformation to form particulate sulphate compounds. Scientific research has been directed primarily at SO_2, although epidemiological studies provide information on sulfur oxide exposure more generally. Sulfur dioxide is a reactive gas that is effectively scrubbed or cleaned from inhaled air in the upper airway. With exercise and a switch to oral breathing as ventilation increases, the inhaled dose of SO_2 increases and more reaches the lung.

Much of the evidence that has driven regulation comes from clinical studies that involve exposure of people with asthma and that show adverse effects without exposures to other pollutants. Asthmatics are particularly sensitive, with some asthmatics having more severe health responses at a particular concentration than others with asthma. With exercise and hyperventilation, some people with asthma respond with increased resistance of the lung to airflow with an associated drop in lung func-

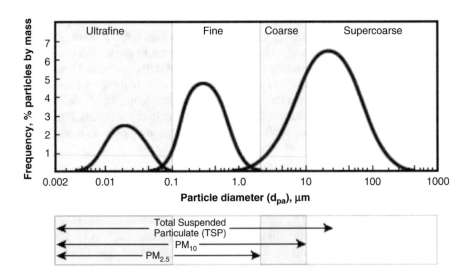

Fig. 6.3 Ambient particulate matter size distribution. (*Source*: Adapted from the EPA)

tion and with respiratory symptoms. Such effects have been demonstrated at concentrations that might be reached in the United States in high-exposure situations and that may be common in some heavily industrialized countries. Epidemiological studies from Hong Kong examined the consequences of a major and rapid reduction in sulfur content in fuels. The investigators found an associated substantial reduction in health effects (childhood respiratory disease and all age mortality outcomes).

Lead   Although lead in gasoline is now phased out in almost all nations, exposure continues from industrial activities, such as smelting, and sometimes results in dangerous exposures for children. Exposure to lead may occur through inhalation and also ingestion in food and water, routes of exposure that have become the most important in high-income countries. A substantial body of epidemiological evidence links lead exposure of children to adverse neurodevelopmental effects, such as lowering the level of intelligence; as a result of that evidence, recommendations as to the acceptable level of lead for children have been lowered progressively. Lead has also been linked to higher blood pressure and cardiovascular disease and to low bone mineral density and osteoporosis.

The Disease Burden from Ambient Air Pollution

Given these adverse health effects and ubiquitous pollution of outdoor air, estimates of the burden of disease caused by air pollution are needed as a basis for priority setting and air quality management. Such estimates of disease burden can also prove useful for motivating action, as they have done recently in China. The conceptual basis for estimating disease burden draws on the conceptual framework of attributable risk, originally proposed in the early 1950s for smoking and lung cancer by Levin (Levin, 1953). He proposed a statistic, the population attributable risk (PAR), which incorporates the prevalence of the exposure of concern (P) and the relative risk (RR) for disease associated with that exposure, as follows: $PAR = (Px[RR - 1])/(1 + Px[RR - 1])$. Thus, the PAR rises as either P or RR increases; in other words, the burden rises as more people are exposed or the risk is increased. For lung cancer, for example, in a population with 40% smokers and a lung cancer RR of 20, the PAR is 0.88, interpreted as 88% of the cases resulting from smoking. The PAR is interpreted based on comparison to a comparison group involving no or lower exposure; for smoking, for example, the comparison is a hypothetical world in which smoking never existed.

For air pollution, estimates of the burden of attributable disease have been made by the World Health Organization and the Institute for Health Metrics and Evaluation (IHME) in the United States, which carries out the Global Burden of Diseases, Injuries, and Risk Factors Study (Cohen et al., 2017). A detailed analysis of the global burden of disease attributable to ambient air pollution was recently reported for the 25 years from 1990 to 2015. With regard to the exposure prevalence (P), $PM_{2.5}$ and ozone concentrations are estimated for smaller level spatial grids that capture variation in urban areas and then aggregated to provide national mean exposures. The IHME estimates for $PM_{2.5}$ include ischemic heart disease (IHD), cerebro-vascular disease, lung cancer, chronic obstructive pulmonary disease (COPD), and lower respiratory infections (LRI), while only COPD is considered for ozone. The RRs for these diseases come from complex analyses of epidemiological data. The counterfactual values for burden estimation are derived from the lowest values considered in the epidemiological studies.

The latest estimates confirm that outdoor air pollution is a leading cause of premature mortality around the world, ranking at the fourth position in 1990 and the fifth in 2015. The total of deaths attributed to $PM_{2.5}$ globally in 2015 is 4.2 million distributed by cause as follows: IHD—1.52 million, cerebrovascular disease—898,000, lung cancer—283,000, COPD—864,000, and LRI—675,000. There is substantial geographic variation globally (Fig. 6.4) with China and India together accounting for more than half of the attributable burden of premature mortality. From 1990 to 2015, estimates have increased in some countries (e.g., India and China), as the population has grown and aged, and air pollution levels have risen. For ozone, the global mortality estimate is much smaller at 254,000.

From a policy perspective, these estimates offer a strong rationale for air quality control and are cautionary in their implications. The estimates of exposure globally describe a stratified target (Fig. 6.4): the high-income countries that have lowered air pollution concentrations over the last half century and the numerous low- and middle-income countries where air pollution has worsened with industrialization and vehicle numbers have increased rapidly. Additionally, many of these countries

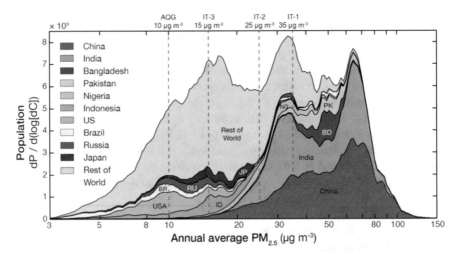

Fig. 6.4 Global and regional distributions of population as a function of annual average ambient $PM_{2.5}$ concentration for the 10 most populous countries in 2013. *Note*: Plotted data reflect local smoothing of bin-width normalized distributions computed over 400 logarithmically spaced bins; equally sized plotted areas would reflect equal populations. Dashed vertical lines indicate World Health Organization Interim Targets (IT) and the Air Quality Guideline (AQG). (*Source*: Brauer, M., et al. (2016). Ambient air pollution exposure estimation for the global burden of disease 2013. *Environmental Science & Technology, 50*(1), 79–88. Figure 3)

still face the added challenge of household air pollution, also a leading cause of premature mortality.

Research Needs

There is a robust body of evidence on the health effects of ambient air pollution, which has both acute and chronic consequences. Given the underlying mechanisms of injury and commonalities among pollution sources, the collective evidence should have general applicability to people around the world. Nonetheless, many nations still lack basic monitoring and air quality management, even as pollution levels have risen in recent decades. In motivating action, locally generated data are likely to be more powerful than the external evidence, particularly if from high-income countries. At the least, monitoring data are needed for $PM_{2.5}$ and other pollutants if relevant to a particular place. Local research could use cross-sectional epidemiological designs, for example, of the respiratory health of school children, and time-series studies of morbidity (e.g., hospitalization counts) and mortality. The Health Effects Institute in the United States has supported time-series studies of air pollution in multiple Asian countries through its Public Health and Air Pollution in Asia (PAPA) Project (Public Health and Air pollution in Asia (PAPA), 2010). The Global Burden of Disease estimates provide a valuable starting point for priority setting around ambient air pollution as a public health issue. The Health Effects Institute has made this information available in a useful form through its State of Global Air/2019 (https://www.stateofglobalair.org/).

Summary and Conclusions

Ambient air pollution is a well-documented threat to global health. Through man's expanding use of fossil fuels for manufacturing, vehicles, and power generation, the numbers of people exposed to risky levels of air pollution has increased progressively; some are exposed at levels historically associated with evident excess mortality. Ambient air pollution now affects large swathes of the world, as contaminated air crosses national boundaries and moves globally. Of course, greenhouse gases represent another form of global pollution. The evidence is sufficiently certain and the burden of disease sufficiently large to motivate action at national and global levels. Evidence-driven regulations have had substantial impact on air quality in the United States and elsewhere, driving down levels of the most prominent pollutants (Fig. 6.3). Unfortunately, in many low- and middle-income countries, outdoor air pollution is a rising problem consequent to population growth, industrialization, and increasing numbers of vehicles—and, in too many countries, insufficient attention to air quality management and regulation. Control of greenhouse gases has the co-benefit of reducing ambient air pollution.

References

Bell, M. L., & Davis, D. L. (2001). Reassessment of the lethal London Fog of 1952: Novel indi-
    cators of acute and chronic consequences of acute exposure to air pollution. *Environmental
    Health Perspectives, 109,* 389–394.
Brimblecombe, P. (1987). *The big smoke: A history of air pollution in London since medieval times*
    (1st ed.). London: Routledge Kegan & Paul.
Cohen, A. J., Brauer, M., Burnett, R., Anderson, H. R., Frostad, J., Estep, K., et al. (2017). Estimates
    and 25-year trends of the global burden of disease attributable to ambient air pollution: An
    analysis of data from the Global Burden of Diseases Study 2015. *Lancet, 389,* 1907–1918.
    https://doi.org/10.1016/s0140-6736(17)30505-6
Di, Q., Wang, Y., Zanobetti, A., Wang, Y., Koutrakis, P., Choirat, C., et al. (2017). Air pollution
    and mortality in the medicare population. *New England Journal of Medicine, 376,* 2513–2522.
    https://doi.org/10.1056/NEJMoa1702747
Gauderman, W. J., Urman, R., Avol, E., Berhane, K., McConnell, R., Rappaport, E., et al. (2015).
    Association of improved air quality with lung development in children. *New England Journal
    of Medicine, 372*(10), 905–913. https://doi.org/10.1056/NEJMoa1414123
IARC (2015). *Outdoor air pollution* (Vol. 109). Lyon, France: International Agency for Research
    on Cancer.
Levin, M. L. (1953). The occurrence of lung cancer in man. *Acta Unio Internationalis Contra
    Cancrum, 9,* 531–541.
Public Health and Air pollution in Asia (PAPA). (2010). *Coordinated studies of short-term
    exposure to air pollution and daily mortality in four cities* (Research Report 154), Boston,
    MA. Retrieved on February 16, 2020 from https://www.healtheffects.org/publication/
    public-health-and-air-pollution-asia-papa-coordinated-studies-short-term-exposure-air-0
U.S. Environmental Protection Agency (2009). *Final report: Integrated science assessment for
    particulate matter* (EPA/600/R-08/139F). Washington, DC: U.S. Environmental Protection
    Agency. Retrieved on February 16, 2020 from https://cfpub.epa.gov/si/si_public_record_
    Report.cfm?dirEntryId=216546
U.S. Environmental Protection Agency (2013). *Integrated science assessment for ozone and
    related photochemical oxidants* (EPA/600/R-10/076F). Research Triangle Park, NC: U.S.
    Environmental Protection Agency. Retrieved on February 16, 2020 from https://cfpub.epa.gov/
    ncea/isa/recorddisplay.cfm?deid=247492

CHAPTER 7
Air Pollution, Oxidative Stress, and Public Health in the Anthropocene

Ulrich Pöschl

Summary Air pollution severely affects air quality, climate, and public health in the Anthropocene, which is the present era of globally pervasive anthropogenic influence on planet Earth. Thus, we need to understand how humanity can best deal with the sources and effects of air pollutants in relation to economic development, human welfare, and environmental preservation. Recent advances in scientific research provide deep insights into the underlying physical, chemical, and biological processes that link air pollution with health effects and reveal the relative importance of different pollutants and sources, including natural and anthropogenic contributions. This knowledge enables the development of efficient strategies and policies to mitigate and counteract the adverse effects of air pollution on the Earth system, climate, and human health ("planetary health"). Building on open access to scholarly publications and data, a global commons of scholarly knowledge in the sciences and humanities will help to augment, communicate, and utilize the scientific understanding. Moreover, public peer review, interactive discussion, and documentation of the scientific discourse on the internet can serve as examples and blueprints for rational and transparent approaches to resolving complex questions and issues ("epistemic web"). With regard to the development, societal communication, and political implementation of appropriate policies for air quality management, it seems worthwhile to emphasize that climate and health effects are two facets of global environmental change that can be and need to be handled together. The Anthropocene notion may help humanity to recognize both rationally and emotionally: We are shaping our planet and environment, so let us get it right.

U. Pöschl (✉)
Multiphase Chemistry Department, Max Planck Institute for Chemistry, Mainz, Germany
e-mail: u.poschl@mpic.de

© The Author(s) 2020   79
W. K. Al-Delaimy, V. Ramanathan, M. Sánchez Sorondo (eds.), *Health of People,
Health of Planet and Our Responsibility*, https://doi.org/10.1007/978-3-030-31125-4_7

Background

Anthropogenic air pollution leads to a massive increase of atmospheric aerosol and oxidant concentrations on local, regional, and global scales. For example, the average mixing ratios of ozone in continental background air have increased by factors of ~2–4 from around 10–20 ppb at the beginning of the nineteenth century to 30–40 ppb in the twenty-first century, and the concentrations of fine particulate matter in polluted urban air are typically a factor of ~10 higher than in pristine air of remote continental and marine regions ($\sim 10-100$ µg m^{-3} vs. $\sim 1-10$ µg m^{-3}). The strong increase of these and other air pollutants is a characteristic feature of global environmental change in the Anthropocene, which is the present era of globally pervasive anthropogenic influence on planet Earth, including the atmosphere, biosphere, land, and oceans (Crutzen, 2002; Foley, Gronenborn, Andreae, et al., 2013; Pöschl & Shiraiwa, 2015; Seinfeld & Pandis, 2016; Waters et al., 2016 and references therein).

Figure 7.1a illustrates the cycling and effects of gas molecules and aerosol particles exchanged between the atmosphere and the Earth's surface, including the biosphere, hydrosphere, and pedosphere/lithosphere (Pöschl, 2005; Pöschl & Shiraiwa, 2015). It involves primary emission from natural and anthropogenic sources as well as secondary formation via oxidation of precursors in the atmosphere. In the atmosphere, gases and aerosols undergo chemical and physical transformation ("aging"), which includes the interaction with solar radiation and atmospheric oxidants as well as cloud droplets and ice crystals (cloud processing). Besides mineral dust, sea spray, and combustion- and traffic-related particles, atmospheric aerosol also contains biological particles such as pollen, fungal spores, bacteria, and viruses, which are important for the functioning of ecosystems as well as the spread of animal, plant, and human diseases (Fröhlich-Nowoisky et al., 2016; Rodriguez-Caballero et al., 2018). The atmospheric cycling of gases, aerosols, clouds, and precipitation is essential for the evolution, current state, and future development of the atmosphere and climate as well as the biosphere and public health or "planetary health." Multiphase chemical reactions, transport, and transformations between gaseous, liquid, and solid matter are key to understanding the Earth system and climate as well as life and health on molecular and global levels, bridging a wide range of spatial and temporal scales from below nanometers to thousands of kilometers and from less than nanoseconds to years and millennia (Fig. 7.1b). From a chemical perspective, life and the metabolism of most living organisms can be regarded as multiphase processes involving gases such as oxygen and carbon dioxide; liquids such as water, blood, lymph, and plant sap; and solid or semisolid substances such as bone, tissue, skin, wood, and cellular membranes. On global scales, the biogeochemical cycling of chemical compounds and elements, which can be regarded as the metabolism of planet Earth, also involves chemical reactions, mass transport, and phase transitions within and across the atmosphere, biosphere, hydrosphere, and pedosphere/lithosphere (Pöschl & Shiraiwa, 2015).

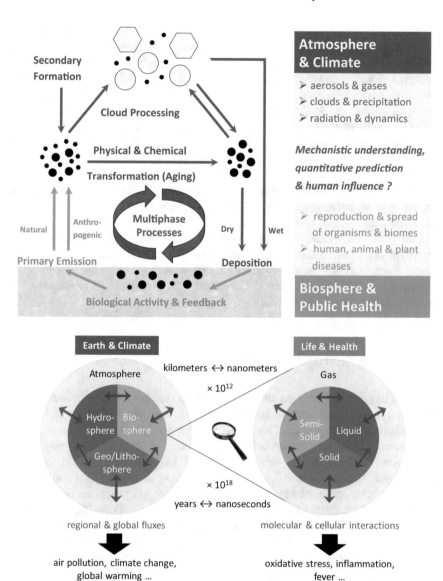

Fig. 7.1 (**a**) Atmospheric cycling of gases and particles, including natural constituents and anthropogenic air pollutants that undergo physical, chemical, and biological interactions. (**b**) Multiphase processes between gaseous, liquid, and solid substances influence the Earth system and climate as well as life and public health from molecular and cellular to regional and global scales (adapted from Pöschl & Shiraiwa, 2015)

Through biogeochemical and metabolic processes, air pollutants such as particulate matter, ozone, and nitrogen oxides can severely affect air quality, climate, and public health. They influence the regional and global distribution of solar radiation, energy, and water; and they enhance mortality, promote cardiovascular, respiratory and allergic diseases, and decrease agricultural crop yields (Pöschl & Shiraiwa, 2015; Shiraiwa, Li, & Tsimpidi, et al., 2017; West et al., 2016; and references therein). Globally, the annual number of premature deaths attributable to ambient air pollution are estimated to exceed 4 million per year, corresponding to over 100 million of disability-adjusted life-years (DALYs) lost every year (Cohen et al., 2017; Landrigan, Fuller, Acosta, et al., 2017; Lelieveld & Pöschl, 2017; Lelieveld, Evans, Fnais, Giannadaki, & Pozzer, 2015; Lelieveld, Haynes, & Pozzer, 2018; Stanaway et al. 2018; and references therein). Recent studies suggest that the mortality rates attributable to air pollution are even higher (up to ~9 million per year; Burnett et al., 2018; Lelieveld et al., 2019). Thus, we need to understand how human activities can create a hazardous atmosphere affecting the health of people and planet Earth, and how humanity can best deal with the sources and effects of air pollution in relation to economic development, human welfare, and environmental preservation in the Anthropocene.

Reactive Oxygen and Nitrogen Species

Reactive oxygen and nitrogen species (ROS/RNS) are central to both atmospheric and physiological chemistry, and their coupling through human airways and epithelia is illustrated in Fig. 7.2a. These chemical species include highly reactive free radicals such as hydroxyl radicals (OH) and nitrate radicals (NO_3) as well as more stable compounds such as hydrogen peroxide (H_2O_2) and nitrous acid (HONO). They play a central role in the adverse health effects of air pollution and are produced in a wide range of atmospheric and physiological processes (Halliwell & Gutteridge, 2015; Pöschl & Shiraiwa, 2015; Shiraiwa, Li, Tsimpidi, et al., 2017; Shiraiwa, Ueda, Pozzer, et al., 2017; Sies et al., 2017; Su et al., 2011; Tong et al., 2017). In the atmosphere, ROS/RNS are generated via photochemical and multiphase reactions involving atmospheric oxidants, aerosols, and clouds. Ozone and related ROS/RNS are among the most noxious components of summer smog, and they are also involved in the secondary formation of air particulate matter by oxidation and condensation of gaseous precursors in the atmosphere (e.g., secondary organic aerosols and sulfate; Cheng et al., 2016; Hallquist et al., 2009). On the other hand, ROS/RNS are crucial for the oxidative self-cleaning of the atmosphere by increasing the water solubility of air pollutants and their removal by precipitation ("washout" or wet deposition). For example, hydroxyl radicals convert nitrogen dioxide into highly soluble nitric acid, which is efficiently absorbed by rain and snow. In fact, hydroxyl radicals are such strong oxidants that they react with and facilitate the removal of most air pollutants and are thus called the "detergent" of the atmosphere (Pöschl & Shiraiwa, 2015; Seinfeld & Pandis, 2016).

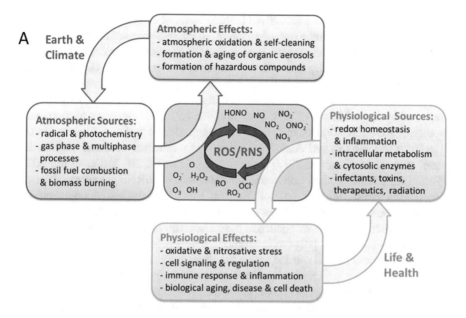

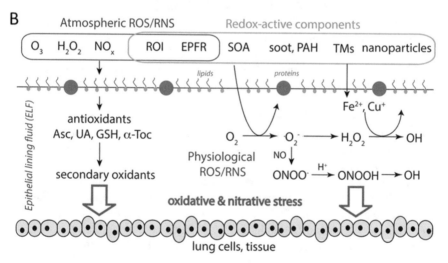

Fig. 7.2 (**a**) Sources, interactions, and effects of reactive oxygen and nitrogen species (ROS/RNS) at the interface of atmospheric and physiological chemistry with feedback loops involving the Earth system, climate, life, and health. (**b**) Interactions of atmospheric and physiological ROS/ RNS with antioxidants (ascorbate, uric acid, reduced glutathione, α-tocopherol) in the epithelial lining fluid (ELF) of the human respiratory tract. Reactive pollutants such as ozone (O_3), hydrogen peroxide (H_2O_2), nitrogen oxides (NO_x), reactive oxygen intermediates (ROI), environmentally persistent free radicals (EPFR), secondary organic aerosol (SOA), soot, quinones, and transition metals can induce ROS/RNS formation in vivo, leading to oxidative stress and biological aging (Pöschl & Shiraiwa, 2015; Reinmuth-Selzle et al., 2017)

Physiological sources of ROS/RNS include metabolic, signaling, and inflammatory processes. In the human immune system, ROS/RNS are used in defense against pathogens by oxidizing relevant biomolecules (DNA, proteins, lipids) and rendering the pathogens biologically less active or inactive. For example, H_2O_2, OH radicals and related ROS are involved in the elimination of pathogenic microbes (Halliwell & Gutteridge, 2015). Under homeostatic conditions, the production of ROS/RNS is balanced by antioxidants. Air pollution, however, can cause excessive production or uptake of ROS/RNS and an imbalance between oxidants and antioxidants (oxidative stress), leading to cell death, biological aging, inflammation, and various diseases (Pöschl & Shiraiwa, 2015; Reinmuth-Selzle et al., 2017; Sies et al., 2017; West et al., 2016). In other words, adequate levels of ROS/RNS are essential for the functioning and self-cleaning of both the atmosphere and the human organism, but excess concentrations of ROS/RNS are detrimental to human health.

Fine particulate matter suspended in the air contains reactive components such as transition metals (TM), polycyclic aromatic hydrocarbons (PAH), semiquinones, nanoparticles, and related substances. In the atmosphere, aerosols undergo transport and transformation, which may lead to a change in their chemical composition and toxicity as well as their optical properties and climate effects depending on temperature, humidity, and particle phase state (Ditas et al., 2018; Mu et al., 2018; Shiraiwa, Ammann, Koop, & Pöschl, 2011; Shiraiwa, Li, Tsimpidi, et al., 2017; Shiraiwa, Ueda, Pozzer, et al., 2017; Socorro et al., 2017; Tong et al., 2017). Upon inhalation and deposition in the human respiratory tract, air pollutants can induce and sustain chemical reactions that produce ROS in the epithelial lining fluid (ELF) covering the airways. As illustrated in Fig. 7.2b, the reactive pollutants and ROS undergo a multitude of chemical reactions in the ELF. A quantitative analysis and assessment of these processes was recently achieved in investigations using a kinetic multilayer model of surface and bulk chemistry in the ELF (KM-SUB-ELF) to obtain chemical exposure–response relations between ambient concentrations of air pollutants and the production rates and concentrations of ROS in the ELF of the human respiratory tract (Lakey et al., 2016).

As illustrated in Fig. 7.3, the total concentration of ROS generated by redox-active substances contained in fine particulate matter (PM2.5) deposited in the ELF can increase by more than a factor of 20 from ~ 10 nmol L^{-1} under clean conditions up to almost ~ 250 nmol L^{-1} under highly polluted conditions. The green-striped horizontal bar indicates ROS concentration levels characteristic of the ELF or bronchoalveolar lavage of healthy humans, respectively, which are around ~ 100 nmol L^{-1}. Compared to this reference level, the ROS concentrations generated by redox-active particulate matter inhaled from pristine marine or rainforest air (PM2.5 < 10 μg m^{-3}) are much lower and appear negligible with regard to airway oxidative stress. In moderately polluted air (PM2.5 $\approx$ 10–50 μg m^{-3}), the particle-generated ROS concentrations can be of similar magnitude or higher than the physiological background level and may thus significantly contribute to oxidative stress depending on aerosol

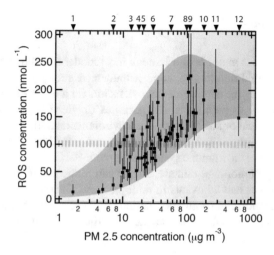

Fig. 7.3 Chemical exposure–response relationships between ambient concentrations of fine particulate matter (PM2.5) and the concentration of reactive oxygen species (ROS) in the epithelial lining fluid (ELF) of the human respiratory tract. The green-striped horizontal bar indicates the ROS level characteristic for healthy humans (~100 nmol L⁻¹). The gray envelope represents the range of aerosol-induced ROS concentrations obtained with approximate upper and lower limit mass fractions of redox-active components observed in ambient PM2.5. The data points represent various geographic locations for which measured or estimated mass fractions are available, including (1) Amazon, Brazil (pristine rainforest air); (2) Edinburgh, UK; (3) Toronto, Canada; (4) Tokyo, Japan; (5) Budapest, Hungary; (6) Hong Kong, China; (7) Milan, Italy; (8) Guangzhou, China; (9) Pune, India; (10) Beijing, China; (11) New Delhi, India; (12) Sumatra, Indonesia (biomass burning/peat fire smoke) (Lakey et al., 2016)

concentration and chemical composition (see error bars and grey-shaded area). In heavily polluted air (PM2.5 > 50 µg m⁻³), the particle-generated ROS concentrations are as high as the ROS concentrations observed in the bronchoalveolar lavage of patients with acute inflammatory diseases in respiratory tract (100–250 nmol L⁻¹). The pathologically high ROS levels calculated for ELF in airways exposed to high ambient aerosol concentrations are consistent with epidemiology-based air quality standards and regulations of the World Health Organization (WHO), aiming at PM2.5 concentrations less than 10 µg m⁻³ averaged over a year (Lakey et al., 2016; WHO, 2013; 2017). Further investigations building on the scientific approach and exposure–response relations outlined above will help to identify key species and processes to be targeted in efficient strategies and policies for air quality control and improvement. For example, they may help to resolve the relative importance and nonlinear interactions of different types and sources of air pollutants such as combustion engine exhaust, brake-wear, and dust emitted or suspended by road or railway traffic.

Allergic Diseases

Allergies constitute a major health issue in most modern societies, and related diseases such as allergic rhinitis and atopic asthma have strongly increased during the past decades. Among the environmental risk factors for allergic diseases are reduced childhood exposure to pathogens and parasites (hygiene hypothesis), nutritional factors, psychological or social stress, and environmental pollutants (diesel exhaust particles, ozone, nitrogen oxides, etc.), which can effectively lead to an over- or under-stimulation of the immune system (Reinmuth-Selzle et al., 2017). As outlined in Fig. 7.4, air pollution and climate change can influence the bioavailability and potency of allergens and adjuvants in multiple ways, including changes in vegetation cover, pollination and sporulation periods, and chemical modification or aggregation. Moreover, air pollutants can act as adjuvants and skew physiological

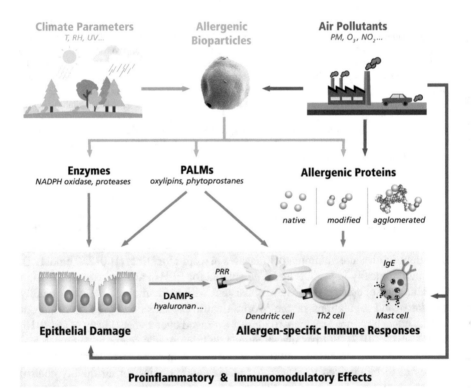

Fig. 7.4 Pathways through which climate parameters and air pollutants can influence the release, potency, and effects of allergens and adjuvants: temperature (T), relative humidity (RH), ultraviolet (UV) radiation, particulate matter (PM), ozone and nitrogen oxides (O_3, NOx), reduced nicotinamide adenine dinucleotide phosphate (NADPH) oxidase, pollen-associated lipid mediators (PALMs), damage-associated molecular patterns (DAMPs), pattern recognition receptors (PRR), type 2 T helper (Th2) cells, immunoglobulin E (IgE), allergenic proteins (green dots), and chemical modifications (red dots) (Reinmuth-Selzle et al., 2017)

processes and the immune system towards the development of allergies, for example, by oxidative stress, inflammation, and disruption of protective epithelial barriers. Environmental allergens are mostly proteins derived from plants, animals, and fungi that can trigger chemical and biological reaction cascades in the immune system leading to allergic sensitization and formation of IgE antibodies. Prominent examples are major allergens of birch pollen (Bet v 1), timothy grass pollen (Phl p 1), ragweed (*Ambrosia*, Amb a 1), molds (*Alternaria alternata*, Alt a 1; *Cladosporium herbarum*, Cla h 1; *Aspergillus fumigatus*, Asp f 1), and dust mites (Der p 1).

Figure 7.5 provides a simplified overview of cellular and molecular interactions that are central to the processes of sensitization and response in IgE-mediated allergies (type I hypersensitivities). Normally, IgE antibodies and related immune reactions are involved in the defense against parasitic infections, and allergic reactions can thus be regarded as a "false alarm" of the immune system. Interactions between protein macromolecules acting as allergens, antibodies, cytokines, or receptors play a key role in the innate and adaptive immune responses involved in the development of allergies. Chemical modification by air pollutants can lead to changes in the molecular structure and interactions of protein macromolecules that trigger such false alarms (Reinmuth-Selzle et al., 2017 and references therein).

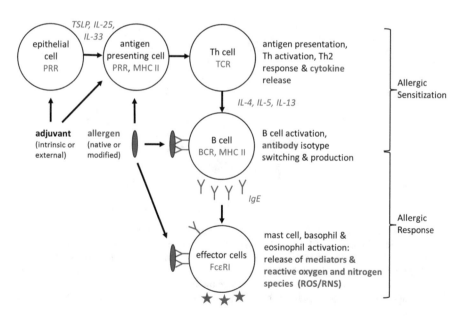

Fig. 7.5 Cellular and molecular interactions involved in allergic sensitization and response of IgE-mediated allergies (type I hypersensitivities): blue color indicates protein macromolecules acting as allergens, antibodies, cytokines, or receptors; red color indicates pro-inflammatory mediators and reactive oxygen or nitrogen species (ROS/RNS) (Reinmuth-Selzle et al., 2017)

Rising concentrations of atmospheric aerosols and oxidants are likely to promote the chemical modification of proteins in the atmosphere as well as in the human body due to elevated ROS/RNS levels and oxidative stress, especially in the airways (ELF) as outlined above. For example, elevated O_3 and NO_2 concentrations can enhance the allergenic potential of proteins through nitration and cross-linking of allergenic proteins through the aromatic amino acid tyrosine as illustrated in Fig. 7.6 (Franze, Weller, Niessner, & Pöschl, 2005; Liu et al., 2017; Reinmuth-Selzle, Ackaert, & Kampf, 2014; Reinmuth-Selzle et al., 2017). Ongoing experimental and theoretical studies provide new insights into the mechanism of these and related processes, which will help to elucidate and counteract the influence of air pollutants on the development and aggravation of allergic diseases.

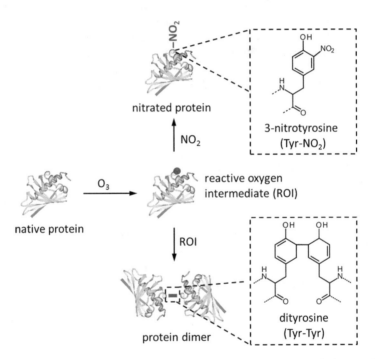

Fig. 7.6 Posttranslational modification of proteins exposed to ozone (O_3) and nitrogen dioxide (NO_2). The initial reaction with O_3 leads to the formation of reactive oxygen intermediates (ROI, tyrosyl radicals), which can further react with each other to form cross-linked proteins (dityrosine) or with NO_2 to form nitrated proteins (nitrotyrosine). The shown protein is Bet v 1.0101, for which nitration and cross-linking were found to influence the immunogenicity and allergenic potential. Red dot indicates a tyrosyl radical; red bar indicates dityrosine cross-link (Reinmuth-Selzle et al., 2017)

Conclusions

In air quality, climate, and health studies, it is often challenging to separate natural and anthropogenic effects and variations. Unperturbed natural conditions are difficult to find or reconstruct from current and recent observations of the environment, which has already undergone massive anthropogenic change on global scales. All the more, it is important to fully unravel and quantify the processes and interactions influencing the current state as well as the history and future development of the Earth system, climate, air quality, and public health. This requires interdisciplinary exchange and collaboration across the Earth, environmental, and life sciences. Recent advances in field observations, laboratory experiments, and mathematical model simulations provide deep insights into the underlying physical, chemical, and biological processes. They enable the development and refinement of target-oriented ways and efficient approaches of effectively mitigating and counteracting the adverse effects of air pollutants. Building on open access to scholarly publications and data, a global commons of scholarly knowledge in the sciences and humanities will help to augment, communicate, and utilize the scientific understanding (Berlin Declaration, 2003; oa2020.org). Moreover, public peer review, interactive discussion, and documentation of the scientific discourse on the internet can serve as examples and blueprints for rational and transparent approaches to resolving complex questions and issues (Pöschl, 2012). Combining the internet with open science will ultimately reflect both what we know and how we know it ("epistemic web"; Hyman & Renn, 2012). In any case, scholars and practitioners in the natural, social, and medical sciences as well as humanities and politics should closely collaborate to transfer the scientific knowledge of air pollutant sources and effects into effective strategies and policies of air quality management and control. This should be pursued in analogy to the successful example of protecting the ozone layer from destruction by chlorofluorocarbons in the late twentieth century rather than the gridlocked struggle for climate protection in the early twenty-first century. In this context, it may be worthwhile to emphasize not only the need to mitigate negative side effects of human activities but also the opportunities for actively using scientific knowledge and technology to protect planet Earth for a sustainable development and healthy future of humanity and its planetary home ("planetary health"). Thus, the Anthropocene notion may help humanity to recognize both rationally and emotionally: We are shaping our planet and environment, so let us get it right.

Acknowledgements This chapter provides an overview and synthesis of recent scientific exchange, studies, and publications of the author with numerous colleagues who are gratefully acknowledged—in particular, P. J. Crutzen, Y. Cheng, H. Su, M. Shiraiwa, P. Lakey, K. Reinmuth-Selzle, C. Kampf, K. Lucas, J. Fröhlich, T. Berkemeier, H. Tong, C. Pöhlker, M. Pöhlker, M. Weller, A. Duschl, I. Bellinghausen, J. Saloga, D. Schuppan, M. O. Andreae, and J. Lelieveld. The chapter is dedicated to Paul J. Crutzen, whose curiosity, ingenuity, and wisdom have helped to prevent ozone layer destruction and nuclear winter, and will help to avoid the perils and grasp the opportunities of the Anthropocene.

References

Berlin Declaration on Open Access to Knowledge in the Sciences and Humanities. (2003). Retrieved on February 16, 2020 from https://openaccess.mpg.de/Berlin-Declaration

Burnett, R., Chena, H., Szyszkowicz, M., Fann, N., Hubbell, B., Pope, I. I. I. C. A., et al. (2018). Global estimates of mortality associated with long-term exposure to outdoor fine particulate matter. *Proceedings of the National Academy of Sciences of the United States of America, 115*, 9592–9597.

Cheng, Y., Zheng, G., Wei, C., Mu, Q., Zheng, B., Wang, Z., et al. (2016). Reactive nitrogen chemistry in aerosol water as a source of sulfate during haze events in China. *Science Advances, 2*. https://doi.org/10.1126/sciadv.1601530

Cohen, A. J., Brauer, M., Burnett, R., Anderson, H. R., Frostad, J., Estep, K., et al. (2017). Estimates and 25-year trends of the global burden of disease attributable to ambient air pollution: An analysis of data from the Global Burden of Diseases Study 2015. *Lancet, 389*, 1907–1918. https://doi.org/10.1016/S0140-6736(17)30505-6

Crutzen, P. J. (2002). Geology of mankind. *Nature, 415*, 23–23.

Ditas, J., Ma, N., Zhang, Y., Assmann, D., Neumeier, M., Riede, H., et al. (2018). Strong impact of wildfires on the abundance and aging of black carbon in the lowermost stratosphere. *Proceedings of the National Academy of Sciences of the United States of America, 115*, E11595–E11603. https://doi.org/10.1073/pnas.1806868115

Foley, S. F., Gronenborn, D., Andreae, M. O., et al. (2013). The Palaeoanthropocene—The beginnings of anthropogenic environmental change. *Anthropocene, 3*, 83–88.

Franze, T., Weller, M. G., Niessner, R., & Pöschl, U. (2005). Protein nitration by polluted air. *Environmental Science and Technology, 39*, 1673–1678. https://doi.org/10.1021/es0488737

Fröhlich-Nowoisky, J., Kampf, C. J., Weber, B., Huffman, J. A., Pöhlker, C., Andreae, M. O., et al. (2016). Bioaerosols in the earth system: Climate, health, and ecosystem interactions. *Atmospheric Research, 182*, 346–376.

Halliwell, B., & Gutteridge, J. M. C. (2015). *Free radicals in biology and medicine*. Oxford: Oxford University Press.

Hallquist, M., Wenger, J. C., Baltensperger, U., Rudich, Y., Simpson, D., Claeys, M., et al. (2009). The formation, properties and impact of secondary organic aerosol: Current and emerging issues. *Atmospheric Chemistry and Physics, 9*, 5155–5236. https://doi.org/10.5194/acp-9-5155-2009

Hyman, M. D., & Renn, J. (2012). Toward an epistemic web. In J. Renn (Ed.), *The globalization of knowledge in history*. Edition Open Access, ISBN: 978-3-945561-23-2.

Lakey, P. S. J., Berkemeier, T., Tong, H., Arangio, A. M., Lucas, K., Pöschl, U., et al. (2016). Chemical exposure-response relationship between air pollutants and reactive oxygen species in the human respiratory tract. *Scientific Reports, 6*, 32916. https://doi.org/10.1038/srep32916.309

Landrigan, P. J., Fuller, R., Acosta, N. J. R., Adeyi, O., Arnold, R., Basu, N., et al. (2017). The Lancet Commission on pollution and health. *Lancet, 391*, 462–512. https://doi.org/10.1016/S0140-6736(17)32345-0

Lelieveld, J., Evans, J. S., Fnais, M., Giannadaki, D., & Pozzer, A. (2015). The contribution of outdoor air pollution sources to premature mortality on a global scale. *Nature, 525*, 367–371. https://doi.org/10.1038/nature15371

Lelieveld, J., Haynes, A., & Pozzer, A. (2018). Age-dependent health risk from ambient air pollution: A modelling and data analysis of childhood mortality in middle-income and low-income countries. *Lancet Planet Health, 2*, E292–E300. https://doi.org/10.1016/S2542-5196(18)30147-5

Lelieveld, J., Klingmüller, K., Pozzer, A., Pöschl, U., Fnais, M., Daiber, A., et al. (2019). Cardiovascular disease burden from ambient air pollution in Europe reassessed using novel hazard ratio functions. *European Heart Journal, 40*, 1590–1596. https://doi.org/10.1093/eurheartj/ehz135

Lelieveld, J., & Pöschl, U. (2017). Chemists can help to solve the air-pollution health crisis. *Nature, 551*, 291–293.

Liu, F., Lakey, P. S. J., Berkemeier, T., Tong, H. J., Kunert, A. T., Meusel, H., et al. (2017). Atmospheric protein chemistry influenced by anthropogenic air pollutants: Nitration and oligomerization upon exposure to ozone and nitrogen dioxide. *Faraday Discussions, 200*, 413–427. https://doi.org/10.1039/c7fd00005g

Mu, Q., Shiraiwa, M., Octaviani, M., Ma, N., Ding, A. J., Su, H., et al. (2018). Temperature effect on phase state and reactivity controls atmospheric multiphase chemistry and transport of PAHs. *Sciences Advances, 4*, eaap7314. https://doi.org/10.1126/sciadv.aap7314

Pöschl, U. (2005). Atmospheric aerosols: Composition, transformation, climate and health effects. *Angewandte Chemie International Edition, 44*, 7520–7540.

Pöschl, U. (2012). Multi-stage open peer review: Scientific evaluation integrating the strengths of traditional peer review with the virtues of transparency and self-regulation. *Frontiers in Computational Neuroscience.* https://doi.org/10.3389/fncom.2012.00033

Pöschl, U., & Shiraiwa, M. (2015). Multiphase chemistry at the atmosphere–biosphere interface influencing climate and public health in the Anthropocene. *Chemical Reviews, 115*, 4440–4475. https://doi.org/10.1021/cr500487s

Reinmuth-Selzle, K., Ackaert, C., Kampf, C. J., Lakey, P. S. J., Lai, S., Liu, F., et al. (2014). Nitration of the birch pollen allergen Bet v 1.0101: Efficiency and site-selectivity of liquid and gaseous nitrating agents. *Journal of Proteome Research, 13*, 1570–1577. https://doi.org/10.1021/pr401078h

Reinmuth-Selzle, K., Kampf, C. J., Lucas, K., Lang-Yona, N., Fröhlich-Nowoisky, J., Shiraiwa, M., et al. (2017). Air pollution and climate change effects on allergies in the anthropocene: Abundance, interaction, and modification of allergens and adjuvants. *Environmental Science and Technology, 51*, 4119–4141. https://doi.org/10.1021/acs.est.6b04908

Rodriguez-Caballero, E., Belnap, J., Budel, B., Crutzen, P. J., Andreae, M. O., Pöschl, U., et al. (2018). Dryland photoautotrophic soil surface communities endangered by global change. *Nature Geoscience, 11*, 185+. https://doi.org/10.1038/s41561-018-0072-1

Seinfeld, J. N., & Pandis, S. N. (2016). *Atmospheric chemistry and physics: From air pollution to climate change*. New York: Wiley.

Shiraiwa, M., Ammann, M., Koop, T., & Pöschl, U. (2011). Gas uptake and chemical aging of semi-solid organic aerosol particles. *Proceedings of the National Academy of Sciences of the United States of America, 108*, 11003–11008.

Shiraiwa, M., Li, Y., Tsimpidi, A. P., Berkemeier, T., Pandis, S. N., Lelieveld, J., et al. (2017). Global distribution of particle phase state in atmospheric secondary organic aerosols. *Nature Communication, 8*, 15002. https://doi.org/10.1038/ncomms15002

Shiraiwa, M., Ueda, K., Pozzer, A., Lammel, G., Kampf, C. J., Fushimi, A., et al. (2017). Aerosol health effects from molecular to global scales. *Environmental Science & Technology, 51*, 13545–13567. https://doi.org/10.1021/acs.est.7b04417

Sies, H., et al. (2017). Oxidative stress. *Annual Review of Biochemistry, 86*, 715–748.

Socorro, J., Lakey, P. S. J., Han, L., Berkemeier, T., Lammel, G., Zetzsch, C., et al. (2017). Heterogeneous OH oxidation, shielding effects and implications for the atmospheric fate of terbuthylazine and other pesticides. *Environmental Science & Technology, 51*, 13749–13754. https://doi.org/10.1021/acs.est.7b04307

Stanaway, J. D., et al. (2018). Global, regional, and national comparative risk assessment of 84 behavioural, environmental and occupational, and metabolic risks or clusters of risks for 195 countries and territories, 1990–2017: A systematic analysis for the Global Burden of Disease Study 2017. *Lancet, 392*, 1923–1994. https://doi.org/10.1016/S0140-6736(18)32225-6

Su, H., Cheng, Y., Oswald, R., Behrendt, T., Trebs, I., Meixner, F. X., et al. (2011). Soil nitrite as a source of atmospheric HONO and OH radicals. *Science, 333*, 1616–1618.

Tong, H. J., Lakey, P. S. J., Arangio, A. M., Socorro, J., Kampf, C. J., Berkemeier, T., et al. (2017). Reactive oxygen species formed in aqueous mixtures of secondary organic aerosols and mineral dust influencing cloud chemistry and public health in the Anthropocene. *Faraday Discussions, 200*, 251–270. https://doi.org/10.1039/c7fd00023e

Waters, C. N., Zalasiewicz, J., Summerhayes, C., Barnosky, A. D., Poirier, C., Galuszka, A., et al. (2016). The Anthropocene is functionally and stratigraphically distinct from the Holocene. *Science, 351*, 6269. https://doi.org/10.1126/science.aad2622

West, J. J., Cohen, A., Dentener, F., Brunekreef, B., Zhu, T., Armstrong, B., et al. (2016). What we breathe impacts our health: Improving understanding of the link between air pollution and health. *Environmental Science and Technology, 50*, 4895–4904.

World Health Organization. (2013). *Review of evidence on health aspects of air pollution—REVI-HAAP project*. Copenhagen, Denmark: WHO Regional Office for Europe.

World Health Organization. (2017). *Evolution of WHO air quality guidelines: Past, present and future*. Copenhagen, Denmark: WHO Regional Office for Europe.

CHAPTER 8
Climate Change, Air Pollution, and the Environment: The Health Argument

Maria Neira and Veerabhadran Ramanathan

Summary There are no aspects of climate and environmental change that are more critical than those that affect health and well-being, and none are more urgent than those that affect the most vulnerable. Air pollution and climate change fit both these categories, and now rank among our greatest contemporary threats to human health.

This is what we know: 92% of the global population breathes air pollution levels that are unsafe. More than seven million lives are lost to indoor and ambient air pollution every year. The major sources of air pollution are the combustion of fossil fuels, the burning of biomass, and agriculture. Global health and welfare losses from air pollution in 2013 were valued at about US $5110 billion, or almost 7% of gross domestic product. However, there is no policy realm in the world that regularly takes into consideration the potential costs and benefits to public health of all decisions, even those that directly produce health-influencing externalities.

Health should be central to discussions around drivers of environmental degradation, such as production methods that pollute, deleterious consumption, distribution patterns and disruption of ecosystems. Moreover, the attainment of health should be promoted as an explicit aim, rather than an afterthought, of decisions in key sectors such as energy, transport, technology, water and sanitation, and urban planning. The health sector must show leadership and work with other sectors to assume its obligations in shaping a healthy and sustainable future. The effects of human actions on the environment are an ethical and human rights issue; they will be felt by future generations and have the most severe impact on the most economically, demographically, and geographically vulnerable populations.

M. Neira (✉)
The World Health Organization, Geneva, Switzerland
e-mail: neiram@who.int

V. Ramanathan
Scripps Institution of Oceanography, University of California at San Diego, La Jolla, CA, USA

© The Author(s) 2020
W. K. Al-Delaimy, V. Ramanathan, M. Sánchez Sorondo (eds.), *Health of People, Health of Planet and Our Responsibility*, https://doi.org/10.1007/978-3-030-31125-4_8

At this point in history, no decision-maker can claim to be ignorant of the adverse health consequences of environmental degradation. Together, we have a major responsibility to drive the transformation needed to make radical changes in policies and behaviors—while there is still time.

Air Pollution: A Major Killer

Known, avoidable environmental risk factors cause at least 13 million deaths every year, which amounts to one quarter of the global burden of disease (Fig. 8.1) Air pollution alone causes about seven million premature deaths a year, or one in eight

Fig. 8.1 (**a**) The big picture: 23% of global deaths are attributable to the environment (*Source*: WHO Infographics, Preventing disease through health environments; http://www.who.int/quantifying_ehimpacts/publications/PHE-prevention-diseases-infographic-EN.pdf?ua=1)

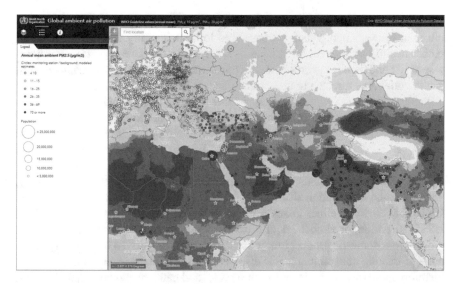

Fig. 8.2 (**a**) Air pollution: 92% of the global population breathes air pollution levels that are not safe. *Source*: Global interactive map of ambient air pollution (http://maps.who.int/airpollution/)

of all deaths, placing it among the top global risks to health. The poorest three billion people are exposed to indoor air pollution due to cooking and heating homes with solid fuels (coal, firewood, and dung and crop residues). Close to four million people die prematurely from illnesses attributable to household air pollution from inefficient cooking practices using polluting stoves paired with these fuels and kerosene. This is an unbearable cost in lives, health, and human development, particularly for the poor (Fig. 8.2).

There is more evidence than ever before about the direct and indirect paths through which a stable and protected environment maintains human health—and, conversely, how failure to manage climate and environmental risk factors are contributing to both communicable and noncommunicable diseases among all populations, from the poorest to the richest. We also have a wealth of evidence on effective interventions to address many of the root drivers of the environmental determinants of health, including action at the local level.

Countries are facing a combination of longstanding, unresolved environmental and health challenges, as well as new ones. These include a lack of universal access to clean household energy; limited access to safe water and sanitation; and consequences from unsustainable development, such as air, water and soil pollution; exposure to hazardous chemicals; more complex, chronic and combined exposure in working and residential settings; ageing infrastructure; stagnating environmental health progress; and increasing inequalities, in all countries.

These challenges result in a triple burden of environmental risks, which include the direct impacts of emergencies, persistent and—in some cases—expanding infectious disease risks, as well as noncommunicable diseases. For noncommunicable diseases, environmental risk factors are now of a comparable magnitude to other well-established risks (such as tobacco consumption, diet, alcohol consumption, and physical inactivity).

Human influences on the global environment continue to grow, and they contribute to climate change, which is considered to be potentially the greatest threat to global health in the twenty-first century. Many countries already suffer significant loss of life and damage to crucial health infrastructure from extreme weather events. These conditions threaten to undermine gains in health and development, and may exacerbate migration and increase social and political tensions within and between countries. In the absence of strong measures to cut carbon emissions and protect populations from the effects of climate change, rising sea levels will submerge extensive and densely populated coastal areas, including some entire small island nations, by the end of this century.

Although the paramount concern is to protect human lives and ensure well-being, environmental degradation also exacts large economic costs from the health sector. Global health and welfare losses from air pollution in 2013 were valued at about US$5110 billion, or almost 7% of gross domestic product. Approximately 10% of global gross domestic product is now spent on health care, driven increasingly by the costs of treating noncommunicable diseases. Failure to manage environmental risks therefore increases the strain on health services and national and household budgets.

The Public Health Response to Climate and Environmental Change

The responsibility and tools to tackle many environmental determinants of health lie beyond those of individuals or the health sector alone. Therefore, we need a wider societal, intersectoral, and population-based public health approach.

A wealth of evidence demonstrates the health impacts of individual exposure to hazards in the environment, for example, to specific chemical or biological contaminants in water. At the same time, there is also strong evidence for the cost-effectiveness of many interventions, from small-scale (e.g., point-of-use water treatment) to large-scale investments (e.g., in sanitation infrastructure).

There have been notable successes in applying such evidence to intersectoral policy. Examples range from removing lead from petrol in many countries to controlling the depletion of the ozone layer and the associated health risks of ultraviolet radiation through the application of measures set out in the Montreal Protocol on Substances that Deplete the Ozone Layer (1987).

In dealing with environmental, climate and other determinants of health, the World Health Organization (WHO) promotes a Health in All Policies approach, including coverage of health in environmental and labor regulations and safeguards, assessment of the health impact of development projects, and tackling several environmental health issues in a single setting, community, or system.

Numerous examples of good practice are available, but such integrated approaches are quite rare and are seldom directed to "upstream" environmental and social determinants, such as more sustainable and equitable resource consumption, climate stabilization, and protection of biodiversity and ecosystem services. This patchy application of integrated approaches drives exposure to hazardous health conditions.

Additionally, many environmental risks to health are interdependent and trans-boundary in nature. These include the transfer of polluting industries, dangerous work processes, and hazardous waste to poorer and less regulated countries; trans-boundary air pollution and radiation risks; and the burning of fossil fuels that drives global climate change. This transfer and extension of risk is occurring against a background of decreased direct investment and a trend of environmental deregulation for some national governments, and an increasing influence of diverse, often politically and economically powerful, multinational, private-sector actors.

At the same time, there is continuous demand for more commonly recognized "downstream" environmental health interventions to deal with the direct and local effects of environmental risks to health—most obviously in the case of responding to emergencies, which can in turn result in or be caused by environmental degradation.

The persistent burden of environmental disease and the evolving range of risks clearly call for a strengthening of primary prevention. A massively scaled-up primary prevention strategy will enormously benefit the health sector and make significant, vital contributions to sustainable development. In fact, it would be a challenge to achieve the 2030 Sustainable Development Agenda without rebalancing the emphases on prevention and healthcare.

Currently, the average engagement of the health sector decreases rapidly. The 97% of global health resources spent on treatment are applied only to the health *effects* at the end of the causal chain. Most of the remaining 3% spent on prevention is applied to addressing specific *exposures* at the individual level.

In order to address the root causes of disease, the health sector needs to engage more effectively to influence decisions that have impacts on the upstream determinants of health. Rather than attempting only to deal with the fallout of having not taken environmental health risks into consideration after the fact, the health sector should aim to reclaim its role at the heart of environment and development policies. We need to ensure that interventions that drive climate and environmental change take account of health considerations.

Bolstering this approach is a growing recognition of opportunities for "co-benefits" of measures that simultaneously protect health and the environment. For example, it is estimated that doubling the share of renewable energy by 2030 would not only reduce air pollution-related disease but also create 24 million jobs and bring a global GDP increase of 1.1%. In another example, improving water and sanitation services benefits public health, increases labor productivity, and reaps an estimated return of between US$5 and 28 per dollar invested. Well-designed policies can maximize synergies and co-benefits: tackling local air pollution to cut greenhouse gas emissions will reduce the economic burden of health problems.

We need to promote more efficient, equitable, and sustainable delivery of Universal Health Coverage, with the reduction of environmental impact of healthcare as a co-benefit, and as an example of the leadership of the health sector.

The health sector should take the opportunity to "lead by example" by promoting environmental sustainability within its own operations, ensuring environmentally sustainable healthcare facilities.

Healthy Opportunities

If implemented and built on, the Paris Agreement on climate change is likely to be remembered as one of the most important public health treaties in history. The public health community should now join forces with the environment community to take the lead in ensuring the Agreement is implemented and increasingly ambitious, and open the door to a green health revolution.

The Paris Agreement should also inspire new and improved agreements to address other threats to global health and well-being, including threats from air and water pollution, waste mismanagement, soil degradation, and diminished biodiversity. In the absence of such agreements, environmental threats to health and well-being are unlikely to be contained.

The 2030 Agenda for Sustainable Development and its associated Sustainable Development Goals provide the integrated framework for the ambitious changes needed to fulfill the commitments made. The Goals and their targets provide the structure to identify and implement actions to safeguard and enhance the upstream determinants of health, and to follow a sustainable pathway to improved, and more equitably distributed, health and well-being.

Health is relevant to all the Goals, not just Goal 3 (Ensure healthy lives and promote well-being for all at all ages). Within the scope of environmental and climate change, there are specific and important opportunities for health gains through shaping the agenda of nutrition (Goal 2, End hunger, achieve food security and improved nutrition and promote sustainable agriculture); water and sanitation (Goal 6, Ensure availability and sustainable management of water and sanitation for all); clean energy (Goal 7, Ensure access to affordable, reliable, sustainable and modern energy for all); decent work (Goal 8, Promote sustained, inclusive and sustainable economic growth, full and productive employment, and decent work for all); sustainable cities (Goal 11, Make cities and human settlements inclusive, safe, resilient, and sustainable); responsible production and consumption (Goal 12, Ensure sustainable consumption and production patterns); and climate change (Goal 13, Take urgent action to combat climate change and its impacts).

The Goals therefore provide an opportunity for the health sector to engage in broad, inclusive and massively expanded primary prevention, effectively bringing together Principle 1 of the Rio Declaration on Environment and Development (1992) ("Human beings are at the centre of concerns for sustainable development. They are entitled to a healthy and productive life in harmony with nature") and Article 1 of the Declaration of Alma-Ata (1978) ("… the attainment of the highest possible level of health … requires the action of many other social and economic sectors in addition to the health sector").

Increasingly, much of this action and many of these decisions will occur in urban settings. Over 50% of the global population now lives in cities, and that number is growing. More than 3.5 billion people living in cities today—half of humanity—suffer from inadequate housing and transport, poor sanitation and waste management, and air quality failing WHO's guidelines. World energy demand is expected to increase by 80% by 2050 (compared to 2010 levels), most of which would be met with fossil fuels if no new energy policies are put in place.

Cities embody the most acute health and environment challenges, from air pollution to sanitation, heat-island effect, chemical contamination, and solid waste management—but the concentration of people, wealth, connectivity, and local leadership through city mayors also provide the greatest opportunities for integrated health, climate and environment programs.

As most future urban growth will take place in cities, urban expansion needs to be planned to make cities a centre of health and well-being. Promoting sustainable transport, access to clean water and energy, waste management, sustainable food production, and healthy urban planning will critically improve health, as well as contribute to the Sustainable Development Goals for Health (3), Energy (7) Water (6) Cities (11) and Climate Change (13).

We need a new framework of health policy leadership positioned to assess and advocate for development that leads to healthier, greener and cleaner cities, and we need a new generation of mayors to lead this healthy development.

Concluding Remarks: The Transformation Needed

In the twenty-first century, health needs to be central to discussions around drivers of environmental degradation, such as production methods that pollute, deleterious consumption, and distribution patterns and disruption of ecosystems. Moreover, attainment of health should be promoted as an explicit aim, rather than an afterthought, of decisions in key sectors such as energy, transport, technology, water and sanitation, and urban planning.

Investment in the capacity of the health sector to engage in policy and to assess and monitor investments that are made in other areas of the economy would support promotion of mutually beneficial measures that simultaneously protect health and the environment. This approach would, in turn, avoid current and future economic costs, allowing for reinvestment in health and sustainable development.

For example, more sustainable urban transport systems that promote public transport, cycling and walking would reduce air pollution, noise and risk of road traffic injuries, and enhance physical activity levels. More generally, it is estimated that placing a price on polluting fuels in line with their health impacts through air pollution would result in a reduction of more than 50% of premature global air pollution deaths, a 20% reduction in greenhouse gas emissions, and the generation of some US$3000 billion in tax revenues every year—more than 50% of current global health spending by governments.

The health sector has the specific responsibility to inform policy-makers and the public of the health impacts of climate and environmental change, because of the importance that populations give to health issues and the generally high quality of, and public trust in, health evidence.

There is a continuous need for evidence on the effectiveness of measures to tackle the environmental root causes of disease burden and on the health impacts of sectoral policies. As such decisions often have wide-ranging effects, there is an associated need for capacity to assimilate, interpret and communicate data and evidence from sources not traditionally used by health policy-makers—evidence that indicates that health-harmful (e.g., fossil) fuels should not be subsidized, for example. Evidence of the impacts on human rights and equity, public acceptance of measures, and information on the socioeconomic and financial costs to individuals and health systems are particularly important.

Implementation must occur not only through influencing other sectors but also within the core functions of the health sector. For example, climate change should be incorporated into risk assessments and preparedness and response plans for health emergencies; climate resilience should be integrated into the building blocks of health systems; and investments should be supported in the provision of energy, water and sanitation for health facilities, as a crucial contribution to universal health coverage.

However, for this approach to work, there needs to be a recognition of its merits that drive institutional change. Major policy and implementation decisions in all sectors are taken at the national and subnational levels. Health actors need to be informed by evidence, connected through institutional mechanisms that allow them to work with other stakeholders (for example, from urban planners to city mayors) and empowered through regulatory frameworks that include health in intersectoral policy-making. Effective regulations and standards should be used to promote energy efficiency and safeguard human health, among other things.

Opportunities for action on health, climate and the environment are rapidly opening up.

Health should be central to the framing of discussions around drivers of environmental degradation and promoted as an explicit aim of decisions in key sectors such as energy, transport, water, sanitation, waste, and urban planning (Figs. 8.3 and 8.4).

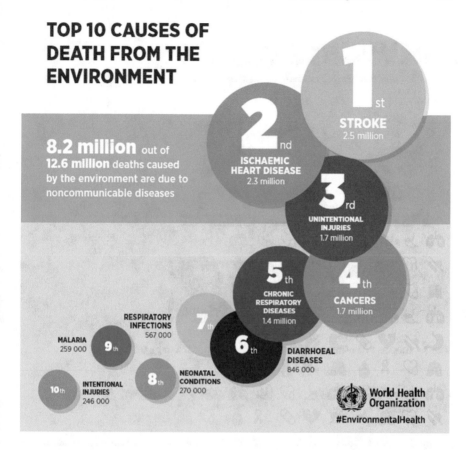

Fig. 8.3 Top ten causes of death from the environment (*Source*: WHO Infographics, Preventing disease through health environments; http://www.who.int/quantifying_ehimpacts/publications/PHE-prevention-diseases-infographic-EN.pdf?ua=1)

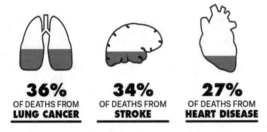

Fig. 8.4 Breathe Life campaign (*Source*: http://breathelife2030.org)

References

Coady, D., Parry, I., Sears, L., & Shang, B. P. (2015). *How large are global energy subsidies?* IMF working paper WP/15/105. International Monetary Fund. Retrieved October 30, 2017, from https://www.imf.org/external/pubs/ft/wp/2015/wp15105.pdf

Government of South Australia; WHO (2017). *Progressing the sustainable development goals through health in all policies: Case studies from around the world.* Adelaide, Australia: Government of South Australia. Retrieved October 30, 2017, from http://www.who.int/social_determinants/publications/hiap-case-studies/en/

OECD (2006). *OECD Health Data 2006, Statistics and Indicators for 30 Countries.* Paris: CD-ROM.

Prüss-Ustün, A., Wolf, J., Corvalán, C., Bos, R., & Neira, M. (2016). *Preventing disease through healthy environments: a global assessment of the burden of disease from environmental risks.* Geneva: World Health Organization. Retrieved October 27, 2017, from www.who.int/quantifying_ehimpacts/publications/preventing-disease/en/. Estimates are based on a combination of quantitative assessment and expert opinion.

WHO (2016). *Global health expenditure database.* Retrieved November 8, 2016., from http://apps.who.int/nha/database

WHO (2018). COP24 Special Report: Health and Climate Change. https://www.who.int/globalchange/publications/COP24-report-health-climate-change/en/

World Bank Group, Institute for Health Metrics and Evaluation (2016). *The cost of air pollution: Strengthening the economic case for action.* Washington, DC: World Bank. Retrieved October 30, 2017, from https://openknowledge.worldbank.org/handle/10986/25013

CHAPTER 9
Reducing Air Pollution: Avoidable Health Burden

Jos Lelieveld

Summary Anthropogenic emissions have transformed atmospheric composition to the extent that biogeochemical cycles, air quality, and climate have changed globally and profoundly. It is estimated that ambient (outdoor) air pollution causes an excess mortality rate of about 4.5 million per year, associated with 122 million years of life lost annually, mostly due to the detrimental health effects of fine particulate matter ($PM_{2.5}$). Globally, the largest source of this pollution is residential energy use for heating and cooking, notably because of its importance in Asia. Agriculture, power production, and road traffic also contribute significantly. If residential energy emissions could be eliminated, up to 1.4 million deaths per year would be avoided. Agriculture, through the release of ammonia, contributes strongly to $PM_{2.5}$. A study of health benefits achieved through European legislation since 1970 indicates that emission controls in the transport and energy sectors have prevented approximately 61,000 deaths per year in Europe and 163,000 per year worldwide, the latter through new technology that penetrated global markets. However, much stronger measures are needed to substantially lower the health burden from air pollution. Clean air is a human right, being fundamental to many sustainable development goals of the United Nations, such as good health, climate action, sustainable cities, clean energy, and protecting life on land and in the sea.

Introduction

Although air pollution episodes have been documented since antiquity by the Greeks and Romans, public health concerns with smog emerged with urbanization and industrialization, such as in London after World War II. The burning of wood and

J. Lelieveld (✉)
Max Planck Institute for Chemistry, Mainz, Germany

Cyprus Institute, Nicosia, Cyprus
e-mail: jos.lelieveld@mpic.de

© The Author(s) 2020
W. K. Al-Delaimy, V. Ramanathan, M. Sánchez Sorondo (eds.), *Health of People, Health of Planet and Our Responsibility*, https://doi.org/10.1007/978-3-030-31125-4_9

coal for power generation, heating, and cooking releases sulfur dioxide (SO_2), nitrogen oxides ($NOx = NO + NO_2$), and carbon-containing compounds from which fine aerosol particles are formed through atmospheric chemistry. London's smog is typically dominated by sulfur compounds, whereas photochemical smog in Los Angeles, for example, is characterized by high levels of ozone (O_3), formed from hydrocarbons and NOx under the influence of sunlight. During the "great acceleration" of the Anthropocene in the middle of the twentieth century (Crutzen, 2002; Steffen, Grinevald, Crutzen, & McNeill, 2011), transboundary air pollution reached regional dimensions, accompanied by the degradation of ecosystems through the deposition of atmospheric sulfur and nitrogen compounds. In recent decades, widespread pollution emissions from traffic, industry, energy production, agriculture, and biomass burning across all continents have emerged into a global challenge.

The chief pollutant that has been related to health outcomes is fine particulate matter, which are aerosols with diameters less than 2.5 μm ($PM_{2.5}$), followed by O_3, which is a strong oxidant. Depending on size, fine particles can penetrate deeply into the lungs, and the smallest ones even reach the bloodstream and affect other organs. Based on emission inventories, satellite observations, measurement data from monitoring networks, and model calculations, it has been unambiguously documented that air quality has degraded on regional and global scales (Akimoto, 2003; Anenberg, Horowitz, Tong, & West, 2010; Lelieveld & Dentener, 2000; Van Donkelaar et al., 2010; Wang & Mauzerall, 2006) (Fig. 9.1). In parallel, evidence that air pollutants adversely impact human health, based on epidemiological studies, has mounted (Burnett et al., 2014; Cohen et al., 2017; Dockery et al., 1993; Lim et al., 2012; Pope et al., 2002). The diseases incited by the long-term exposure to air

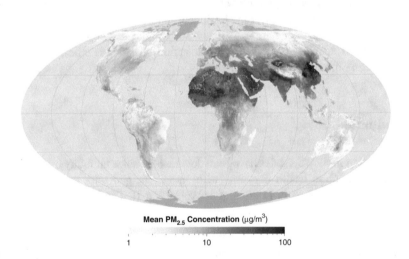

Mean $PM_{2.5}$ Concentration (μg/m^3)

1   10   100

Fig. 9.1 Global mean $PM_{2.5}$ in the period 2010–2012 derived from satellite observations (NASA). The World Health Organization (WHO) estimates that 90% of people on Earth are exposed to $PM_{2.5}$ concentrations higher than the air quality guideline of 10 μg/m^3 (see also Brauer et al., 2016)

pollution include acute lower respiratory infections, chronic obstructive pulmonary disease, cerebrovascular disease, ischemic heart disease, and lung cancer. This chapter uses a global model to estimate the health burden from air pollution, and especially how it can be avoided by reducing emissions.

Methods: Earth System Chemistry Model

We used the ECHAM/MESSy Atmospheric Chemistry (EMAC) computer model that comprehensively simulates meteorological and atmospheric chemical processes worldwide. It has been developed at the Max Planck Institute for Chemistry in Mainz, building on the climate model of the Max Planck Institute for Meteorology in Hamburg (Roeckner et al., 2006). EMAC couples atmosphere, ocean, and land processes to simulate biogeochemical cycles of reactive compounds. Therefore, it can be seen as an Earth system chemistry model, which can cover a wide range of space and time scales, typically up to decades or centuries (Pozzer, Jöckel, Kern, & Haak, 2011). Currently, the EMAC development is pursued by an international consortium (see http://www.messy-interface.org for details and references). The model can be applied at various horizontal resolutions (between about 50 and 250 km grid spacing), and covers the lower and middle atmosphere (Jöckel et al., 2006, 2010). Here, we use EMAC to investigate the global impact of growing air pollution emissions in recent years and also to determine possible chemical states of the atmosphere up to the middle of the twenty-first century, applying a grid spacing of ~100 km latitude/longitude. Air pollution emissions have been adopted from the Emission Database for Global Atmospheric Research (EDGAR) (http://edgar.jrc.ec.europa.eu/).

Model calculations have been extensively tested against measurement data of gases and particles from ground-based air quality networks and global observations from satellites (e.g., Karydis, Tsimpidi, Pozzer, Astitha, & Lelieveld, 2016; Karydis et al., 2017; Lelieveld, Gromov, Pozzer, & Taraborrelli, 2016; Pozzer et al., 2012a, 2012b; Tsimpidi, Karydis, Pandis, & Lelieveld, 2016). To compute the public health impacts attributable to air pollution, the model simulations have been combined with integrated exposure–response (IER) functions that use annual mean pollution concentrations to assess long-term health outcomes, following Anenberg et al. (2010), Lim et al. (2012), Burnett et al. (2014), and Lelieveld, Evans, Fnais, Giannadaki, and Pozzer (2015). We applied the updated IER coefficients according to the Global Burden of Disease (GBD, 2017) and Cohen et al. (2017) to compute the relative risk of mortality by cardiovascular and respiratory diseases and lung cancer (Lelieveld, 2017). Data sets used as inputs, such as country-level baseline mortality rates for the different disease categories and population data, have been adopted from the WHO Global Health Observatory (www.who.int/gho/database/en/), being representative of the year 2015.

Multi-pollutant Index (MPI)

Considering that human health is affected by multiple gas and aerosol species, it is helpful to combine them into an air quality metric, accounting for the so-called criteria pollutants SO_2, NO_2, carbon monoxide (CO), O_3 and $PM_{2.5}$, for which many countries have implemented control measures based on health standards. To assess air quality changes globally, regionally and nationally, and also account for the number of people subjected to poor air conditions, we adopted a Multi-pollutant Index, MPI (Pozzer et al., 2012b). The index is population weighted to represent the impact on public health. The scale extends from -1 to infinity, whereby levels above zero indicate poor air quality according to the pollution thresholds of the WHO for the criteria pollutants. Figure 9.2 compares the air quality development in seven regions since preindustrial times.

Figure 9.2 illustrates the differences between regions and the time development in the Anthropocene, that is, since the preindustrial period. For example, in the Middle East, air quality has been relatively low even without human-induced emissions, related to the high concentration of airborne desert dust, which has a significant impact on human health (Giannadaki, Pozzer, & Lelieveld, 2014). Figure 9.2 also includes results from future projections with the EMAC model according to a

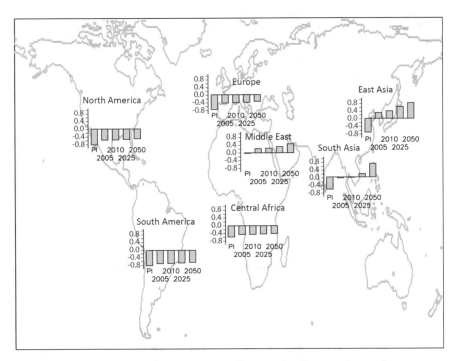

Fig. 9.2 Population-weighted multi-pollutant index for seven regions and five periods: preindustrial times (PI), 2005, 2010, 2025, and 2050. An MPI of -0.8 indicates good, whereas $+0.8$ indicates poor air quality

business-as-usual (BaU) emission scenario for the years 2025 and 2050 (Pozzer et al., 2012b). BaU means that only current legislation is being considered in the modeled emission scenario. Therefore, the projections represent the present trajectory into the future.

The largest changes in the population-weighted MPI occurred during industrialization in the twentieth century, that is, between the preindustrial period and 2005. This is related to the rapidly growing pollution emissions, especially since the middle of the twentieth century (the great acceleration), which strongly impacted air quality for a large fraction of the world population. Although the changes between 2005 and 2050 are smaller, they partly occur rather abruptly. In Europe, the Americas and Africa, the anticipated MPI differences are incremental; however, in economically emerging regions, such as East Asia, very rapid increases have taken place, and may continue in the near future, especially in South Asia (Fig. 9.3). On a global scale, the population-weighted MPI has increased from −0.63 (PI) to −0.13 in 2005, and an additional strong increase up to +0.18 is projected until 2050. This implies that the BaU scenario will lead to worldwide violation of WHO air quality standards.

Figure 9.3 presents the projected population-weighted MPI increase per country. Under the BaU scenario, India may experience the strongest air quality degradation, followed by Bangladesh and China. Figure 9.3 also includes countries in the Middle East where both air pollution and population growth are rapid; hence, many more

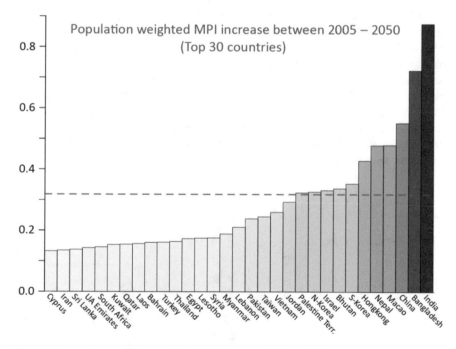

Fig. 9.3 Increasing population-weighted MPI between 2005 and 2050 under a business-as-usual scenario. The red dashed line shows the global mean MPI increase in this period

people may become exposed to poor air quality. The large, densely populated countries with high levels of air pollution, notably in Asia, relatively strongly affect the global mean decline of air quality, and thus adversely influence the health conditions of the "average citizen." Note that Asia is projected to continually dominate the world population (about four of globally seven billion in 2010, and about five of globally nine billion in 2050). Further, it is expected that the urban population will grow relatively rapidly from 3.6 billion in 2010 to 5.2 billion in 2050.

Health Impacts of Air Pollution

The model results have been combined with integrated exposure–response (IER) functions. We computed the relative risk of mortality by acute lower respiratory infections (ALRI) in childhood, chronic obstructive pulmonary disease (COPD), cerebrovascular disease (CEV) leading to mortality by ischemic and hemorrhagic strokes (CEVI and CEVH, respectively), ischemic heart disease (IHD) leading to mortality by heart attacks, and lung cancer (LC). These diseases are related to $PM_{2.5}$, while COPD is additionally related to O_3 (Fig. 9.4). Our methodology follows that of the Global Burden of Disease for 2010 (Lim et al., 2012), being updated for the year 2015 (GBD, 2017).

The minimum risk exposure level distribution for annual mean $PM_{2.5}$, applied in the IER functions, is 2.4–5.9 $\mu g/m^3$, which is lower than the previously assumed range of 5.8–8.8 $\mu g/m^3$ for the Global burden of Disease 2010, and that of O_3 is 33.3–41.9 ppbv (unchanged). The recent downward adjustment of the "safe threshold" for $PM_{2.5}$ is the main reason for the higher mortality rates presented here compared to previous estimates for 2010, since a larger fraction of the population is exposed to hazardous levels of air pollution. In particular, the mortality rates for COPD are significantly higher than reported previously. Thus, the state-of-the-science is that only very low mean $PM_{2.5}$ concentrations under 4.2 ± 1.8 $\mu g/m^3$ do not significantly contribute to morbidity and mortality. The global results presented here agree with the GBD for the year 2015, on average within a few percent.

Fig. 9.4 Global excess mortality rates attributed to air pollution due to different disease categories

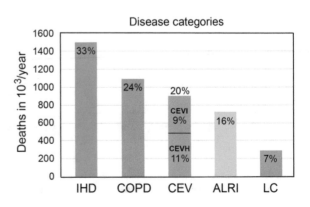

The updated calculations indicate a total excess mortality rate attributable to ambient (outdoor) air pollution of 4.5 million persons per year, which is about a third higher than estimated previously by Lelieveld et al. (2015) and the GBD for the year 2010 (Lim et al., 2012). The statistical uncertainty range is estimated at 3.4–5.6 million per year (95% confidence interval), hence approximately ±25%, being dominated by uncertainties in the IER functions and to a lesser extent to estimates of annual mean pollution concentrations (Kushta, Pozzer, & Lelieveld, 2018). Note that the IER functions for very high levels of $PM_{2.5}$ are associated with relatively large uncertainty, as epidemiological data for such conditions are scarce.

The estimated mortality attributable to air pollution is about 265,000 per year in the EU-28, 373,000 per year for Europe as a whole (not including Russia), and 120,000 per year for the United States. The global total number of years of life lost (YLL) is about 122 million per year (92–148 million per year), which implies that each individual who dies from diseases incited by air pollution loses about 28 ± 7 years of his or her life. The life expectancy reduction is greatest for ALRI as it also affects children, especially in countries where nutrition and medical care are inadequate, being on average 54 years, and 20–25 years for the other disease categories. Among the leading GBD risk factors, ambient particulate matter pollution ranks fifth, after high blood pressure, tobacco smoking, diabetes, and being overweight (GBD, 2017). Poor air quality is the largest environmental risk factor worldwide.

To put the problem in perspective, the WHO estimates that the number of people who die annually from malaria and HIV/AIDS together is about 1.5 million, which is roughly one third compared to outdoor air pollution. The additional mortality attributable to household (indoor) air pollution is about 2.9 million per year (GBD, 2017). Even though the global implementation of the WHO air quality guideline for $PM_{2.5}$ of 10 $\mu g/m^3$ would significantly reduce mortality attributable to air pollution by more than 50% (Giannadaki, Pozzer, & Lelieveld, 2016), this threshold is still too high, especially because the minimum risk exposure level has decreased to 4.2 ± 1.8 $\mu g/m^3$ (Lelieveld & Pöschl, 2017). While the control of emissions in polluted regions such as South and East Asia and Eastern Europe has a high potential to reduce mortality, air quality improvements at relatively low concentrations still have major health benefits because of the nonlinear response to exposure (Cohen et al., 2017).

Table 9.1 ranks the countries where mortality exceeds 30,000 individuals per year. China and India lead, partly related to their large population size because comparatively large numbers of people are exposed to poor air quality. The attribution to source sectors in Table 9.1 has been adopted from Lelieveld et al. (2015), calculated by omitting the respective emissions in the EMAC model. These results represent the elimination of source categories one at a time, which to a large extent represents the efficiency of $PM_{2.5}$ reductions by controlling these sources, as mortality attributable to O_3 represents only 6% of the total. On a global scale, residential energy use for heating and cooking is the largest source category (31%), notably because it is the major one in Asia. In eastern Asia and Europe, agricultural emissions are important, such as ammonia (NH_3) from animal husbandry and fertilizer

Table 9.1 Top 20 ranked countries of mortality attributable to outdoor air pollution in 2015 (the 95% confidence interval is about ±25%), and contributions of seven main source categories (in %), the leading ones in bold

Country	Deaths (×10³)	Residential energy	Agriculture	Natural	Power generation	Industry	Biomass burning	Road traffic
China	1,44	**32**	29	9	18	8	1	3
India	967	**50**	6	11	14	7	7	5
Russia	203	7	**43**	1	22	8	8	11
Nigeria	158	14	1	**77**	0	0	8	0
Pakistan	133	31	2	**57**	2	2	2	3
USA	120	6	29	2	**31**	6	5	21
Indonesia	90	**60**	2	0	5	4	27	2
Japan	84	12	**38**	0	17	18	5	10
Bangladesh	77	**55**	10	0	15	7	7	6
Ukraine	70	6	**52**	0	18	9	5	10
Sudan[a]	58	1	0	**92**	1	0	6	0
Germany	49	8	**45**	0	13	13	1	20
Egypt	47	1	3	**92**	2	1	0	1
Brazil	41	9	4	1	2	6	**71**	7
Vietnam	34	**51**	12	0	13	8	12	4
Iran	32	1	6	**81**	4	3	1	4
Italy	32	7	**39**	6	15	14	2	17
Turkey	31	9	**29**	15	19	11	6	11
N-Korea	31	21	**38**	0	20	12	4	5
Poland	30	15	**43**	0	16	10	3	13
World	4,55	**31**	20	18	14	7	5	5

[a]Including South Sudan

use, because this gas is an effective precursor of $PM_{2.5}$ (Pozzer, Tsimpidi, Karydis, de Meij, & Lelieveld, 2017).

World Avoided by Improving Technology in the EU

To illustrate the effectiveness of air pollution legislation and the consequent public health improvement in the past, an emission scenario was adopted, which accounts for trends in gaseous and particulate pollution sources in the period 1970–2010 (Crippa et al., 2016). The scenario focuses on the EU where the EURO norms for light- and heavy-duty vehicles have been introduced. The EU additionally imposed standards on fuel quality and flue gas desulfurization in the power production sector. The measures taken in the EU not only impacted European air pollution but also worldwide, because the EURO standards penetrated global markets through European and Japanese manufacturers, thus effectively exporting EU regulations. The scenario assesses end-of-pipe reduction measures in the EU by comparing actual emissions to historical fuel consumption along with a stagnation of technology, by applying constant emission factors since 1970, that is, without abatement measures imposed by European emission standards. It includes three regulated sectors—power generation, manufacturing industry and road traffic—which together account for 28% of mortality attributable to air pollution (Table 9.1).

The scenario calculations indicate that, without emission regulation, the European sources of SO_2 and NOx would have been about 70% and $PM_{2.5}$ would have been a factor of 2.3 higher in 2010 compared to 1970, mainly due to end-of-pipe controls (Crippa et al., 2016). The model calculations suggest that regulation of the road transport sector has made a relatively large contribution to emission control of SO_2, NOx, non-methane hydrocarbons (NMHC), and $PM_{2.5}$ (including black and organic carbon, BC and OC), while power production also contributed strongly, particularly to the reduction of SO_2 and $PM_{2.5}$. In Europe, the 2010 emissions by industry were higher than from power production, as abatement strategies for this sector have been less stringent. Globally, European emission controls in power production have played a significant role in reducing SO_2 and BC, and road traffic in the reduction of NMHC, NOx, and BC emissions. Power production was still the largest contributor to SO_2 emissions, both in 1970 and 2010. Nevertheless, without EU emission controls, global SO_2 emissions would have been 30% higher in 2010.

Emissions that have been avoided as a consequence of EU legislation have led to a reduction of annual mean European concentrations of $PM_{2.5}$ by about 5 $\mu g/m^3$ and of O_3 by about 2.5 ppbv, and in Asia 1–2 $\mu g/m^3$ and ppbv, respectively. Comparing the scenario with uncontrolled EU emissions to the reference calculations suggests that the mortality rate attributable to air pollution in the EU has been reduced by about 61,000 persons per year, which means that 18–19% have been avoided. Globally, the reduction has been about 163,000 persons per year, that is, 3–4% were avoided; in China and Japan, approximately 2% and 7% were avoided, respectively. Within the EU, relatively large reductions have been achieved in Western Europe, such as in Germany (21%), France (23%), and the Netherlands (27%).

Discussion and Conclusions

Based on recently updated data on air pollution, disease and population statistics, the excess mortality rate attributable to outdoor air pollution is estimated at 4.5 million individuals per year, associated with about 122 million years of life lost in the year 2015. This represents mortality through cardiovascular and respiratory ailments that could have been avoided when annual average $PM_{2.5}$ would be below 4.2 ± 1.8 $\mu g/m^3$ and O_3 below 37.6 ± 4.3 ppbv. The fractional contribution of $PM_{2.5}$ to mortality is about 94%, and of O_3 about 6%; hence, fine particulates constitute the main health risk. Recent downward adjustment of the minimum risk exposure level for $PM_{2.5}$ to ~4.2 $\mu g/m^3$ (from ~7.3 $\mu g/m^3$) is a main reason for the 30% higher mortality presented here for 2015 compared to previous estimates that referred to the year 2010 (Lelieveld et al., 2015; Lim et al., 2012), since a much larger fraction of the population is exposed to hazardous levels of air pollution.

In view of the revised minimum risk exposure level of ~4.2 $\mu g/m^3$, air quality standards should be adjusted accordingly, currently being 12 $\mu g/m^3$ in the USA and 25 $\mu g/m^3$ in the EU, and higher or absent in Asian and African countries. The air quality guideline by the WHO of 10 $\mu g/m^3$ for $PM_{2.5}$ should also be reevaluated. While the global implementation of this WHO guideline would reduce mortality attributable to air pollution by more than 50% (Giannadaki et al., 2016), further reduction will be necessary. About 75% of the global mortality occurs in Asia, and more than 50% in China and India (together 2.4 million per year). While the latter countries lead the ranking of total mortality, the per capita mortality rate is highest in Eastern Europe (e.g., Russia) with 1.6 deaths per 1000 inhabitants annually, being more than twice that in India and four times compared to North America.

The most rapid increase in mortality attributable to air pollution worldwide occurs in South Asia, particularly in India. Although the leading source category is residential energy use, the recent increase is largely related to fossil fuel consumption, notably of coal, which doubled in the past decade (IEA, 2016). Worldwide, approximately 31% of mortality is due to residential energy use, amounting to about 1.4 million per year. This source category is also a main cause of household (indoor) air pollution-related mortality, which adds about 2.9 million per year (GBD, 2017). In South and Southeast Asia, residential energy use contributes >50% to mortality attributable to outdoor air pollution. Globally, approximately 18% (~800,000 deaths per year) is related to natural processes and is difficult to avoid, mostly due to eolian desert dust in Africa and Asia.

Pozzer et al. (2017) found that strong reductions of $PM_{2.5}$ can be achieved by controlling agricultural ammonia (NH_3) emissions from animal husbandry and fertilizer use. The absolute impact of NH_3 on $PM_{2.5}$ is largest in East Asia, even for relatively small emission decreases. In Europe and North America, the formation of $PM_{2.5}$ is not directly limited by NH_3, but strong emission controls could nevertheless substantially decrease $PM_{2.5}$ and associated mortality, especially when the NH_3 limited regime is reached at about 50% emission reduction. In other regions, including South Asia, the influence of NH_3 emissions on public health is relatively minor. Based on the fractional contribution of agricultural emissions to $PM_{2.5}$, it is estimated that a 50% reduction of global NH_3 emissions could avoid about 350,000 deaths per year. A theoretical reduction of 100% could avoid about 900,000 deaths per year.

A study of European technology development illustrates the public health benefit achieved by emission controls (Crippa et al., 2016). An emission scenario shows the actual world avoided by EU legislation since 1970 through the EURO norms for light- and heavy-duty vehicles, and standards on fuel quality and flue gas desulfurization in the power production sector. The EU measures have influenced pollution emissions in Europe and worldwide because the EURO standards penetrated global markets. The scenario calculations suggest that mortality attributable to air pollution in the EU has decreased by about 61,000 persons per year, which means that about 18–19% mortality have been avoided. Globally, the reduction has been about

163,000 persons per year, which means 3–4% were avoided. It should be noted that these calculations do not account for illegitimate diesel-related NOx emissions. Anenberg et al. (2017) estimated that about a third of heavy-duty and more than half of light-duty diesel vehicle emissions have been in excess of certification limits, associated with about 38,000 deaths globally in 2015.

Zhang et al. (2017) investigated the relationship between exports of goods and services and the exposure to poor air quality. In effect, by exporting goods and services, countries with emerging economies in Asia and eastern Europe "import" air pollution-related mortality. These countries trade monetary revenue against health expenditures through emissions embodied in exports. It is likely that some polluting industries have been transferred internationally, such as from the EU and the USA to regions with more permissive environmental regulations. To prevent such practices, international norms and agreements are needed. It would be an important signal if the UN adopts a sustainable development goal on "clean air." It can be argued that actions to mitigate climate change and improve air quality have co-benefits. Furthermore, clean air contributes to other sustainable development goals that aim at good health, sustainable cities, clean energy, taking climate action, and protecting life on land and in the water.

References

Akimoto, H. (2003). Global air quality and pollution. *Science, 302*, 1716–1719.

Anenberg, S. C., Horowitz, L. W., Tong, D. Q., & West, J. J. (2010). An estimate of the global burden of anthropogenic ozone and fine particulate matter on premature human mortality using atmospheric modeling. *Environmental Health Perspectives, 118*, 1189–1195.

Anenberg, S. C., Miller, J., Minjares, R., Du, L., Henze, D. K., Lacey, F., et al. (2017). Impacts and mitigation of excess diesel-related NOx emissions in 11 major vehicle markets. *Nature, 445*, 467–471.

Brauer, M., Freedman, G., Frostad, J., van Donkelaar, A., Martin, R. V., Dentener, F., et al. (2016). Ambient air pollution exposure estimation for the global burden of disease 2013. *Environmental Science & Technology, 50*, 79–88.

Burnett, R. T., Pope III, C. A., Ezzati, M., Olives, C., Lim, S. S., Mehta, S., et al. (2014). An integrated risk function for estimating the Global Burden of Disease attributable to ambient fine particulate matter exposure. *Environmental Health Perspectives, 122*, 397–403.

Cohen, A. J., Brauer, M., Burnett, R., Anderson, H. R., Frostad, J., Estep, K., et al. (2017). Estimates and 25-year trends of the global burden of disease attributable to ambient air pollution: An analysis of data from the Global Burden of Diseases Study 2015. *The Lancet, 389*, 1907–1918.

Crippa, M., Janssens-Maenhout, G., Dentener, F., Guizzardi, D., Sindelarova, K., Muntean, M., et al. (2016). Regional policy-industry interactions with global impacts. *Atmospheric Chemistry and Physics, 16*, 3825–3841.

Crutzen, P. J. (2002). Geology of mankind. *Nature, 415*, 23.

Dockery, D. W., Pope II, C. A., Xu, X., Spengler, J. D., Ware, J. H., Martha, E. F., et al. (1993). An association between air pollution and mortality in six US cities. *The New England Journal of Medicine, 329*, 1753–1759.

Giannadaki, D., Pozzer, A., & Lelieveld, J. (2014). Modeled global effects of airborne desert dust on air quality and premature mortality. *Atmospheric Chemistry and Physics, 14*, 957–968.

Giannadaki, D., Pozzer, A., & Lelieveld, J. (2016). Implementing the US air quality standard for PM2.5 worldwide can prevent millions of premature deaths per year. *Environmental Health, 15*, 88. https://doi.org/10.1186/s12940-016-0170-8

Global Burden of Disease (GBD)—Risk Factors Collaborators (2017). Global, regional, and national comparative risk assessment of 79 behavioural, environmental and occupational, and metabolic risks or clusters of risks in 188 countries, 1990–2013: A systematic analysis for the Global Burden of Disease Study 2013. *The Lancet, 388*, 1659–1724.

Jöckel, P., Tost, H., Pozzer, A., Brühl, C., Buchholz, J., Ganzeveld, L., et al. (2006). The atmospheric chemistry general circulation model ECHAM5/MESSy: Consistent simulation of ozone from the surface to the mesosphere. *Atmospheric Chemistry and Physics, 6*, 5067–5104.

Jöckel, P., Kerkweg, A., Pozzer, A., Sander, R., Tost, H., Riede, H., et al. (2010). Development cycle 2 of the Modular Earth Submodel System (MESSy2). *Geoscientific Model Development, 3*, 717–752.

Karydis, V. A., Tsimpidi, A. P., Bacer, S., Pozzer, A., Nenes, A., & Lelieveld, J. (2017). Global impact of mineral dust on cloud droplet number concentration. *Atmospheric and Physics, 17*, 5601–5621.

Karydis, V. A., Tsimpidi, A. P., Pozzer, A., Astitha, M., & Lelieveld, J. (2016). Effects of mineral dust on global atmospheric nitrate concentrations. *Atmospheric Chemistry and Physics, 16*, 1491–1509.

Kushta, J., Pozzer, A., & Lelieveld, J. (2018). Uncertainties in estimates of mortality attributable to ambient PM2, 5 in Europe. *Environmental Research Letters, 13*, 064029.

Lelieveld, J. (2017). Clean air in the Anthropocene. *Faraday Discussions, 200*, 693–703.

Lelieveld, J., & Dentener, F. J. (2000). What controls tropospheric ozone? *Journal of Geophysical Research, 105*, 3531–3551.

Lelieveld, J., Evans, J. S., Fnais, M., Giannadaki, D., & Pozzer, A. (2015). The contribution of outdoor air pollution sources to premature mortality on a global scale. *Nature, 525*, 367–371.

Lelieveld, J., Gromov, S., Pozzer, A., & Taraborrelli, D. (2016). Global tropospheric hydroxyl distribution, budget and reactivity. *Atmospheric Chemistry and Physics, 16*, 12477–12493.

Lelieveld, J., & Pöschl, U. (2017). Chemists can help to solve the air pollution health crisis. *Nature, 551*, 291–293.

Lim, S. S., Vos, T., Flaxman, A. D., Danaei, G., Shibuya, K., Adair-Rohani, H., et al. (2012). A comparative risk assessment of burden of disease and injury attributable to 67 risk factors and risk factor clusters in 21 regions, 1990–2010: A systematic analysis for the Global Burden of Disease Study 2010. *The Lancet, 380*, 2224–2260.

Pope III, C. A., Burnett, R. T., Thun, M. J., Calle, E. E., Krewski, D., Ito, K. et al. (2002). Lung cancer, cardiopulmonary mortality, and long-term exposure to fine particulate air pollution. *Journal of the American Medical Association, 287*, 1132–1141.

Pozzer, A., Jöckel, P., Kern, B., & Haak, H. (2011). The atmosphere-ocean general circulation model EMAC-MPIOM. *Geoscientific Model Development, 4*, 771–784.

Pozzer, A., Tsimpidi, A., Karydis, V., de Meij, A., & Lelieveld, J. (2017). Impact of agricultural emissions on fine particulate matter and public health. *Atmospheric Chemistry and Physics, 17*, 12813–12826.

Pozzer, A., de Meij, A., Pringle, K. J., Tost, H., Doering, U. M., van Aardenne, J. et al. (2012a). Distributions and regional budgets of aerosols and their precursors simulated with the EMAC chemistry-climate model. *Atmospheric Chemistry and Physics, 12*, 961–987.

Pozzer, A., Zimmermann, P., Doering, U. M., van Aardenne, J., Tost, H., Dentener, F. et al. (2012b). Effects of business-as-usual anthropogenic emissions on air quality. *Atmospheric Chemistry and Physics, 12*, 6915–6937.

Roeckner, E., Brokopf, R., Esch, M., Giorgetta, M., Hagemann, S., Kornblüh, E., et al. (2006). Sensitivity of simulated climate to horizontal and vertical resolution in the ECHAM5 atmosphere model. *Journal of Climate, 19*, 3771–3791.

Steffen, W., Grinevald, J., Crutzen, P., & McNeill, J. (2011). The Anthropocene: conceptual and historical perspectives. *Philosophical Transactions of the Royal Society A: Mathematical, Physical and Engineering Sciences, 369*, 842–867.

Tsimpidi, A. P., Karydis, V. A., Pandis, S. N., & Lelieveld, J. (2016). Global combustion sources of organic aerosols: Model comparison with 84 AMS factor analysis data sets. *Atmospheric Chemistry and Physics, 16*, 8939–8962.

van Donkelaar, A., Martin, R. V., Brauer, M., Kahn, R., Levy, R., Verduzco, C. et al. (2010). Global estimates of ambient fine particulate matter concentrations from satellite-based aerosol optical depth: Development and application. *Environmental Health Perspectives, 118*, 847–885.

Wang, X., & Mauzerall, D. L. (2006). Evaluating impacts of air pollution in China on public health: Implications for future air pollution and energy policies. *Atmospheric Environment, 40*, 1706–1721.

Zhang, Q., Jiang, X., Tong, D., Davis, S. J., Zhao, H., & Geng, G. (2017). Transboundary health impacts of transported global air pollution and international trade. *Nature, 543*, 705–709.

Part III
Climate Change and Health: Sustainability and Vulnerable Populations and Regions

CHAPTER 10
Vulnerable Populations and Regions: Middle East as a Case Study

Wael K. Al-Delaimy

Summary There are published analyses predicting that the Middle East and North Africa (MENA) region will be the most affected by climate change and global warming. This region has a multitude of factors that makes it more vulnerable to climate change. It is the first region in the world expected to run out of fresh water. It has diversity in income levels between one country and another, but most low-income countries rely on farming and agriculture that is rain-dependent, and there is poor governance and lack of resources to address the impact of climate change through adaptation. There is genuine interest by the MENA countries to address climate change through submission of Intended Nationally Determined Contributions by many countries in the region, but there is a lack of implementation with regard to climate change action plans and no public awareness within countries in the region. The heat waves, sand storms, declines in agriculture and food security, declines in water access, and associated health outcomes in this region are going to drive more migration within and outside the specific countries. The health impact from these climate change scenarios will be overwhelming in many regions in MENA, even if the target 2 °C global minimum temperature increase is achieved. There is a dearth of data from the MENA region regarding climate change and its impact on health.

Historical and Religious Relevance

The MENA region is identified differently by different entities, but for the purpose of this chapter it is represented by the predominately Arab countries (Morocco, Algeria, Tunisia, Libya, Egypt and Sudan in North Africa, and Syria, Lebanon, Palestine, Jordan, Iraq, Saudi Arabia, Yemen, Oman, Kuwait, Qatar, United Arab Emirates-

W. K. Al-Delaimy (✉)
University of California San Diego, La Jolla, CA, USA
e-mail: waldelaimy@ucsd.edu

© The Author(s) 2020
W. K. Al-Delaimy, V. Ramanathan, M. Sánchez Sorondo (eds.), *Health of People, Health of Planet and Our Responsibility*, https://doi.org/10.1007/978-3-030-31125-4_10

121

UAE, and Bahrain in the Middle East) as well as Iran, based on the World Bank criteria. This area is inhabited by predominantly Arab ethnic groups, of whom the majority (91.2%) are Muslims (Pew Research Center, 2011). The origin of the Arabs in the region goes back to nomadic tribes roaming the region in search of pasture and water for their livestock through regular seasonal migration in search of these resources. The Arabs have been in the desert in the Arabian Peninsula or in farming rural communities in Iraq, Syria, Palestine, Egypt, and coastal Mediterranean countries. With increasing affluence in the past century, most of the migration has been internal from rural areas to cities, which offer more secure and higher paying jobs. This trend has changed recently with increasing migration out of the region because of conflicts and political instability, and is likely exacerbated by global warming impacts.

The Quran, Islam's Holy Scripture, documents a clearly identifiable historical severe drought in great detail, which is corroborated by other Biblical and Judaic Scripture. Prophet Joseph, son of Prophet Jacob who is revered by Muslims, Christians, and Jews, lived around 600 B.C. and relayed his experiences of a 7-year drought that affected all of today's Egypt, Palestine, Israel, Lebanon, Syria, and Jordan in the MENA region. Many lessons are provided by the story of Joseph and the drought that apply to the region today, which includes good governance, adaptation, and conservation. This concept is typically overlooked by the public, who read the verses of the Quran about the story of Joseph but do not relate it to today's climate change, drought, and needed adaptation measures. Other teachings of Islam through the Holy text in the Quran that clearly point to the role of mankind in destroying the planet include Chapter 30 verse 41: *Corruption has appeared in the land and the sea on account of what the hands of men have wrought, that He may make them taste a part of that which they have done, so that they may return.* The teaching of Prophet Mohamed has many examples relevant to protecting the planet as a duty (He said, *Don't waste water even if you are on the bank of a flowing river*; He also said, *if the end of the world happened while you are planting a tree, continue planting it*—to overemphasize the importance of planting trees). This teaching in Islam is supportive of an anthropogenic origin of climate change, as well as the means to mitigate it and to protect the resources of the planet.

Extreme Heat Wave Events

The occurrence of frequent heat waves is a direct manifestation of climate change and has impacted many countries in recent years. Extreme heat waves are events with record temperatures beyond the usual average heat wave event. Extreme heat waves are expected to cause several countries in the Arabian/Persian Gulf area to exceed the capacity of humans to be outdoors during summer periods (Lelieveld, et al., 2016; Pal & Eltahir, 2016; World Bank, 2014a). The Warm Spell Duration Index (WSDI) is defined as the annual or seasonal count of days with at least 6 consecutive days of a warm spell. During this warm spell, the maximum temperature each day is above the 90th percentile of past daily temperatures that occurred at the same time of the year during 1979–2009. In a 4 °C scenario, 80% of summer months are projected to be hotter than 5-standard deviations above the average (unprecedented heat extremes) by 2100, and about 65% are projected to be hotter than 5-standard deviations above the average during the 2071–2099 period (see Fig. 10.1). In a 2 °C scenario, the annual number of hot days with exceptionally high temperatures and high thermal discomfort is expected to increase in several capital cities, from 4 to 62 days in Amman (Jordan), from 8 to 90 days in Baghdad (Iraq), and from 1 to 71 days in Damascus (Syria). The greatest increase is expected in Riyadh (Saudi Arabia), where the number of these extremely hot days is pro-

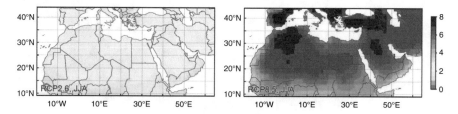

Fig. 10.1 Multi-model mean temperature anomaly for RCP2.6 (2 °C world, left) and RCP8.5 (4 °C world, right) for the months of June–July–August for the Middle East and North African region (temperature anomalies in degrees Celsius are averaged over the time period 2071–2099 relative to 1951–1980) (From World Bank, 2014a)

Table 10.1 Mean WSDI (Warm Spell Duration Index) for capital cities in the MENA region for different levels of global warming based on regional climate model projections (World Bank, 2014a)

	Period of observation	Observed	1.5	2.0	3.0	4.0
Abadan	1951–2000	6	43	82	99	134
Amman	1959–2004	4	31	62	84	115
Ankara	1926–2003	7	44	67	111	128
Athens	1951–2001	1	40	61	121	166
Baghdad	1950–2000	8	47	90	113	162
Beirut	–	–	47	93	126	187
Belgrade	1951–2010	9	39	39	76	113
Cairo	–	–	32	53	80	94
Damascus	1965–1993	1	36	71	98	129
Istanbul	1960–2010	0	26	41	78	113
Jerusalem	1964–2004	7	26	46	73	102
Kuwait	–	–	45	87	123	167
Nicosia	1975–2001	6	25	58	81	162
Riyadh	1970–2004	3	81	132	157	202
Sofia	1960–2010	1	40	49	88	136
Tehran	1956–1999	5	48	92	122	159
Tirana	1951–2000	6	49	71	125	168
Tripoli	1956–1999	3	13	22	33	59

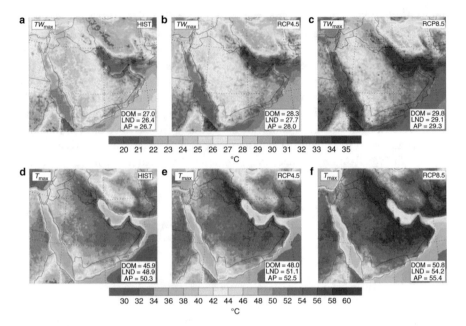

Fig. 10.2 Spatial distributions of extreme wet bulb temperature and extreme temperature. (**a–f**) Ensemble average of the 30-year maximum TWmax (**a–c**) and Tmax (**d–f**) temperatures for each GHG scenario: historical (**a**, **d**), RCP4.5 (**b**, **e**) and RCP8.5 (**c**, **f**). Averages for the domain excluding the buffer zone (DOM), land excluding the buffer zone (LND) and the Arabian Peninsula (AP) are indicated in each plot. TW_{max} and T_{max} are the maximum daily values averaged over a 6-h window (From Pal & Eltahir, 2016 reprinted with permission)

jected to rise from 3 to 132 days per year. In a 4 °C world, the average number of such days is projected to exceed 115 days per year in all of these cities (see Table 10.1 and Fig. 10.2) (Pal & Eltahir, 2016).

Heat waves have been shown to be related to increased mortality across many countries (Guo, et al., 2017) and with some recent heat waves receiving wide media attention coverage for their high death tolls, such as the 2003 European heat wave that caused an estimated 70,000 deaths (Robine, Cheung, Le Roy et al., 2008) and the 2010 Russian heat wave that caused an estimated 11,000 deaths (Shaposhnikov et al., 2014). Heat stroke, exhaustion, and hyperthermia are the most common causes of mortality, but other causes of mortality from respiratory, cardiovascular and diabetic complications have also been demonstrated (Ye et al., 2012). However, there is limited data describing the impact of heat waves on health in the MENA region. Given the extreme temperatures expected from the business-as-usual scenario leading to a 4 °C increase in temperature, the resulting mortality from high temperatures in the MENA is going to be the highest in history because 80% of summer is going to have temperatures that are 5-standard deviations above the average temperature. The fact that temperatures have never reached 60 °C, but are expected to reach such levels by the end of this century in the Gulf Region, means the health impact will be devastating. While the European heat wave claimed so many lives when reaching 37–41 °C for 9 consecutive days, the number of deaths will be much higher if reaching 60 °C. Unfortunately, there are no health-related studies regarding heat wave-related mortality from the MENA region to better quantify the risk. There are also no clear action plans to proactively deal with such heat waves within capital cities in the MENA.

Therefore, the consequences of climate change, even with a 2 °C scenario, are rather serious for the MENA region and will be felt well before the end of this century. Record temperatures are already being recorded in several countries.

Desert Sand and Dust Storms

The MENA region is a major source of desert dust storms travelling to Europe and Asia (Fig. 10.3). The higher temperatures affecting the region are leading to lower soil moisture and more arid conditions, which are exacerbated by lower precipitation and longer drought periods. Aridity is expected to increase by 50% throughout the entire coastal region if the business-as-usual model is followed (World Bank, 2014a). These conditions support more desert dust storms in the region. However, local conflicts played an important role in desertification and dust storms from the desert. The 1991 Gulf War and use of tanks and heavy equipment disrupted the desert ecology in Iraq, Kuwait, and Saudi Arabia (in areas which are now major sources of desert dust storms) with the topsoil becoming suspended as fine particles at higher levels in the air. There have been more frequent desert dust storms affecting the region in the last two decades. In Iraq, 122 dust storms were recorded in a single year in 2010 and are expected to more than double in incidence to around 300 dust storms a year in 2020 (UNEP, 2016). The size and type of dust particles also

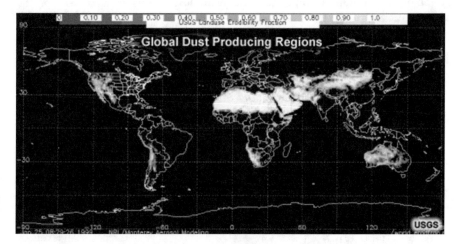

Fig. 10.3 The National Geospatial Program of the US Geological Survey map of origins of dust storms from the Middle East and North Africa region

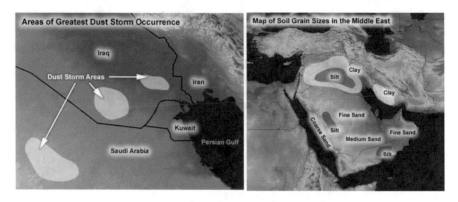

Fig. 10.4 Types of particles from the dust storms in the Middle East and the specific places with most frequent dust storms (From the Comet Program by NOAA National Weather Service)

have an impact on human health, where finer particles that are more common in the Arabian Peninsula (Fig. 10.4) reach deeper in the lung and cause more damage.

Frequent desert dust storms cause increased emergency visits for those with respiratory illnesses. We have demonstrated in a study from Japan that desert dust storms from the Taklamakan Desert in China led to a significant increase in asthma admission among children in Japan (Kanatani et al., 2010). The best-fit model adjusting for climate and other variables demonstrated an odds ratio (OR) of 1.71 (95% CI 1.18–2.48; $p = 0.005$). OR was higher for boys at 2.32 (95% CI 1.10–4.87; $p = 0.005$) and for 6- to 12-year-olds at 3.33 (95% CI 1.02–10.92). The impact of these desert dust storms will continue to inflict damage on these countries, but the level of health impact from these frequent dust storms is unknown for the MENA region.

There are other important health impacts of dust storms, including poor visibility, leading to more road traffic accidents. Dust storms have also been shown to carry bacteria and viruses for thousands of miles across the globe, with the bacteria types varying by region. Studies have documented microorganisms in dust from Tunisia (Hezbri et al., 2016), Lebanon (Itani & Smith, 2016), Kuwait (Leski, Malanoski, Gregory, Lin, & Stenger, 2011), and Iraq (Al-Dabbas, Ayad Abbas, & Al-Khafaji, 2012). This has new implications and understandings for what might trigger certain medical conditions. Toxins and allergens are also carried on dust particles in dust storms (Abuduwailil, Zhaoyong, & Fengqing, 2015; Kameda et al., 2016).

Climate change will specifically impact the health of populations in the MENA region through an increase in dust storms that are of higher frequency and wider geographical spread, affecting millions of people every year. There is a need for active adaptation approaches based on scientific evidence of who is at risk and what is the magnitude of that risk.

Water Scarcity

All countries in the MENA region suffer from water shortage. The increase in drought and high temperatures with lower precipitation and increased demand will lead to scarcity in access to water. The MENA region will be the first region in the world to run out of fresh water. It is expected that water availability will decrease in most parts of the MENA region by 15% in a 2 °C scenario and up to 45% in parts of the region in a 4 °C scenario (World Bank, 2014a). Water access is mostly dependent on underground water or rain. The Gulf Countries have developed desalination projects to cover the supply of water, although this is at high financial and environmental costs.

Some of the countries with the poorest water supplies in the world are in the MENA region, namely Jordan and the Palestinian Territories under occupation of Israel. According to the Israeli Information Center for Human Rights in the Occupied Territories, Palestinians get as little as 20–50 L/day per capita compared to 287 L/day per capita for Israelis, while the WHO recommendation is to have a minimum of 100 L per capita per day (B'Tselem, 2016). The lack of water access is going to be exacerbated in that area because of overuse of underground water, lower precipitation, longer droughts and higher temperatures. Dwindling resources can trigger further conflict in an area already suffering from ongoing violence. It is estimated that the withdrawal-to-availability ratio for underground water in Jordan, Yemen, Libya, and most of the Gulf countries is more than 100%, meaning that withdrawal is not being replenished and the water is going to be completely depleted (World Bank, 2014a). This is going to lead to lower local agricultural production and food insecurity.

Yemen is another country that is at the forefront of the water scarcity situation in the MENA region. It is the least able to adapt; has a growing population; lacks government control of water sources, which are dominated by local tribes; and has an ongoing conflict, which is preventing any policies or approaches to ration water or address the shortage (World Bank, 2014b). There is a 1.4 billion cubic meter shortfall in water supply per year, which is being met by water pumped up with modern tube wells or boreholes, depleting reserves of underground water. In rural areas, when wells run dry, social tensions escalate into local conflicts. Mass displacement from water scarcity causes migration and fuels the risk of wider conflicts (World Bank, 2014b). With Yemen being mostly an agricultural country, crops in Yemen are dying out because of water shortage leading to food insecurity, lower incomes, higher unemployment and more displacement. Although southern parts of Yemen are expected to have an increase in precipitation, this increase will be through extreme rain events that are likely to cause considerable damage to cropland and will not be enough to replenish underground water (World Bank, 2014a).

The above examples within the region give an indication of the volatility of the water situation in MENA and by no means cover all countries and challenges related to water scarcity. Water scarcity will bring with it a multitude of negative impacts in terms of agriculture and income, but also sanitation, violence and mass migration. Droughts will be most severe in the coastal areas of North Africa, leading to decreases in precipitation and a decline in rain-dependent agriculture and farming activities.

Climate Change and Political Instability in Relation to Refugees

Mass migration from flooding is expected to reach two billion at the global level by the end of this century. More people will move inland as a result of coastal flooding, becoming environmental refugees (Geisler & Currens, 2017). In the MENA region, mass migration has largely been the result of violence and economic reasons.

There are attempts to link the refugee crises in Syria to global warming, citing the drought that preceded the Syrian war in 2009. This is championed by notable reporters such as Thomas Friedman and several other researchers (Kelley, Mohtadi, Cane, Seager, & Kushnir, 2015); however, it might be an oversimplification of complex geopolitical interests by regional and worldwide powers to control the region. The evidence for climate change being responsible for the Arab Spring or the Syrian war is only correlational in nature, and there is no evidence it was responsible for this uprising. Such oversimplification ignores the history and political situation of Syria and these countries. This connection between environment and war is not new and originated from Malthus in the eighteenth century (Malthus, 1798). This concept has been applied elsewhere in the MENA region such as Darfur, but there are strong arguments against such an association (Verhoeven, 2011). Could the drought in Syria and rapid urbanization by its rural populations have exacerbated the uprising? It might have, but this cannot be said with certainty. There is anecdotal evidence that frustration over lack of water and drought by farmers made it easier to join a rebellion (Wendle, 2016). A more balanced explanation for the association between climate change and the Syrian war and other conflicts would be to label it as an eco-social or risk-multiplier (Butler, 2017), but the conflict would have happened regardless. The nature of this conflict goes back to decades of oppression; the rebellion against dictatorships and oppression started in one place and then spread elsewhere.

However, the risk is real regarding climate change leading to exacerbation or triggering mass movements and future conflicts that result from such crises. According to Brigadier General Stephen Cheney, a member of the US Department of State's Foreign Affairs Policy Board and CEO of the American Security Project, "Climate change could lead to a humanitarian crisis of epic proportions. We are already seeing migration of large numbers of people around the world because of food scarcity, water insecurity and extreme weather, and this is set to become the new normal. Climate change is acting as an accelerant of instability in parts of the world on Europe's footsteps, including the Middle East and Africa" (Carrington, 2017).

Health of Refugees

Even though the initial trigger of the conflicts that led to the current refugee crises are not necessarily related to the environment, the environment does exacerbate the suffering of refugees. Challenges of extreme heat and cold, limited water access, unemployment, rising food prices because of agricultural land degradation and other factors relevant to climate change increase the suffering of refugees in their host transitional countries such as Jordan, Lebanon, Egypt and Turkey. It is worth noting that Jordan and Lebanon have the highest number of refugees per capita of their respective populations. Therefore, the burden from migration is to a large extent kept within the region. There was a massive refugee influx from transition countries to Europe, mainly through Turkey. This in turn encouraged other refugees coming from Afghanistan, Iraq, and sub-Saharan Africa to join the exodus and mass migration by foot and boats, as smugglers established routes of illegal migration to Europe. The Mediterranean Sea has become the graveyard for thousands who have drowned on their path to reach Italy or other European countries. Others who survived the journey suffer the trauma of losing loved ones in the war or during their dangerous path to Europe.

The health of refugees has been undermined by the lack of access to health care for basic medical needs and the limited resources available in transitional countries where they settle, such as Jordan, Lebanon, and Egypt. In Turkey, refugees generally have better access to health care. Refugees can also spread infectious diseases through their mass movement across countries and because of living in unsanitary conditions with no treatment, and limited or no access to health care (Vittecoq et al., 2014).

One of the most important and highly neglected health needs of refugees is recognition and treatment of mental illnesses among both adults and children. Historically, the region has had a stigma against mental illnesses and has a very limited number of psychiatrists, estimated to be in the order of one psychiatrist per 100,000–200,000 people. Trauma from the conflicts and displacement is a long-term cause of mental illnesses such as depression, anxiety, PTSD and suicide. While the focus of NGOs and countries that care for refugees is on food and shelter, mental illnesses are neglected and generations of children from Syria and other countries suffer silently from loss and displacement.

We have shown through our own research among Somali refugees in San Diego that mothers who have a history of severe trauma from before being resettled in the USA continue to suffer mental illnesses and are more likely to have children who suffer from depression or poor adjustment (East, Gahagan, & Al-Delaimy, 2017). The trauma continues and might not heal like a physical illness. Therefore, poor mental health is going to be a major health impact of climate change in the MENA region among displaced populations, whether resulting from climate impacts on the environment and livelihood of populations in the region or from ongoing civil unrest and violence.

Solutions and Level of Preparedness

There is, to some extent, an understanding of the consequences of climate change on the welfare of the countries in the MENA region by the governments of 11 countries that have submitted Intended Nationally Determined Contributions (INDC) as of September 2017 (Algeria, Bahrain, Egypt, Jordan, Saudi Arabia, Sudan, Qatar, UAE, Palestine, Tunisia, and Morocco). Submissions vary in details of actions and policies; the most comprehensive was Jordan's, which took a holistic approach and developed a climate action plan. These countries plan mitigation and adaptation strategies to a variable degree of detail. This is a positive indication, although there is an equal number of countries from the region that still have not submitted any INDC.

Some countries are generating specific data to support adaptation policies. One such example is a recent study from Morocco, which demonstrated a tripling in water deficit by 2050 due to increased demand, 30% less precipitation, and decreased surface and underground water reserves (Seif-Ennasr et al., 2016). The authors proposed 38 adaptation measures to address the impact of climate change and demonstrated that the most cost-effective measure is water preservation and management. In Lebanon, there is a new direction towards conservation agriculture as an alternative to intensive conventional agriculture (Chalak et al., 2017). UAE is focusing on renewable energy and proposed the development of renewable energy and nuclear energy as partial replacement energy sources to reach 25% of its energy expenditure by 2020 (Paul, Al Tenaiji, & Braimah, 2016). In Saudi Arabia, the focus is on water conservation by developing desalination plants, expanding water recycling processes and infrastructure, transitioning from domestic agriculture to outsourcing food products, and taking the reins on solar development to phase out fossil fuels. Saudi Arabia is doing just about everything it can to ensure domestic water resources will be available and accessible, and that it will be able to sustain these practices (DeNicola, Aburizaiza, Siddique, Khwaja, & Carpenter, 2015).

Emergency heat wave adaptation plans should be implemented in each of the MENA countries by utilizing several such plans already in place elsewhere in the world. The same should be done for dust storms that are becoming more frequent; these storms require preparation and prevention for their health impacts on vulnerable populations through education and collection of data.

Solutions to decrease the impact of this major water shortage are possible but require much effort, better governance, and a shift in thinking towards new approaches and water conservation. Because the teaching of Islam indicates that not wasting water is part of the faith, this can be put into the context of climate change consequences for the predominately Muslim followers in MENA. Recycling water is another important approach that can help prepare populations to deal with water scarcity. Grey water (GW), which is wastewater generated from domestic activities such as laundering, dishwashing, and bathing used to meet rising water demands, comprises 50–75% of residential wastewater and is distinct from black wastewater, which is collected from toilets (Eriksson, Auffarth, Henze, & Ledin, 2002; Friedler, 2004). Adopting GW recycling practices on a large scale may lower freshwater demands and decrease groundwater extraction rates by up to 30–50%; this can result in a decrease in the risk of aquifer salinization (Jeppesen, 1996). It has been found by our survey and review of the literature that the main limitation for use of GW is cost (Leas, Dare & Al-Delaimy 2014). There is general cultural acceptability in MENA to use GW and no health-related impacts are reported, although more epidemiological studies are needed to document its safety for use. Further, scholars of Islam have already addressed the type of uses for GW that are allowed in Islam, including its use for irrigation (Leas, Dare & Al-Delaimy 2014).

Conclusions

The MENA region is the global ground zero for climate change. It will be suffering catastrophic heat waves and, especially during several months in summer, will be beyond human capacity to be outdoors. It will be the first region to run out of fresh water, it has a population growth that will increase demand for resources, and it is suffering ongoing wars and civil unrest leading to displacement of mass populations. All of these have serious health consequences for the populations living in the region. Lower-income populations will suffer the most, as they are less likely to be protected by infrastructure or the means to adapt to this new environment brought by climate change. Water scarcity and frequent desert dust storms bring many adverse health consequences. There is also lack of awareness and education by the general population of these impending catastrophes in this century and local media coverage is very limited.

There is a need to support and encourage individual countries to aggressively address mitigation and adaptation measures for the sake of current and future generations living in this region. Civil unrest and wars will be a major barrier towards such efforts. However, as was done elsewhere in Latin America and Africa regarding vaccination programs to create peace, the climate change problem can become a unifying source in finding ways to work together to address the impact of climate change as the leading public health problem of the century. Religious bodies and entities in the region need to be informed and educated about the association between the Muslim faith and protection of the planet to prevent and reduce the health consequences of climate change. There is much at stake for the livelihood of the millions living in the MENA region, but also in the rest of the world, as witnessed by the mass displacements and refugee crises that have affected other regions.

References

Abuduwailil, J., Zhaoyong, Z., & Fengqing, J. (2015). Evaluation of the pollution and human health risks posed by heavy metals in the atmospheric dust in Ebinur Basin in Northwest China. *Environmental Science and Pollution Research, 22*, 14018–14031.

Al-Dabbas, M. A., Ayad Abbas, M., & Al-Khafaji, R. M. (2012). *Arabian Journal Geoscience, 5*, 121.

B'Tselem. (2016, September 27). *The Israeli Information Center for human rights in the occupied territories.* Retrieved October 19, 2017, from http://www.btselem.org/water/discrimination_in_water_supply

Butler, C. (2017). Limits to growth, planetary boundaries, and planetary health. *Current Opinion in Environmental Sustainability, 25*, 59–65.

Chalak, A., Irani, A., Chaaban, J., Bashour, I., Seyfert, K., Smoot, K., et al. (2017). Farmers' willingness to adopt conservation agriculture: New evidence from Lebanon. *Environmental Management, 60*, 693–704. https://doi.org/10.1007/s00267-017-0904-6

Damian Carrington. (2017, December 1). Climate change will stir unimaginable refugee crisis, says military. *The Guardian.*

DeNicola, E., Aburizaiza, O. S., Siddique, A., Khwaja, H., & Carpenter, D. O. (2015). Climate change and water scarcity: The case of Saudi Arabia. *Annals of Global Health, 81*, 342–353.

East, P. L., Gahagan, S., & Al-Delaimy, W. K. (2017). The impact of refugee mothers' trauma, posttraumatic stress, and depression on their children's adjustment. *Journal of Immigration and Minority Health, 20*(2), 271–282. https://doi.org/10.1007/s10903-017-0624-2

Eriksson, E., Auffarth, K., Henze, M., & Ledin, A. (2002). Characteristics of grey wastewater. *Urban Water, 4*, 85–104.

Friedler, E. (2004). Quality of individual domestic greywater streams and its implication for on-site treatment and reuse possibilities. *Environmental Technology, 25*, 997–1008.

Geisler, C., & Currens, B. (2017). Impediments to inland resettlement under conditions of accelerated sea level rise. *Land Use Policy, 66*, 322–330.

Guo, Y., Gasparrini, A., Armstrong, B. G., Tawatsupa, B., Tobias, A., Lavigine, E., et al. (2017). Heat wave and mortality: A multicountry, multicommunity study. *Environmental Health Perspectives, 125*, 087006.

Hezbri, K., Ghodhbane-Gtari, F., Montero-Calasanz, M. D., Nouioui, I., Rohde, M., Spröer, C., et al. (2016). Geodermatophilus pulveris sp. nov., agamma-radiation-resistant actinobacterium isolated from the Sahara desert. *The International Journal of Systematic and Evolutionary Microbiology, 66*, 3828–3834.

Itani, G. N., & Smith, C. A. (2016). Dust rains deliver diverse assemblages of microorganisms to the Eastern Mediterranean. *Scientific Reports, 6*, 22657.

Jeppesen, B. (1996). Domestic greywater re-use: Australia's challenge for the future. *Desalination, 106*, 311–315.

Kameda, T., Azumi, E., Fukushima, A., Tang, N., Matsuki, A., Kamiya, Y., et al. (2016). Mineral dust aerosols promote the formation of toxic nitropolycyclic aromatic compounds. *Scientific Reports, 6*, 24427.

Kanatani KT, Ito I, Al-Delaimy WK, Adachi Y, Mathews WC, Ramsdell JW; Toyama Asian Desert Dust and Asthma Study Team. (2010) Desert dust exposure is associated with increased risk of asthma hospitalization in children. *American Journal of Respiratory and Critical Care Medicine, 182*, 1475–1481.

Kelley, C. P., Mohtadi, S., Cane, M. A., Seager, R., & Kushnir, Y. (2015). Climate change in the Fertile Crescent and implications of the recent Syrian drought. *Proceedings of the National Academy of Sciences of the United States of America, 112*, 3241–3246.

Leas, E. C., Dare, A., & Al-Delaimy, W. K. (2014). Is gray water the key to unlocking water for resource-poor areas of the Middle East, North Africa, and other arid regions of the world? *Ambio, 43*, 707–717.

Lelieveld, J., Proestos, Y., Hadjinicolaou, P., Tanarhte, M., Tyrlis, E., & Zittis, G. (2016). *Climatic Change, 137*, 245–260.

Leski, T. A., Malanoski, A. P., Gregory, M. J., Lin, B., & Stenger, D. A. (2011). Application of a broad-range resequencing array for detection of pathogens in desert dust samples from Kuwait and Iraq. *Applied and Environmental Microbiology, 77*, 4285–4292.

Malthus, T. R. (1798). *An essay on the principle of population.* London. Retrieved October 7th 2019 from http://www.esp.org/books/malthus/population/malthus.pdf

Pal, J. S., & Eltahir, E. A. B. (2016). *Nature Climate Change, 6*, 197–200, London. Retrieved October 7th 2019 from http://www.esp.org/books/malthus/population/malthus.pdf.

Paul, P., Al Tenaiji, A. K., & Braimah, N. (2016). A review of the water and energy sectors and the use of a nexus approach in Abu Dhabi. *International Journal of Environmental Research and Public Health, 13*, 364. PMC.

Pew Research Center (2011, January 27). *Religion and public life.* Retrieved October 19, 2017 from http://www.pewforum.org/2011/01/27/future-of-the-global-muslim-population-regional-middle-east/

Robine, J. M., Cheung, S. L., Le Roy, S., et al. (2008). Death toll exceeded 70,000 in Europe during the summer of 2003. *Comptes Rendus Biologies, 331*, 171–178.

Seif-Ennasr, M., Zaaboul, R., Hirich, A., Caroletti, G. N., Bouchaou, L., El Morjani, Z. E., et al. (2016). Climate change and adaptive water management measures in Chtouka Aït Baha region (Morocco). *Science of Total Environment, 573*, 862–875.

Shaposhnikov, D., Revich, B., Bellander, T., Bedada, G. B., Bottai, M., Kharkova, T., et al. (2014). Mortality related to air pollution with the moscow heat wave and wildfire of 2010. *Epidemiology, 25*, 359–364. https://doi.org/10.1097/EDE.0000000000000090

UNEP (2016, June 14). Retrieved October 19, 2017, from https://www.unenvironment.org/news-and-stories/story/more-action-needed-sand-and-dust-storms

Verhoeven, H. (2011). Climate change, conflict and development in Sudan: Global neo-Malthusian narratives and local power struggles. *Development and Change, 42*, 679–707.

Vittecoq, M., Thomas, F., Jourdain, E., Moutou, F., Renaud, F., et al. (2014). Risks of emerging infectious diseases: Evolving threats in a changing area, the Mediterranean basin. *Transboundary and Emerging Diseases, 61*, 17–27.

Wendle, J. (2016). Syria's climate refugees. *Scientific American, 314*, 50–55.

World Bank. (2014a). *Turn down the heat: Confronting the new climate normal.* Washington, DC: World Bank.

World Bank. (2014b). *Future impact of climate change visible now in Yemen.* Retrieved October 19, 2017, from http://www.worldbank.org/en/news/feature/2014/11/24/future-impact-of-climate-change-visible-now-in-yemen

Ye, X., Wolff, R., Yu, W., Vaneckova, P., Pan, X., & Tong, S. (2012). Ambient temperature and morbidity: A review of epidemiological evidence. *Environmental Health Perspectives, 120*, 19–28.

CHAPTER 11
Climate Change Risks for Agriculture, Health, and Nutrition

Joachim von Braun

Summary The stability of global, national, and local food systems is at risk under climate change. Climate change affects food production, availability of and access to food, food quality, food safety, diet quality, and thus people's nutrition and health. Climate change may further slow progress towards a world with food security for all. Climate change impacts will exacerbate food shortages, especially in areas that already show a high prevalence of food insecurity. Climate change will affect good nutrition through complex indirect pathways, such as income shocks when droughts or floods occur, loss of employment opportunities, health effects resulting from air pollution and changed water systems. A conceptual framework for a food systems analysis is presented here, and the linkages within the system and expected changes among them are elaborated—some on a global scale and some on a micro scale. Policy actions are proposed and research gaps are identified.

Introduction

The global food system is malfunctioning, leaving large segments of the population undernourished or malnourished and causing large environmental damage. Climate change adds to the challenge. Climate change and agriculture are in a reciprocal relationship: Greenhouse gas emissions related to land use and animal production make agriculture an important cause of climate change. In turn, climate change and its related extreme events affect agriculture and the food system through changing production conditions and complex environmental and economic externalities, some of which have serious impacts on human health and nutrition.

It is well understood that food security has strong linkages to agriculture with its multiple functions for economy, ecology, and people's livelihoods. The more spe-

J. von Braun (✉)
University of Bonn, Bonn, Germany
e-mail: jvonbraun@uni-bonn.de

cific linkages between agriculture, food security, health, and nutrition have received more research attention recently (Fan & Pandya-Lorch, 2012; von Braun, Ruel, & Gillespie, 2012). However, further research is needed as the complex linkages between food security and health are dynamic in the context of global change processes, including climate change, urbanization, and market integration (Ruddiman, Ellis, Kaplan, & Fuller, 2015). The linkages are not easily quantifiable which, however, does not imply they are weak (Masset, Haddad, Cornelius, & Isaza-Castro, 2012; Webb & Kennedy, 2014).

The focus of this chapter lies on some of these linkages in particular, with an emphasis on additional risks for agriculture, food security, and related nutrition and health problems emerging from climate change. Climate change may slow down progress towards a world with food security for all. The stability of whole food systems may be at risk under climate change, through shocks and increased variability in markets. On a global scale, a pattern of the impacts of climate change on crop productivity, and thereby on food availability, emerges. Unless public policy actions are taken, these climate change impacts will exacerbate food shortages, especially in areas that already show a high prevalence of food insecurity. Climate change will also affect access to food and high-quality nutrition through indirect pathways such as income shocks, when droughts or floods occur, burden on (women's) time use, and loss of employment opportunities. Health effects are caused by air pollution (indoor and outside) and possibly through changed water systems.

In view of the complex links through which climate change may have an impact on global food security, a food systems approach is needed. A conceptual framework for a food systems analysis is presented first, then linkages within the system and expected changes among them are elaborated—some on a global scale and some (if widely prevalent) on a micro scale. Finally, policy actions are proposed and research gaps that warrant our attention are identified.

Conceptual Relationships of Agriculture, Food Security, and Health Under Climate Change

Agriculture is inherently sensitive to climate variability and change, whether due to natural causes or human activities. Climate change due to emissions of greenhouse gases is expected to directly impact crop production systems for food, feed, or fodder; to affect livestock health; and to alter the pattern and balance of trade of food and food products. These impacts will vary with the degree of warming and associated changes in rainfall patterns, and from one location to another, but underpinning these impacts are a number of direct effects on the physiology of crops grown for food, feed, fuel, and fiber as well as on the pests and diseases of livestock and crops.

A conceptual framework is put forward here. The concept takes a broad perspective on agriculture, comprising crops, animal production, and connected value chains as well as the natural resource base of land and water use, and the technological

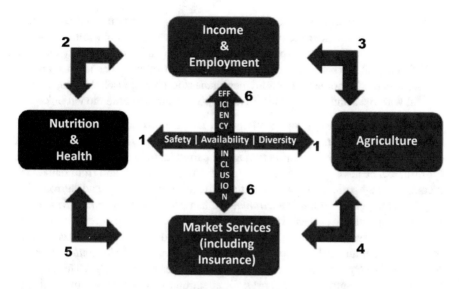

Fig. 11.1 Conceptual framework—a food systems perspective (Adapted from von Braun, 2017)

foundations of agriculture. Institutions, information, and behavior are interdependent issues that influence linkages in all of the domains that describe the framework. The linkages of food security and agriculture with health can be broadly grouped into six domains as depicted in Fig. 11.1, and all of these are influenced by climate change in various ways:

Link 1: Agricultural production- and processing-related health linkages are central, especially in the long run. They entail, for example, diet-, nutrition-, and health-sensitive agriculture, paying attention to food supply and access to nutrients with diverse diets, quality and safety of foods, and production conditions (Pontifical Academy of Sciences and Global Alliance for Improved Nutrition, 2018). To the extent that climate risks add instability in production, the direct and indirect linkages to food security and health are adversely impacted. Technology is critical for these links.

Links 2 and 3: Income- and employment-related linkages greatly determine access to nutrition and health, especially by marginal farm households and agricultural laborers (Gatzweiler & von Braun, 2016). If, under a climate shock, rural labor markets contract for services and landless laborers, there may be adverse implications for health and nutrition. For instance, health services may become unaffordable and often school fees are no longer paid. Agriculture and income and employment links are fundamental for onward linkages to nutrition.

Links 4, 5, and 6: Linkages with markets and services drive nutrition and health via income, but also directly via food price formation and forward and backward market linkages. If services and infrastructures are malfunctioning under climate stress, the adverse impacts of climate change increase. Markets and services are critical for delivering affordable nutritious food as well as nutrition and health-related services. Under increased climate risks, investments in health and nutrition services will need to expand to enhance the adaptation capacities of the poor. Efficiency of markets and inclusive services with insurance mechanisms are key forces behind income and employment generation and households' capacity to deal with risks and shocks. This, in turn, drives many aforementioned linkages.

Like any framework, simplifications and abstractions are implicit in order to focus on key causal linkages. The linkages include various *feedbacks among them,* such as from health to production through impaired labor productivity (which may, for instance, be adversely affected by stress from heat waves), and from markets to income through prices and consumption effects. Overarching, and surrounding Fig. 11.1, are *agricultural and environmental as well as macroeconomic framework conditions.* Related linkages exist on a global scale, such as links related to greenhouse gas emissions and, on a local scale, water and sanitation in the context of irrigated agriculture. All the links operate with *diverse dynamics under short- or long-term time lags,* which require attention in policies and programs. This outlined framework also demonstrates the need for attention to the complex causal linkages and behavior change in food systems that need to be considered in modelling climate change impacts on populations and wellbeing. Neglecting these linkages and indirect effects may over- or underestimate climate change threats for people.

The following sections of the chapter address selected components of the framework in which food security, agriculture, and health/nutrition linkages may be affected by climate change.

Climate Risks for the Stability of the Entire Global Food System

Before reviewing key linkages within the food system, the system as a whole is considered because the stability of whole food systems may be at risk under climate change. Important issues here are extreme events, especially in production, crop and animal pests, and markets. Climate change is likely to increase food market volatility from the production and supply side. Demand-side shocks are also possible in a political economy context, such as in the case of aggressive bio-energy subsidies and quota policies. Policy shifts of the last decade in the USA and EU were partly motivated by energy security and partly by climate mitigation objectives. The resulting destabilization effects in food markets, which contributed to large food security problems, were therefore partly related to climate change policies. Climate change may be one trigger of effects for broader destabilizations of food security, including the risk of *high and volatile food prices,* which temporarily limit poor people's food consumption (Kalkuhl, Braun, & Torero, 2016); *financial and economic shocks,* which lead to job loss and credit constraints; and the risks that *political disruptions and failed political systems* pose for food security, such as conflicts over land and water, which are increasing under climate change. These complex system risks can assume a variety of patterns and can become high-risk combinations.

Landmark studies on climate change and agriculture since 1994 include those by Parry, Rosenzweig, Iglesias, Livermore, and Fischer (2004), Cline (2007), and the World Bank (2010). These studies focus mainly on production effects and, in some cases, on the expected changes of average prices. More recent studies highlight the roles of mitigation and adaptation interventions for the reduction of climate change impacts (Richardson, Lewis, Wiltshire, & Hanlon, 2018). Baldos and Hertel (2015) quantify the crucial role of a well-functioning global trade system for the reduction of climate change impacts on food security. Extreme weather events need to be considered in comprehensive climate change impact analyses (Lesk, Rowhani, & Ramankutty, 2016). In a recent global modelling study, this is embedded in the determinants of risks for global crop production for maize, wheat, rice, and soybeans over the period 1961–2013 (Haile, Wossen, Tesfaye, & von Braun, 2017). Using seasonal production data, price change and price volatility information at the country level, as well as future climate data from 32 global circulation models, it is projected that climate change could reduce global crop production by 9% in the 2030s and by 23% in the 2050s. Climate change is expected to lead to 1–3% higher annual fluctuations of global crop production over the next four decades. Furthermore, Haile et al. (2017) found:

- A strong, positive and statistically significant supply response to changing prices for all four crops. However, output price volatility, which signals risk to producers, reduces the supply of these key global agricultural staple crops—especially for wheat and maize.
- Climate change has significant adverse effects on production of the world's key staple crops.
- In particular, weather extremes—in terms of shocks in both temperature and precipitation—during crop growing months have detrimental impacts on the production of the abovementioned crops. Weather extremes also exacerbate the year-to-year fluctuations of food availability, and thus may further increase price volatility with its adverse impacts on production and poor consumers.

The inclusion of volatility in the production and price developments (i.e., the inclusion of risks associated with climate change) is the new aspect of this research, which gives further insights. Combating climate change using both mitigation and adaptation technologies is crucial for global production and hence food security on a global scale. The projected effects of climate change on production (Fig. 11.2) use the national climate data from country specific GCMs as forecasts with national resolution to capture the heterogeneity of future climate change in the study countries. The climate change impacts are more severe in the 2050s. Future climate change also increases the risk associated with crop production: climate change increases the variance of crop production by 1.4% and 2.8% in the 2030s and 2050s, respectively.

Projected average crop production shows positive but small changes for countries such as the Russian Federation, Turkey, and Ukraine in the 2030s, whereas production changes are negative and more pronounced for all countries in the 2050s.

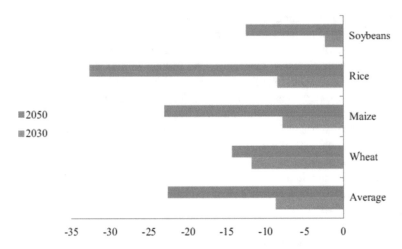

Fig. 11.2 Projected effect of climate change on food production (%) (*Source*: Haile et al., 2017)

These effects on the climate-induced average food production changes have significant implications for global food security for two reasons: (1) wheat, rice, maize, and soybeans make up about three-quarters of the food calories of the global population; (2) our study countries produce more than 85% of the global production of these crops. Technological advances for sustainable production systems and trade for balancing regional and local deficits may need to expand to cope with the risks for food security.

Functioning agricultural markets are important for access to food and services and, thereby, also for income and health (Kalkuhl et al., 2016). However, volatility in prices, even in efficiently functioning markets, may adversely impact consumption capabilities and health. Price volatility is not a new phenomenon, though some of its causes are. Several studies (e.g., Abbott, Hurt, & Tyner, 2011) have identified the drivers of price upsurges, such as biofuel demand, speculation on future commodity markets, public stockholding, trade restrictions, macroeconomic shocks to money supply, exchange rates, and economic growth. Tadesse, Algieri, Kahkuhl, and von Braun (2014) found that a food crisis is more closely related to extreme price spikes, whereas long-term volatility is more strongly aligned with general price risks. In 2007–2008, the prices for almost all food commodities increased significantly, inducing negative effects especially for low-income countries and the part of their population that spends the biggest share of their income on staple foods. To the extent to which climate shocks result in price spikes, health and nutrition consequences can be serious for the poor (Kalkuhl et al., 2016).

Large parts of the world, where negative impacts of climate change on crop production are expected, coincide with countries that currently have high levels of food insecurity. There is a robust and coherent pattern at the global scale of the impacts of climate change on crop productivity, and most likely on food availability. These impacts will exacerbate food insecurity in areas that currently show

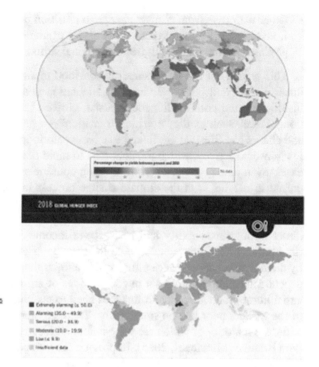

Impacts of climate change on the productivity of food crops in 2050
Source: World Bank Publishers
World bank Development report 2010
http://wdronline.worldbank.org/

Global Hunger Index 2018
Source: Welthungerhilfe and Concern Worldwide 2018
https://www.globalhungerindex.org/res
ults/

Fig. 11.3 Climate variability and change, and food security (*Sources*: see inside the figure)

a high prevalence of hunger (Fig. 11.3) as identified by the Global Hunger Index (Welthungerhilfe and Concern International, 2018).

Agricultural Production- and Processing-Related Health Linkages Under Climate Stress

Utilization of food depends much upon behavior and the health environment (including water and sanitation), so any impact of climate change on the health environment also affects this dimension of and pathway toward food security. Research on these links and climate change is rather weak, but some meaningful extrapolations can be derived from related research (Wheeler & von Braun, 2013):

- Links with drinking water may be the most obvious when climate variability further strains clean drinking water availability. Hygiene may be equally affected by extreme weather events (e.g., floods) in environments where sound sanitation is absent.
- Additionally, the uptake of micronutrients is affected negatively by diarrheal diseases, which strongly correlate with temperature.

- Other pathways from climate change to nutrition potentially work through risks from climate change for diet quality or, for example, increased storage costs and pest attacks that may result from ecological shifts.

This brings out the importance of non-food inputs in food security, including infrastructure and health services for adaptation and coping. While problems of insufficient and poor-quality food persist, global changes are creating new nutritional issues such as the "nutrition transition"—a process by which urbanization and changes in lifestyle are linked to excess caloric intake, poor-quality diets, and low physical activity, which together lead to rapid rises in obesity and chronic diseases, even among the poor in developing countries. The nutrition transition will unfold in parallel with the climate change process in coming decades. Very little research on the potentially re-enforcing effects of the two has been conducted.

As pointed out in the framework above, agriculture-related health linkages are direct through availability and indirect via income-related effects. In particular, rural populations in low- and middle-income countries are significantly influenced by the linkages between agriculture and health, as their economies are often highly dependent on agriculture as a major provider of employment and income. Direct agricultural production–health linkages need to take into account that the majority of the world´s poor live on small farms. There are about 570 million farms worldwide, the majority of which are very small and located in China, India, and Africa (von Braun & Mirzabaev, 2015). Improving the productivity and incomes of smallholder farmers can help significantly to reduce poverty. There are several ways how this can be achieved. First, commercialization and technological innovations among smallholders can help reduce poverty (Gatzweiler & von Braun, 2016).

Diversification from solely staple crop production to the production of commercially high-value crops is another avenue for reducing poverty among smallholders. Diversification typically may also involve specialization at the farm level, with growing variety of production and processing at the regional level. Efficient storage and food processing plants are as important as improving agricultural systems to provide a diverse and safe diet (Keding, Schneider, & Jordan, 2013).

Part of the solution is nutrition-sensitive agriculture, which focuses on the causes of malnutrition and is more long-term oriented (Balz, Heil, & Jordan, 2015). The approach encourages production and consumption of a variety of foods, recognizes and emphasizes the nutritional value of food, and considers the significance of food production for the agricultural sector as well as for rural livelihoods (Thompson & Amoroso, 2014). The provision of information for consumers is essential to the success of this kind of intervention.

A host of studies are now available on communities and households on the microlevel being exposed to climate shocks. These approaches tend to capture more adaptation capabilities than macro-models, such as assets draw down, job-switching migration, social policy responses, and collective action, as shown in detail by Rakib and Matz (2016) for rural Bangladesh. These studies should be a basis to expand analyses into health effects. Taken together, macro modeling and micro-level analyses of climate change linkages to food security are complementary.

Income- and Employment-Related Linkages of Climate Change

Food access and utilization of foods may be impacted by climate change through indirect but rather strong pathways. Access to food is largely a matter of household and individual-level income, of capabilities and rights as well as of behaviors. Food access issues have been studied through two types of approaches: top-down by models that attempt to link macro-shocks related to climate change to responses and adaptation outcomes (Nelson et al., 2010; van Meijl et al., 2018); and by community- and household-level studies that try to assess climate change effects bottom-up. Specific findings of top-down approaches are heavily dependent on the assumptions made about future income and population growth, but in general clear linkages between economic growth and resilience to climate change can be shown. Springmann et al. (2016) modelled the relationships between climate change, diet change, and related health outcomes at the global scale and found adverse health effects.

Agricultural practices can also be the source of health hazards through the use of toxic substances, such as pesticides and fertilizers. Mycotoxins are another case in point, which evolve, for example, due to improper crop handling or use of inferior seeds (Smith, Stoltzfus, & Prendergast, 2012). More unpredictable flooding and strong rains are likely to increase the risks of mycotoxins.

In addition, there are several foodborne diseases that can occur through the mis-handling of agricultural products. For this reason, effective food safety schemes are just as important to promote health as are policies to reduce food insecurity (Pontifical Academy of Sciences and Global Alliance for Improved Nutrition, 2018). Heat waves make food handling more difficult in low-income countries. Cold chains may in the long run become more accessible for the poor due to decentralized (solar) energy systems, but that is still in a distant future for many low-income rural households. Facilitating access to safe foods by such means as cold chains should increasingly be considered a matter of public health investment, rather than consumer subsidies.

Linkages of Climate Change and Health via Markets and Services

A particularly complex set of agriculture-health linkages lies within the broader ecological and public health context of farming communities. An important subset of these is the environmental health linkages of partly agriculture-related water quality and quantity problems interacting with farm populations' sanitation and hygiene behaviors. The importance of water quality for health is well known. The impact of agriculture on both the water quality and quantity with respect to health issues, though, is not well researched. The health costs of sanitation and water problems are large. In addition, in peri-urban settings, farmers often have neither the choice to use other water than wastewater for irrigation, nor do they have the knowledge on the potential risks related to that resource (Qadir et al., 2010)—including harm for human health and the environment through the introduction of pathogens or contamination with chemical substances. There is also a clear linkage between these health, environmental, and behavioral issues on the one hand and the nutritional status and farming on the other hand (Usman, Gerber, & Braun, 2018).

The Water-Energy-Food Security Nexus surrounds the energy decision-making process of rural households. The rural energy system and its impacts on environment, agriculture and health are connected in complex ways to climate change. Attention is drawn here to just one of the complex relationships: The majority of rural households in developing countries continue to depend on unreliable, inefficient, and harmful household energy technologies. As a consequence, many of the poorest people are exposed to household indoor air pollution from cooking and heating with dirty energy and inappropriate technology. This results in large mortality and disease rates, mainly among children and women of poor households who cook and spend more time indoors than men (Smith et al., 2014). The problem is old, but it may change further with increased climate risks. The link to climate and environmental change is indirect but can be particularly strong for the poor, as it is often based on collected fuel wood (mainly by women), cow dung cakes (South Asia), and charcoal, with adverse effects on ecologies. Women's time allocation for energy production (such as fuel wood collection or producing dung cakes), for food and for home goods production is a critical element of the linkages. When, for instance, distance to collectable fuel wood increases due to deforestation and land-use change—both being important forces of greenhouse gas emissions—women's time investments for collecting fuel wood increases, which may come at the cost of sustainable local food systems. Development of village energy systems (e.g., solar or bio-based) is important to facilitate the needed cooking, lighting, power, and irrigation of the village community in sustainable ways, and this cannot be achieved by individual household initiatives. Also, clean cooking stoves fit financially, technically, and according to the perceptions of communities, have a potential to reduce the adverse health effects of indoor pollution from smoke in low-income communities in rural areas.

Policy Implications and Research Challenges

Action needs to be taken by policy-makers and practitioners confronted with the prospect of climate change impacts on food security, nutrition, and health primarily regarding these threats:

1. Climate change-related production and price volatilities will increase risks and uncertainties within the global food system, and impact food and nutrition security through food availability and price effects.

2. Climate change impacts on food security as well as related health and nutrition aspects will be worst in countries already suffering from high levels of hunger, and they will aggravate over time. Doing nothing in response to climate change will have potentially large consequences for global undernutrition and malnutrition, and this will worsen over time.
3. Inequalities and locations of food insecurity may change. People and communities who are currently vulnerable to the effects of extreme weather may become more vulnerable in the future, as they are less resilient to climate shocks.

Action needs to be directed towards what could be termed a "climate-smart food system"—not just climate smart agriculture—that addresses climate change impacts on all dimensions of food and nutrition security. Research initiatives need to be part of the action agenda (Campbell et al., 2016). Governance of food systems matters too. Exploiting the positive linkages between agricultural production and health is particularly constrained by intersectoral and disciplinary bias that impedes collaboration (von Braun et al., 2012). A policy study of Africa found that intersectoral collaboration for nutrition is enhanced when there is inclusive programming and jointly agreed-upon and compatible monitoring and evaluation in place (Malabo Montpellier Panel, 2017). Much more of that is needed under increased climate risks. Considerable investment in adaptation and mitigation actions is needed to slow the impacts of climate change on the progress of global hunger. There is a wide range of potential adaptation and resilience options. These need to address food security in its broadest sense and be integrated into the development of agriculture worldwide.

Strengthening agricultural resilience through science, change in technology and production systems is a key part of this. However, it is not sufficient on its own for global food security. The whole food system needs to adjust with strong attention to science-based insights, trade, stocks, nutrition and health, and social policy options, as outlined by the InterAcademy Partnership (2018). Related priorities for research include:

1. Climate change impacts on food access, nutrition, and health should be more comprehensively considered.
2. Much of the climate change impacts on food security are rather indirect, but that does not mean they are small. Understanding these indirect impacts requires more comprehensive analytical approaches and sophisticated modelling, including their political economy linkages.
3. Better integration of research on the human dimensions, including understanding behavior and collective actions, together with biophysical changes will be central to understanding and identifying policy options in support of food security and healthy living under climate change-related risks.

Important is the strengthening of data basis and monitoring. The "Lancet Countdown" aims to track the health impacts of climate hazards; health resilience and adaptation; health co-benefits of climate change mitigation; economics and finance; and political and broader engagement (Watts et al., 2017). This could also strengthen capacities to integrate health in climate modelling.

References

Abbott, P. C., Hurt, C., & Tyner, W. E. (2011). *What's driving food prices in 2011?* Issue report, July 2011. Oak Brook, IL: Farm Foundation, NFP. Retrieved on February 16, 2020 from http:// farmfoundation.org/news/articlefiles/1742-FoodPrices_web.pdf

Baldos, U. L. C., & Hertel, T. W. (2015). The role of international trade in managing food security risks from climate change. *Food Security, 7,* 275–290. https://doi.org/10.1007/ s12571-015-0435-z

Balz, A. G., Heil, E. A., & Jordan, I. (2015). Nutrition-sensitive agriculture: New term or new concept? *Agriculture & Food Security, 4,* 6. https://doi.org/10.1186/s40066-015-0026-4

Campbell, B. M., Vermeulen, S. J., Aggarwal, P. K., Corner-Dolloff, C., Girvetz, E., Loboguerrero, A. M., et al. (2016). Reducing risks to food security from climate change. *Global Food Security, 11,* 34–43. https://doi.org/10.1016/j.gfs.2016.06.002

Cline, W. R. (2007). *Global warming and agriculture: Impact estimates by country.* Washington, DC: Center for Global Development, Peterson Institute for International Economics.

Fan, S., & Pandya-Lorch, R. (2012). *Reshaping agriculture for nutrition and health. An IFPRI 2020 Book.* Washington, DC: International Food Policy Research Institute (IFPRI). Retrieved on February 16, 2020 from http://ebrary.ifpri.org/cdm/ref/collection/p15738coll2/id/126825

Gatzweiler, F. W., & von Braun, J. (Eds.). (2016). *Technological and institutional innovations for marginalized smallholders in agricultural development.* Cham: Springer. https://doi. org/10.1007/978-3-319-25718-1

Haile, M. G., Wossen, T., Tesfaye, K., & von Braun, J. (2017). Impact of climate change, weather extremes, and price risk on global food supply. *Economics of Disasters and Climate Change, 1,* 1–17. https://doi.org/10.1007/s41885-017-0005-2

Kalkuhl, M., von Braun, J., & Torero, M. (Eds.). (2016). *Food price volatility and its implications for food security and policy.* Cham: Springer. Retrieved on February 16, 2020 from http://link. springer.com/10.1007/978-3-319-28201-5

Keding, G. B., Schneider, K., & Jordan, I. (2013). Production and processing of foods as core aspects of nutrition-sensitive agriculture and sustainable diets. *Food Security, 5,* 825–846. https://doi.org/10.1007/s12571-013-0312-6

Lesk, C., Rowhani, P., & Ramankutty, N. (2016). Influence of extreme weather disasters on global crop production. *Nature, 529,* 84–87. https://doi.org/10.1038/nature16467

Masset, E., Haddad, L., Cornelius, A., & Isaza-Castro, J. (2012). Effectiveness of agricultural interventions that aim to improve nutritional status of children: Systematic review. *BMJ, 344*:d8222.

Nelson, G. C., Rosegrant, M. W., Palazzo, A., Gray, I., Ingersoll, C., Robertson, R., et al. (2010). *Food security, farming, and climate change to 2050: Scenarios, results, policy options.* Washington, DC: IFPRI. Retrieved on February 16, 2020 from http://www.ifpri.org/cdmref/ p15738coll2/id/127066/filename/127277.pdf

Malabo Montpellier Panel. (2017, August). *Nourished: How Africa can build a future free from hunger and malnutrition,* Dakar.

Parry, M. L., Rosenzweig, C., Iglesias, A., Livermore, M., & Fischer, G. (2004). Effect of change on global food production under SRES emissions and socio economic scenarios. *Global Environmental Change, 14,* 53–67. https://doi.org/10.1016/j.gloenvcha.2003.10.008

InterAcademy Partnership (2018). *Opportunities for future research and innovation on food and nutrition security and agriculture: The InterAcademy Partnership's global perspective.* Trieste: Schaefer Druck und Verlag GmbH. Retrieved on February 16, 2020 from http://www.interacademies.org/File.aspx?id=49085&v=b6788f3d

Pontifical Academy of Sciences and Global Alliance for Improved Nutrition (GAIN) (2018). Final statement of the workshop on food safety and healthy diets. *The Vatican.* Retrieved on February 16, 2020 from http://www.pas.va/content/accademia/en/events/2018/food/statement.html

Qadir, M., Wichelns, D., Raschid-Sally, L., McCornick, P. G., Drechsel, P., Bahri, A., et al. (2010). The challenges of wastewater irrigation in developing countries. *Agricultural Water Management, 97*, 561–568. https://doi.org/10.1016/j.agwat.2008.11.004

Rakib, M., & Matz, J. A. (2016). The impact of shocks on gender-differentiated asset dynamics in Bangladesh. *The Journal of Development Studies, 52*, 377–395. https://doi.org/10.1080/0022 0388.2015.1093117

Richardson, K. J., Lewis, K. H., Krishnamurthy, P. K., Kent, C., Wiltshire, A. J., & Hanlon, H. M. (2018). Food security outcomes under a changing climate: Impacts of mitigation and adaptation on vulnerability to food insecurity. *Climatic Change, 147*, 327–341. https://doi. org/10.1007/s10584-018-2137-y

Ruddiman, W. F., Ellis, E. C., Kaplan, J. O., & Fuller, D. Q. (2015). Defining the epoch we live in. *Science, 348*, 38–39. https://doi.org/10.1126/science.aaa7297

Smith, K. R., Bruce, N., Balakrishnan, K., Adair-Rohani, H., Balmes, J., Chafe, Z., et al. (2014). Millions dead: How do we know and what does it mean? Methods used in the comparative risk assessment of household air pollution. *Annual Review of Public Health, 35*, 185–206. https:// doi.org/10.1146/annurev-publhealth-032013-182356

Smith, L. E., Stoltzfus, R. J., & Prendergast, A. (2012). Food chain mycotoxin exposure, gut health, and impaired growth: A conceptual framework. *Advances in Nutrition, 3*, 526–531. https://doi.org/10.3945/an.112.002188

Springmann, M., Mason-D'Croz, D., Robinson, S., Garnett, T., Godfray, H. C. J., Gollin, D., et al. (2016). Global and regional health effects of future food production under climate change: A modelling study. *The Lancet, 387*, 1937–1946. https://doi.org/10.1016/ S0140-6736(15)01156-3

Tadesse, G., Algieri, B., Kahkuhl, M., & von Braun, J. (2014). Drivers and triggers of international food price spikes and volatility. *Food Policy, 47*, 117–128. https://doi.org/10.1016/j. foodpol.2013.08.014

Thompson, B., & Amoroso, L. (Eds.). (2014). *Improving diets and nutrition: Food-based approaches*. Croydon, UK: FAO and CABI.

Usman, M. A., Gerber, N., & von Braun, J. (2018). The impact of drinking water quality and sanitation on child health: Evidence from rural Ethiopia. *The Journal of Development Studies, 55*, 2193–2211. https://doi.org/10.1080/00220388.2018.1493193

van Meijl, H., Havlik, P., Lotze-Campen, H., Stehfest, E., Witzke, P., Pérez Domínquez, I., et al. (2018). Comparing impacts of climate change and mitigation on global agriculture by 2050. *Environmental Research Letters, 13*. https://doi.org/10.1088/1748-9326/aabdc4

von Braun, J. (2017). Agricultural change and health and nutrition in emerging economies. In P. Pingali & G. Feder (Eds.), *Agriculture and rural development in a globalizing world*. Earthscan Food and Agriculture Series (pp. 273–291). London: Routledge.

von Braun, J., & Mirzabaev, A. (2015). *Small farms: Changing structures and roles in economic development*. ZEF-Discussion Papers on Development Policy, 204. Retrieved on February 16, 2020 form https://www.zef.de/uploads/tx_zefportal/Publications/zef_dp_204.pdf

von Braun, J., Ruel, M. T., & Gillespie, S. (2012). Bridging the gap between the agriculture and health sectors. In S. Fan & R. Pandya-Lorch (Eds.), *Reshaping agriculture for nutrition and health. An IFPRI 2020 book 2020* (pp. 183–190). Washington, DC: International Food Policy Research Institute (IFPRI).

Watts, N., Adger, W. N., Ayeb-Karlsson, S., Bai, Y., Byass, P., Campbell-Lendrum, D., et al. (2017). The Lancet countdown: Tracking progress on health and climate change. *The Lancet, 389*, 1151–1164. https://doi.org/10.1016/S0140-6736(16)32124-9

Webb, P., & Kennedy, E. (2014). *Impact pathways from agricultural research to improved nutrition and health: Literature analysis and research priorities*. Rome: Food and Agriculture Organization of the United Nations (FAO) and World Health Organization (WHO).

Welthungerhilfe and Concern International. (2018). *Global Hunger Index*, Bonn and Dublin. Retrieved on February 16, 2020 from https://www.globalhungerindex.org/results/

Wheeler, T., & von Braun, J. (2013). Climate change impacts on global food security. *Science, 341,* 508–513. https://doi.org/10.1126/science.1239402

World Bank. (2010). *World Bank development report 2010: Development and climate change.* Washington, DC: World Bank.

CHAPTER 12
Sustaining Water Resources

Paolo D'Odorico and Ignacio Rodriguez-Iturbe

"As the rain and the snow
come down from heaven,
and do not return to it
without watering the earth
and making it bud and flourish, …"

(Isaiah 55:10)

Summary Life depends on water and there are no substitutes. Fundamental to the functioning of the Earth's system, water is a renewable natural resource that is available in finite amounts. Human appropriation of freshwater resources is mainly used for agriculture. To prevent loss of natural habitat and biodiversity, the growing demand for agricultural products will likely need to be met by enhancing crop yields in currently cultivated land rather than through farmland expansion. Because of local limitations in water availability, irrigation can be sustainably expanded only in part by the currently rain-fed farmland. Any further withdrawal from water bodies would deplete environmental flows or groundwater stocks. Local water deficits (i.e., gaps between availability and demand, including water for food) are often compensated for by food imports from water-rich regions and the associated transfers of "virtual water." Thus, water is a strategic resource that has become strongly globalized and controlled by a few countries, while the rest of the world is in conditions of trade dependency. Water-saving strategies—based on food waste reduction, moderation of diets, use of more suitable crops, or increased water use efficiency—need to be adopted in order to meet human needs without compromising water sustainability.

P. D'Odorico (✉)
University of California, Berkeley, CA, USA
e-mail: paolododo@berkeley.edu

I. Rodriguez-Iturbe
Texas A&M University, College Station, TX, USA

© The Author(s) 2020
W. K. Al-Delaimy, V. Ramanathan, M. Sánchez Sorondo (eds.), *Health of People, Health of Planet and Our Responsibility*, https://doi.org/10.1007/978-3-030-31125-4_12

149

Background

Water is essential for life on Earth and plays a crucial role in determining the global distribution of biomes as well as habitats for aquatic and terrestrial species (e.g., Whittaker, 1975). It also shapes the land surface through erosion and deposition processes; transports sediments, nutrients, and pollutants; and drives the dispersal of a number of species and pathogens (e.g., Dunne & Leopold, 1978). Water is also an important determinant of the Earth's climate because of heat storage in water bodies, the impact of evaporation and transpiration on the surface energy balance, and the effect of water vapor on atmospheric radiative transfer. Water availability affects the productivity of terrestrial ecosystems, the rate of global biogeochemical cycles, and the composition of plant, microbial, and animal communities (e.g., Rodriguez-Iturbe & Porporato, 2005). The impact of water resources on human societies is evidenced by the critical role of water in a number of human activities, such as agriculture, manufacturing, energy generation, and mining.

The majority of terrestrial ecosystems and most human activities rely on freshwater, which accounts for about 3% of the global water resources (with the remaining 97% being in the oceans). Most of the freshwater on Earth is in ice caps ($\approx$69%) and groundwater ($\approx$30%), while only about 1% is stored in surface water bodies and vegetation.

Atmospheric water reaches the land surface as precipitation and then either flows downhill as surface runoff or infiltrates into the ground, thereby contributing to root zone soil moisture and groundwater recharge (Fig. 12.1). Water retained by capillary

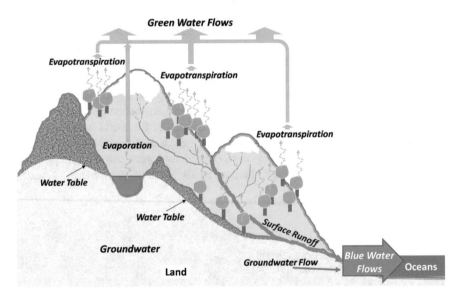

Fig. 12.1 Blue water flows (i.e., runoff) and green water flows (i.e., evapotranspiration) out of land masses

forces in shallow unsaturated soils is often termed "green water" because it is largely used by plants and soil microorganisms that can exert the suction required for extraction. The majority of terrestrial plants relies on green water. Conversely, groundwater (i.e., in the saturated portion of below-ground water stocks) and surface water are often referred to as "blue water" (Falkenmark & Rockström, 2004).

Water leaves land masses either in the liquid phase as blue water flows, or as water vapor fluxes associated with evapotranspiration (i.e., green water flows) (Fig. 12.2). Globally, blue water flows account for about 40% and green water flows for 60% of the water fluxes from land masses. These proportions, however, vary across geographic regions and have been altered by human action.

The two major anthropogenic disturbances to the water cycle are associated with land use change and irrigation. These phenomena have opposite effects on hydrological processes. Land use change—typically in the form of deforestation and land conversion to agriculture—tends to reduce evapotranspiration and increase runoff (e.g., Bonan, 2008). In other words, it reduces green water flows and increases blue water flows. Conversely, irrigation, which accounts for the majority ($\approx 70\%$) of water withdrawals for human activities, uses blue water from streams, lakes, and aquifers to enhance agricultural yields by increasing crop evapotranspiration (i.e., turning blue water into green water flows). In many regions of the world, the latter effect is predominant and explains the observed reduction in river flows in major agricultural regions. Indeed, some rivers (e.g., the Rio Grande, Colorado, Nile, and Yellow Rivers) are so strongly depleted that they do not reach the ocean anymore. Particularly emblematic are the cases of endorheic basins, such as Lake Chad, The Dead Sea, The Aral Sea, and Lake Urmia, which are drying out because of water withdrawals from their tributaries (e.g., Stone, 2015). Likewise, many aquifers around the world are strongly depleted because withdrawals (often for irrigation or municipal uses)

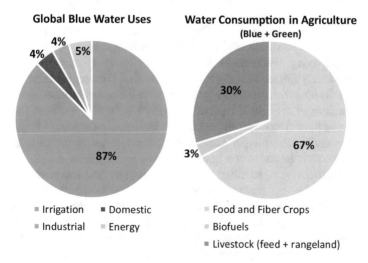

Fig. 12.2 Global blue water consumption (left) and total (blue and green) water consumed in agriculture (data from Table 12.1 for years around 2000)

are occurring at rates exceeding those of groundwater recharge, a phenomenon often known as "overpumping" (Konikow, 2011; Wada et al., 2010). Moreover, agriculture may affect the water cycle through the impacts of irrigation and land use change on precipitation, evapotranspiration and runoff (e.g., Bonan, 2008).

Changes in the water cycle are also emerging as a result of climate warming, which causes a slight intensification of the water cycle with a 2–3% increase both in global precipitation and evapotranspiration per Celsius degree of global tempera- ture increase. These global averages, however, do not show how local precipitation is expected to either substantially decrease or increase depending on the location. As a general pattern, humid regions (e.g., at the equator and midlatitudes) are expected to become wetter, while drier regions (e.g., the subtropics) drier (Held & Soden, 2006). The impact on crop production, however, should also account for the effect of increased atmospheric CO_2 on photosynthesis and the nutritional proper- ties of crops (Myers et al., 2014).

Thus, water is a renewable resource that is conserved within the earth system. Some water stocks (e.g., glaciers, aquifers, and endorheic lakes), however, are undergoing depletion. Likewise, in some areas, water quality is degraded by pollu- tion from chemicals, suspended sediments, and pathogens. Despite its renewability, water is available on Earth only in limited amounts. Therefore, some human activi- ties are constrained by the pace of the water cycle, and competition often exists between human uses and environmental needs.

Human Appropriation of Water Resources

Human societies use "blue" water resources for agriculture, industrial, energy, and household needs. The majority of these uses are related to crop production, in the form of irrigation water (Fig. 12.2). Therefore, contrary to common perception, the global water crisis is all about finding solutions to mitigate and prevent hunger rather than thirst (e.g., Falkenmark & Rockström, 2004). Thus, problems of water scarcity are closely related to important constraints on food production. While eco- nomic access is often recognized as the main cause for undernutrition, limited avail- ability in periods of drought is typically the driver overlying spikes in food prices and associated food crises. Insufficient caloric intake may cause acute malnutrition (weight loss and/or inability to gain weight), a major cause for child mortality. Insufficient protein intake in the first 1000 days in the life of a child can cause chronic malnutrition (i.e., stunting) with permanent physical and cognitive impair- ment in children. Overall, malnutrition affects about 800–900 million people world- wide. Even though protein-rich food products tend to require three to six times more water per calorie than plant food, this chapter shows how the water needed to eradi- cate undernutrition can be sustainably met by the global water resources of the planet.

In this context, there is a difference between water withdrawals from water bod- ies and water consumption (i.e., evapotranspiration) by agriculture. Only a fraction of the water that is withdrawn from water bodies and applied to cropland is actually evapotranspired. The difference between the two eventually contributes to return flow to nearby water bodies and is available downstream for human uses or environmental need. Conversely, the water lost through evapotranspiration (or "consumptive use") will become available again after precipitation at a rate determined by the pace of the water cycle.

The energy sector is also a major water consumer, though at much smaller rates than that of agriculture. Almost all forms of energy production require water, both for the extraction of energy resources and for power generation. It has been estimated that energy production (2014 data) accounts for 400 km^3/year of blue water withdrawals and 50 km^3/year of consumptive use (IEA, 2016). These estimates include blue water for primary energy source extraction and power generation (Table 12.1) but do not include green water consumption for energy production from biomass (e.g., wood, bioethanol, and biodiesel) and losses from reservoirs for hydropower generation. The total water consumption for biofuel production is roughly 200 km^3/year (Rulli, Bellomi, Cazzoli, Carolis, & D'Odorico, 2016), which is about 2.4% of the water use in agriculture (Table 12.1).

Environmental Impacts of Water Withdrawals

The impacts of overpumping differ depending on whether it affects aquifers or surface water bodies. Many regions of the world, including the North-American Southwest, Northern Africa, the Indo-Gangetic Plain, the North China Plain, and the Arabian Peninsula, are affected by groundwater depletion (Konikow, 2011). Globally, groundwater depletion accounts for 20–33% of global groundwater withdrawals (Table 12.1).

The case of overuse of surface water resources is different because it raises concerns about the loss of aquatic habitat. Indeed, freshwater ecosystems exhibit high levels of biodiversity but are susceptible to much greater extinction rates (about five times) than those of other terrestrial ecosystems (Postel & Richter, 2003). To maintain biodiverse freshwater ecosystems, we need to ensure that suitable minimum flow and water quality requirements are met and that the timing, frequency, and magnitude of both high and low flows are not disrupted. Ecologists have defined minimum flow requirements (or "environmental flows") to place a cap on sustainable water use (e.g., Pastor, Ludwig, Biemans, Hoff, & Kabat, 2014; Richter, Davis, Apse, & Konrad, 2012).

Water Use for Agriculture

Most of the agricultural production on Earth is rain-fed and does not rely on irrigation. Roughly 18% of the global cultivated land is presently irrigated (Fig. 12.3) and contributes to about 40% of global agricultural production (e.g., Rosa et al., 2018).

Table 12.1 Global water flows and demands in km³/year

	Annual flow (km³ year⁻¹)	Year	Source
Precipitation over land	120,000		Chow (1988)
– Evapotranspiration from land (green water flows)	72,000		
– Global runoff (blue water flows)	48,000		
Planetary boundaries of blue water	4000		Rockström et al. (2009)
Water withdrawal for irrigation	2560	2000	Sacks et al. (2009)
Water consumption for irrigation	1280	2000	Siebert and Döll (2010)
	899	1996–2005	Mekonnen and Hoekstra (2011)
Unsustainable water consumption from irrigation	336	2000	Rosa et al. (2018)
Groundwater consumption for irrigation	540	2000–2010	Siebert and Döll (2010)
Groundwater withdrawals	730	2000	Wada et al. (2010)
Groundwater depletion	140	2001–2008	Konikow (2011)
	200	2000	Wada et al. (2010)
Water consumption for crop production (green + blue water)	6670	1996–2005	Mekonnen and Hoekstra (2011)
	11,800	2010	Carr et al. (2013)
Green water consumption for food production	5771	1996–2005	Mekonnen and Hoekstra (2011)
	6150	2000	Rosa et al. (2018)
Water consumption in rangelands and pastures	910	1996–2005	Mekonnen and Hoekstra (2011)
Water for livestock production (blue and green)	2260	1996–2005	Hoekstra and Mekonnen (2012)
Freshwater used for biofuel production	220	2013	Rulli et al. (2016)
– Green water for biofuel crops	170		
– Blue water for biofuel crops	11		
Virtual water trade (food only)	2810	2010	Carr et al. (2013)
Water grabbing	450	2013	Rulli et al. (2013)
Freshwater withdrawals for energy production	400		
– Primary energy source extraction	50	2016	IEA (2016)
– Power generation	350		
Freshwater consumption for energy production	50		
– Primary energy source extraction	34	2016	IEA (2016)
– Power generation	16		
Freshwater withdrawals from domestic sector	400–450	2000–2010	Wada et al. (2016)
Freshwater consumption from domestic sector	42	1996–2005	Hoekstra and Mekonnen (2012)

"Year" denotes the period considered in each study (see also D'Odorico et al., 2018)

Fig. 12.3 Current irrigated land. Areas with less than 5% irrigation (based on data by Siebert, Henrich, Frenken, & Burke, 2013) and where more than 90% of crop water needs are met by rainfed agriculture have been masked. According to these criteria, irrigated areas account for 2.5×10^6 km², which is about 18% of the global cultivated land. (Based on Rosa et al., 2018)

Rain-fed agriculture relies only on green water resources, while irrigated agriculture uses both green and blue water.

Green water is often taken for granted because it is directly acquired with the land without requiring investments in infrastructure for its capture or institutional arrangements to define the right to use it. The choice to plant certain crops may strongly affect evapotranspiration losses, with an indirect impact also on runoff. These effects of agriculture on croplands' hydrologic response are often ignored.

Sustainable and Unsustainable Water Use for Irrigation

Estimates of global water consumption for irrigation (Table 12.1) vary between 900 and 1280 km³/year (Mekonnen & Hoekstra, 2012; Siebert & Döll, 2010), while withdrawals range between 2410 (Jägermeyr, Pastor, Biemans, & Gerten, 2017) and 2560 km³/year (Sacks et al., 2009). Even though these values are well within the sustainability limits (or "planetary boundaries") identified by some authors (≈ 1100–4500 km³/year, Gerten et al., 2013; Rockström et al., 2009) and are way below the global estimates of blue water flows (4800 km³/year, see Table 12.1), there are numerous instances of locally unsustainable water uses.

As a major form of human appropriation of blue water resources, agriculture is the main reason why environmental flows have been strongly depleted in many rivers worldwide. About 41% of water withdrawals for irrigation (Jägermeyr et al., 2017) and 40% of irrigation water consumption (Rosa et al., 2018) are unsustainable because they occur at the expenses of environmental flows and/or groundwater stocks (Fig. 12.3).

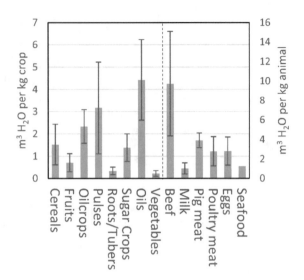

Fig. 12.4 The water footprint of major crops and animal products (based on data in Davis et al., 2016)

Livestock Production

About 30% of agricultural water consumption is used for livestock production, mostly (>90%) in the form of green water for feed, fodder, and rangeland grazing (Table 12.1). It takes much more water to produce a calorie of food from animal products than from crops. The consumption of meat-rich diets remains a water-intensive (and carbon-intensive) lifestyle (e.g., Davis et al., 2016; Tilman, Balzer, Hill, & Befort, 2011). Thus, the observed trend of decreasing cereal consumption and increasing reliance on animal products with increasing levels of affluence (Tilman et al., 2011) is expected to enhance the human pressure on global freshwater resources (e.g., Davis et al., 2016). Are these hydrologic effects of dietary changes on water consumption of the same order of magnitude as those of demographic growth? Using a GDP-based projection of dietary change, Davis et al. (2016) estimated that by 2050 the average per capita water consumption for food production will increase by 19% from the present level, while population growth is expected to increase by 35%, reaching 9.8 billion people by 2050. Thus, the hydrologic impact of dietary shifts in the burgeoning developing societies is expected to be of the same order of magnitude as demographic growth (Fig. 12.4).

Projections of Future Water Needs for Food Production

To meet the food needs of the increasing global population while eradicating under-nourishment, humanity will likely need to reduce food waste and the consumption of resource-intensive products (e.g., Foley et al., 2011). In fact, there are limits to the sustainable increase in food production, and some of these limits are associated with water resources (Falkenmark & Rockström, 2004).

Recent studies have stressed the danger of increasing food production by expanding agriculture because it would result in habitat destruction, biodiversity loss and CO_2 emissions. Therefore, scientists have been advocating for the intensification of agriculture in land that is presently cultivated (or "land sparing") (Foley et al., 2011). Intensification typically requires the use of fertilizers, irrigation, and other modern technology to close the yield gap, which is the difference between actual and maximum potential yields. Yield gap estimates are typically developed without evaluating to what extent the local hydrologic conditions allow for a sustainable development of irrigation. Indeed, large tracts of agricultural land are not suitable for sustainable irrigation, while in 26% of currently rain-fed areas it is possible to irrigate without depleting environmental flows or groundwater stocks (Rosa et al., 2018). If at the same time current unsustainable irrigation practices (Fig. 12.3) were eliminated, sustainable irrigation expansion would allow for a 24% increase in global food production that could feed an additional 1.8 billion people (Rosa et al., 2018). Collectively, these results demonstrate the existence of limits to the sustainable appropriation of water resources for agriculture and to humanity's ability to sustainably increase food production. Further needs for water resources for food production will need to result from water conservation (D'Odorico et al., 2018).

Pathways to Water Conservation

A more sustainable approach to water and food security looks at a reduction of consumption as an alternative to increasing production. The ongoing pattern of overconsumption can be broken by reducing food waste, promoting the shift to diets that are less resource intensive, and decreasing the amount of food intake in overnourished populations.

Waste Reduction

About 24% of the water used for food production is lost through food waste (Kummu et al., 2012). These losses occur at different stages along the "farm to fork" supply chain, depending on the type of food, access to refrigeration and transportation facilities, and household habits. In the last few years, there has been a push toward a circular economy of food to reduce resource consumption and waste accumulation by repurposing and reusing food products (e.g., Stahel, 2016). In some countries, the destruction of unsold food by retailers or distributors is avoided through timely donations to food charities. Some food types can be reused to produce other food commodities, or utilized as feed, compost or substrate for biogas production. Likewise, nutrient-rich wastewater can be reused for irrigation, biogas production, or to recover critical nutrients such as phosphorus (Cordell, Drangert, & White, 2009).

Changes in Diets

A reduction in human demand for water resources could result from a switch to less meat-intensive diets. For instance, the shift to vegetarian or pescatarian diets would save 18–21% of the water used for food production and improve human health (Davis et al., 2016; Tilman & Clark, 2014). Human diets are the result of historical habits and cultural heritages and are therefore hard to change (e.g., Wellesley, Happer, & Froggatt, 2015). A positive sign can be seen in the ongoing *livestock transition*, the replacement of resource-intensive bovine meat with pork and poultry (Davis et al., 2016; Thornton, 2010).

Crop Water Management

While transpiration is a productive water loss from terrestrial ecosystems, evaporation does not contribute to plant growth and can even negatively affect soil properties by increasing salinity. Therefore, to produce more food with less water, agricultural practices need to reduce evaporation and maximize transpiration (Falkenmark & Rockström, 2004). Improvement in water productivity based on soil water management could reduce current water withdrawals for irrigation by about 1200 km^3/year (Jägermeyr et al., 2016).

Improving Crop Suitability by Replacing Crops

Improvements in water use efficiency to produce "more crop per drop" can be attained either by using new cultivars or by changing crop distribution. The current distribution of crops often makes a suboptimal use of water resources. Water-demanding crops are often planted in arid and semiarid regions. Perennial plants are often cultivated in climates affected by episodic droughts that could kill what took years to grow. Alternative crop distributions can maximize production and feed about 825 million additional people, while saving 12% of irrigation water without sacrificing crop diversity, nutritional value, or land (Davis, Rulli, Seveso, & D'Odorico, 2017). This approach does not require major investments in modern technology that small-scale subsistence farmers in the developing world might not be able to afford.

Water Savings through Trade

Another option to save water is through trade, that is, by producing crops in more suitable areas and exporting them to other regions of the world. Indeed, the current patterns of trade contribute to water savings for about 350 km³/year (Hoekstra & Chapagain, 2008). Other studies found smaller savings and documented their growth over time, from roughly 50 km³/year in 1986 to 240 km³/year in 2008 (Dalin, Konar, Hanasaki, Rinaldo, & Rodríguez-Iturbe, 2012). Further savings could be attained by reshaping trade patterns and future trade intensification.

Globalization of Water

When the withdrawal and consumption of freshwater resources are examined at the global scale, human uses are dwarfed by the magnitude of global runoff and no major environmental impact would seem to emerge from human action (Table 12.1). At a local scale, however, water resources can be strongly stressed by agricultural, municipal, and industrial uses. Several regions of the world are in a state of chronic water deficit because their population by far exceeds the number of people the local land and water resources are able to feed. The existence of countries that are strongly water and food deficient appears to be paradoxical, as one would expect that these conditions would trigger mass emigrations, social unrest, and conflict over the scarce water resources. However, in general this is not the case because trade allows societies to indirectly rely on water resources existing in other regions of the world (Allan, 1998). In other words, the trade of food commodities is associated with a virtual transfer of the water resources required for the production of those goods. Known as "virtual water trade," this mechanism contributes to the globalization of water resources.

The international trade of virtual water accounts for about 25% of water use in agriculture, including both green and blue water (Mekonnen & Hoekstra, 2012). Thus, while 10–16% of the evapotranspiration by the terrestrial biosphere is contributed by agroecosystems, roughly one-fourth of it is virtually traded (Table 12.1). In the last two decades and a half, virtual water trade has more than doubled in volume and the trade network has become more interconnected. Net exports are mostly contributed by a handful of countries, such as Brazil, the USA, Argentina, Thailand, Canada and Australia (Carr, D'Odorico, Laio, & Ridolfi, 2013). These results suggest that water is a strategic resource that is strongly globalized and controlled by few countries, while the rest of the world is in conditions of trade dependency.

Water is also globalized though transnational land investments, a phenomenon that has intensified over the last decade and a half. Since 2002, about half a million square kilometers of land (i.e., 1.7 times the area of Italy) have been acquired for agriculture and forestry (The Land Matrix, 2017). Land investors can be domestic, foreign, or mixed domestic-foreign ventures that operate as private agribusiness corporations, government-owned companies, or retirement funds (Cotula, 2013). Investors often target both the land and the water that comes with it. Thus, land acquisitions often become opportunities to secure access to both land and water resources abroad (Dell'Angelo, Rulli, & D'Odorico, 2018; Rulli, Saviori, & D'Odorico, 2013). "Water grabbing" is defined as the appropriation of water resources by powerful investors at the expenses of local communities (Dell'Angelo et al., 2018). Appropriation mechanisms may include the redefinition or reallocation of water rights, river diversions, dam constructions, and virtual appropriation of water resources through land acquisitions (Mehta, Veldwisch, & Franco, 2012; Rulli et al., 2013). Interestingly, this phenomenon exceeds the rate of water grabbing from future generations through groundwater depletion (Table 12.1).

Conclusion

Freshwater is crucial to sustain ecological function, food security, energy production, and investments in a variety of businesses. With challenges to water sustainability emerging from growing societal pressures, humanity is facing the trilemma of allocating water to food, energy, or environmental needs (D'Odorico et al., 2018).

How did we reach this point? There have been some major revolutions in the food–water system. The industrial revolution brought about new sources of energy and power, which allowed for the withdrawal and transport of unprecedented amounts of water from surface water bodies and aquifers and the expansion of irrigation to areas not suitable for gravity-driven distribution systems. The green revolution in the decades following World War II led to a major growth in agricultural production with an incredible increase in crop yields. The early phases of the green revolution hinged on the invention of nitrogen fertilizers (e.g., Erisman, Sutton, Galloway, Klimont, & Winiwarter, 2008). At that point, water remained a major limiting factor for crop production, which led to a renewed interest in irrigation infrastructure and improvements in irrigation efficiency.

More recently, the water and food system have been transformed by a "trade revolution." The intensification of trade has allowed some populations to rely on water resources existing elsewhere (Allan, 1998).

A "fourth food revolution" (D'Odorico et al., 2018) is associated with large-scale land acquisitions, a phenomenon that has rapidly accelerated in the last few years, likely in response to recent food crises (Cotula, 2013; Rulli et al., 2013). After some trade-dependent countries experienced conditions of food insufficiency, several agribusiness corporations started to look at regions of the developing world where yields are low because of the lack of investments in modern agricultural technology. This phenomenon contributes to the transition from subsistence or small-scale farming to large-scale commercial agriculture and is often associated with large-scale investments in agricultural land.

All these "revolutions" led to major changes in agriculture, through industrialization, new technology, globalization, and transition in farming systems. With these changes, humanity has been trying to meet the increasing needs of human societies by sustaining ongoing growth trends in agriculture. Sadly, these changes are frequently accompanied by unsustainable practices and acute social injustices working against the poorest part of the populations. There is the need for new approaches that focus on a reduction in consumption. The pathways to water saving reviewed in this chapter could substantially reduce human consumption of water resources.

References

Allan, J. A. (1998). Virtual water: A strategic resource. *Ground Water, 36*, 545–547.

Bonan, G. B. (2008). *Ecological climatology* (p. 550). New York: Cambridge University Press.

Carr, J. A., D'Odorico, P., Laio, F., & Ridolfi, L. (2013). Recent history and geography of virtual water trade. *PLoS One, 8*, e55825.

Chow, V. T. (1988). *Applied hydrology*. New York: McGraw-Hill Education.

Cordell, D., Drangert, J. O., & White, S. (2009). The story of phosphorus: Global food security and food for thought. *Global Environmental Change, 19*, 292–305.

Cotula, L. (2013). The new enclosures? Polanyi, international investment law and the global land rush. *Third World Quarterly, 34*, 1605–1629.

D'Odorico, P., Davis, K. F., Rosa, L., Carr, J. A., Gephart, J., Seekell, D. A., et al. (2018). The food-energy-water nexus. *Reviews of Geophysics, 56*, 456–531.

Dalin, C., Konar, M., Hanasaki, N., Rinaldo, A., & Rodríguez-Iturbe, I. (2012). Evolution of the global virtual water trade network. *Proceedings of the National Academy of Sciences of the United States of America, 109*, 5989–5994.

Davis, K. F., Gephart, J. A., Emery, K. A., Leach, A. M., Galloway, J. N., & D'Odorico, P. (2016). Meeting future food demand with current agricultural resources. *Global Environmental Change, 39*, 125–132.

Davis, K. F., Rulli, M. C., Seveso, A., & D'Odorico, P. (2017). Increase in food production and reduction in water use through optimized crop distribution. *Nature Geosciences, 10*, 919–924. https://doi.org/10.1038/s41561-017-0004-5

Dell'Angelo, J., Rulli, M. C., & D'Odorico, P. (2018). The global water grabbing syndrome. *Ecological Economics, 143*, 276–285.

Dunne, T., & Leopold, L. B. (1978). *Water in environmental planning*. San Francisco: W.H. Freeman.

Erisman, J. W., Sutton, M. A., Galloway, J., Klimont, Z., & Winiwarter, W. (2008). How a century of ammonia synthesis changed the world. *Nature Geoscience, 1*, 636.

Falkenmark, M., & Rockström, J. (2004). *Balancing water for humans and nature: The new approach in ecohydrology*. London: Earthscan.

Foley, J. A., et al. (2011). Solutions for a cultivated planet. *Nature, 478*, 337–342.

Gerten, D., Hoff, H., Rockström, J., Jägermeyr, J., Kummu, M., & Pastor, A. V. (2013). Towards a revised planetary boundary for consumptive freshwater use: Role of environmental flow requirements. *Current Opinion in Environmental Sustainability, 5*, 551–558.

Held, I. M., & Soden, B. J. (2006). Robust responses of the hydrological cycle to global warming. *Journal of Climate, 19*, 5686–5699.

Hoekstra, A. Y., & Chapagain, A. (2008). *Globalization of water*. New York: Wiley.

Hoekstra, A. Y., & Mekonnen, M. M. (2012). The water footprint of humanity. *Proceedings of the National Academy of Sciences of the United States of America, 109*, 3232–3237.

IEA (2016). *World Energy Outlook 2016*. International Energy Agency.

Jägermeyr, J., Gerten, D., Schaphoff, S., Heinke, J., Lucht, W., & Rockström, J. (2016). Integrated crop water management might sustainably halve the global food gap. *Environmental Research Letters, 11*, 025002.

Jägermeyr, J., Pastor, A., Biemans, H., & Gerten, D. (2017). Reconciling irrigated food production with environmental flows for sustainable development goals implementation. *Nature Communications, 8*, 15900. https://doi.org/10.1038/ncomms15900

Konikow, L. F. (2011). Contribution of global groundwater depletion since 1900 to sea-level rise. *Geophysical Research Letters, 38*, L17401. https://doi.org/10.1029/2011GL048604

Kummu, M., de Moel, H., Porkka, M., Siebert, S., Varis, O., & Ward, P. J. (2012). Lost food, wasted resources: Global food supply chain losses and their impacts on freshwater, cropland, and fertiliser use. *Science of the Total Environment, 438*, 477–489.

Mehta, L., Veldwisch, G. J., & Franco, J. (2012). Introduction to the special issue: Water grabbing? Focus on the (re) appropriation of finite water resources. *Water Alternatives, 5*, 193–207.

Mekonnen, M. M., & Hoekstra, A. Y. (2011). National water footprint accounts: The green, blue and grey water footprint of production and consumption. Value of Water Research Report Series No. 48. UNESCO-IHE, Delft, the Netherlands.

Mekonnen, M. M., & Hoekstra, A. Y. (2012). A global assessment of the water footprint of farm animal products. *Ecosystems, 15*, 401–415.

Myers, S. S., et al. (2014). Increasing CO_2 threatens human nutrition. *Nature, 510*, 139–142.

Pastor, A. V., Ludwig, F., Biemans, H., Hoff, H., & Kabat, P. (2014). Accounting for environmental flow requirements in global water assessments. *Hydrology and Earth System Sciences, 18*, 5041–5059.

Postel, S., & Richter, B. (2003). *Rivers for life: Managing water for people and nature*. Washington, DC: Island Press.

Richter, B. D., Davis, M. M., Apse, C., & Konrad, C. (2012). A presumptive standard for environmental flow protection. *River Research and Applications, 28*, 1312–1321.

Rockström, J., et al. (2009). A safe operating space for humanity. *Nature, 461*, 472–475.

Rodriguez-Iturbe, I., & Porporato, A. (2005). *Ecohydrology*. New York: Cambridge University Press.

Rosa, L., Rulli, M. C., Davis, K. F., Chiarelli, D. D., Passera, C., & D'Odorico, P. (2018). Sustainable yield gap closure through irrigation intensification. *Environmental Research Letters, 13*, 104002.

Rulli, M.C., Bellomi, D., Cazzoli, A., De Carolis, G., & D'Odorico, P. (2016). The water-land-food nexus of first-generation biofuels. *Scientific Reports, 6*, 22521. https://doi.org/10.1038/srep22521

Rulli, M. C., Saviori, A., & D'Odorico, P. (2013). Global land and water grabbing. *Proceedings of the National Academy of Sciences of the United States of America, 110*, 892–897.

Sacks, W. J., Cook, B. I., Buenning, N., Levis, S., & Helkowski, J. H. (2009). Effects of global irrigation on the near-surface climate. *Climate Dynamics, 33*, 159–175. https://doi.org/10.1007/s00382-008-0445-z

Siebert, S., & Döll, P. (2010). Quantifying blue and green virtual water contents in global crop production as well as potential production losses without irrigation. *Journal of Hydrology, 384*, 198–217.

Siebert, S., Henrich, V., Frenken, K. & Burke, J. (2013). *Global map of irrigation areas version 5*. Bonn, Germany/Rome, Italy: Rheinische Friedrich-Wilhelms-University/FAO.

Stahel, W. R. (2016). Circular economy: A new relationship with our goods and materials would save resources and energy and create local jobs. *Nature, 531*, 435–439.

Stone, R. (2015). Saving Iran's great salt lake. *Science, 349*, 1044–1047.

The Land Matrix. (2017). *The Land Matrix Global Observatory*. Retrieved December 1, 2019 from www.landmatrix.org.

Thornton, P. K. (2010). Livestock production: Recent trends, future prospects. *Philosophical Transactions of the Royal Society of London B: Biological Sciences, 365*, 2853–2867.

Tilman, D., Balzer, C., Hill, J., & Befort, B. L. (2011). Global food demand and the sustainable intensification of agriculture. *Proceedings of the National Academy of Sciences of the United States of America, 108*, 20260–20264.

Tilman, D., & Clark, M. (2014). Global diets link environmental sustainability and human health. *Nature, 515*, 518–522.

Wada, Y., van Beek, L. P., van Kempen, C. M., Reckman, J. W., Vasak, S., & Bierkens, M. F. (2010). Global depletion of groundwater resources. *Geophysical Research Letters, 37*. https://doi.org/10.1029/2010GL044571

Wada, Y., Flörke, M., Hanasaki, N., Eisner, S., Fischer, G., Tramberend, S., et al. (2016). Modeling global water use for the 21st century: Water Futures and Solutions (WFaS) initiative and its approaches. *Geoscientific Model Development, 9*, 175–222.

Wellesley, L., Happer, C., & Froggatt, A. (2015). *Changing climate, changing diets: Pathway to lower meat consumption*, Chatham House Report, The Royal Institute of International Affairs, London.

Whittaker, R. H. (1975). *Chapter 5: Communities and ecosystems*. New York: McMillan Press.

CHAPTER 13
Health Climate Justice and Deforestation in the Amazon

Virgilio Viana

Summary Health is a dimension of tropical deforestation that deserves greater attention in light of the current trends in climate change, environmental degradation and national and subnational development policies. Forest fires are becoming more frequent and larger in scale globally (Baccini et al., 2017). Deforestation fuels global warming, and reducing forest fires is essential to meet the Paris Agreement targets (UNEP, 2018).

Climate change is affecting rainfall regimes, including the length of dry seasons, which fuels forest fires and thus creates a feedback loop mechanism. Increased fire frequency results in air pollution, which is associated with a number of health problems. Ambient air pollution accounts for an estimated 4.2 million deaths per year due to stroke, heart disease, lung cancer, and chronic respiratory diseases. Around 91% of the world's population lives in places where air quality levels exceed World Health Organization (WHO) limits (WHO, 2019).

The Amazon is home to 34 million people who are exposed to air pollution associated with forest fires. This includes more than 380 indigenous groups of peoples who suffer from one of the greatest climate injustices: despite having had a very low carbon footprint, they are facing the consequences of climate change, including health problems associated with air pollution from forest fires. Reducing deforestation and forest fires is relevant both for local populations and for mitigating global climate change.

This chapter analyzes the linkages between health, air pollution and forest fires in the Amazon. The goal is to highlight the importance of having an understanding of the complexity of these linkages to encourage further research and policy approaches.

V. Viana (✉)
Sustainable Amazon Foundation (FAS), Cambridge, MA, USA
e-mail: virgilio.viana@fas-amazonas.org

Deforestation and Fires in the Amazon

The Amazon is the largest tropical rainforest in the world, covering 6.7 million hectares, spreading across eight countries and one province: Brazil (64.3%), Peru (10%), Bolivia (6.2%), Colombia (6.2%), Venezuela (5.8%), Guyana (2.8), Suriname (2.1%), Ecuador (1.5%), and French Guiana (1.1%) (RAISG, 2015). While Brazil has 71.5% of the Amazon Basin, it has only 64.3% of the Amazon Biome. The Amazon Basin represents 66% of the Brazilian territory, while the Amazon Biome represents 58.8% of the country area. By contrast, Suriname and French Guiana have practically no area in the Amazon Basin, while having 100% of their territory in the Amazon Biome (Table 13.1). The Amazon is extremely diverse in both biophysical and human dimensions. Therefore, this heterogeneity and complexity has to be considered and generalizations for the region should be taken with caution. This chapter focuses on Brazil, but most of the data and conclusions are also relevant to other Amazonian countries.

The Amazon plays a vital role for the earth's climate as a mega-provider of ecosystem services, including storage of live carbon (176.297 million tons/25% of global storage) (FAO, 2011), conservation of biodiversity (over 10% of total species) and a water pump to the atmosphere (over 20 billion tons of water every day) (Nobre, 2014). The Amazon generates approximately half of its own rainfall by recycling moisture five to six times as air masses move from the Atlantic across the basin to the west (Salati et al., 1979).

Deforestation continues at an alarming rate: 7.900 km^2 in 2018 in Brazil alone, an increase of 13.7% from 2017 (INPE, 2019). Forest fires have increased similarly. The Amazon has warmed, reaching 0.6–0.7 °C over the last 40 years, with 2016 as the warmest year since at least 1950 (0.9 °C + 0.3 °C). The dry-season length increased and more climate extremes affect the region with impacts on carbon sequestration and storage, river streamflow, and biodiversity, even in the large forested areas designated to conservation and protection (Marengo et al., 2018).

Recent studies have demonstrated that interactions between deforestation, fire, and severe drought potentially lead to losses of carbon storage and changes in regional precipitation patterns and river discharge (Davidson et al., 2012). A study in the state of Rondonia, Brazil—a region that has been continuously deforested since the 1970s—indicates that in areas highly deforested, the beginning of the rainy season has significantly shifted to, on average, 11 days (and up to 18 days) later in the year over the last three decades. However, in areas that have not been heavily deforested, that did not happen. Delayed onset of rains may be a result of land use change, and this signal may strengthen in future (Butt et al., 2011).

Amazon forest's susceptibility to fires during extreme events is not only driven by climate, but also the previous disturbance history, and modulated by forest structure. Heavily degraded forests with low canopy cover and biomass suffer water stress and increased flammability earlier in the drought period than old-growth forests (Longo et al., 2018).

Table 13.1 Land and environmental profile of Amazon countries

Country	Total area (million km²)	% of total Amazon basin found in each country	% of Amazon basin for each country	% of total Amazon biome found in each country	% of Amazon biome for each country	% forest loss
Brasil	8,516,000	71.5	66	64.3	58.8	17.6
Bolivia	1,099,000	9.2	66	6.1	43.2	7.3
Colombia	1,142,000	4.4	30	6.2	42.3	9.9
Equador	283,560	1.8	51	1.5	41.1	10.7
Guyana	214,970	0.1	5	2.8	100	2.5
French Guiana	83,534	0	0	1.1	100	2.8
Peru	1,285,000	12.3	75	10.1	60.9	9.1
Suriname	163,821	0	0	2.1	100	4.2
Venezuela	916,445	0.7	6	5.8	49.5	3.3
Total	13,704,330	100		100		6.2

Source: SDSN-Amazonia at https://www.sdsn-amazonia.org/pt-plataforma

Amazon Tipping Point

A suite of general circulation model projections generally agrees, showing robust cross-model drying in the Amazon and other regions (Cook et al., 2014). This can generate feedback loops given the fact that the Amazon generates approximately half of its own rainfall by recycling moisture five to six times as air masses move from the Atlantic across the basin to the west (Salati et al., 1979).

Negative synergies between deforestation, climate change, and widespread use of fire indicate a tipping point for the Amazon system to flip to non-forest ecosystems in eastern, southern, and central Amazonia at 20–25% deforestation (Lovejoy & Nobre, 2018). However, with the combined stress of rising temperature, deforestation, and increased forest fires, the tipping point may be reached significantly earlier. To date, almost 18% of the Amazonian basin has been deforested (Gaffney et al., 2018). While there is scientific uncertainty about the tipping point (Fig. 13.1), the precautionary principle recommends setting a goal for zero net deforestation no later than 2025—preferably earlier.

There are already signs of a transition to a disturbance-dominated regime in the Amazon, driven by interactions between deforestation, fire, and drought, potentially leading to losses of carbon storage and changes in regional precipitation patterns and river discharge (Davidson et al., 2012).

The Amazon biodiversity is highly vulnerable to climate change. The Amazon contains one of Earth's richest assortments of biodiversity with recent compilations indicating at least 40,000 plant species, 427 mammals, 1294 birds, 378 reptiles, 427 amphibians, 3000 fishes, and likely over a million insect species (Case, 2006). Even a 2 °C rise would make the new average temperature hotter than previous extremes, and would threaten more than one-third of species in all groups in the absence of

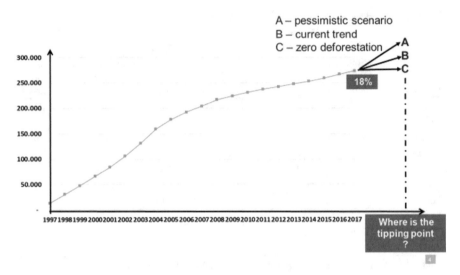

Fig. 13.1 Cumulative deforestation in the Brazilian Amazon (km²). Based on official deforestation data (http://www.inpe.br)

dispersal. A business-as-usual scenario would see this figure increase to around two-thirds (WWF, 2018).

Forest Fires and Health

Forest fires increase air pollution, and this results in health problems. Evidence on the nexus between forest fires, air pollution and health is not new. Human health impacts of forest fires include biophysical effects, psychosocial impacts, occupational exposure issues among fire crew members, and visibility impairment (Fowler, 2003). A recent literature review of public health issues caused by forest fires between 1990 and 2018 presented evidence on mortality, respiratory and cardiovascular morbidity as well as other possible impacts on human health (Cancedo-González & Viqueira, 2018).

A study of 13.5 years of data including 48 days affected by wildfire smoke in Sydney, Australia, demonstrated a significant increase in mortality associated with smoke-affected days (Johnston et al., 2011). An earlier study of mortality in Sydney, using 8 years of data, found an increase in mortality associated with wildfire-related PM10 (Morgan et al., 2010). An analysis of forest fires that blanketed the Indonesian islands of Kalimantan and Sumatra in late 1997 found that individuals who were exposed to haze experienced greater increases in difficulty with activities of daily living and that haze had a negative impact on respiratory and general health (Frankenberg et al., 2005). A meta-analysis of data from 2003 to 2010 in 10 cities in southern Europe found increases in cardiovascular mortality associated with

PM10 that were stronger on smoke-affected days than on non-affected days (Faustini et al., 2015).

Large-scale intentional forest fires in Indonesia cause haze in neighboring countries. The causal impact of these fires on Singaporean health outcomes had estimated partial health and avoidance costs of US$184 million from September 2012 to August 2017 (Sankaran & Sheldon, 2018). A study on prenatal exposure to the 1997 Indonesian forest fires based on child nutritional outcomes found that mean exposure to air pollution during the prenatal stage was associated with a 0.5 standard deviation decrease in height-for-age z score at age 17. Because adult height is associated with income, this implies a loss of 4% of average monthly wages for approximately one million Indonesian workers born during this period (Tan-Soo & Pattanayak, 2019).

An analysis of hospitalizations due to respiratory diseases in people over 60 years of age who were residents of Cuiabá in the Brazilian Amazon during 2012 found significant correlations between exposure to these contaminants and hospitalizations. There was a hospitalization risk increase of 31.8%, with an increase of 3.5 $\mu g/m^3$ of PM2.5 concentrations and an increase of 188 in the total number of hospitalizations (Machin et al., 2019).

The spatial variability of aerosols over Amazon Basin shows that large areas of three to five million km^2 are affected by high aerosol concentrations. The concentration of trace gases and aerosols from forest fires in the dry season increase by factor of 2–8 in large areas (Artaxo et al., 2006).

A 10 $\mu g/m^3$ increase in the level of exposure to particulate matter in the Amazon was associated with increases of 2.9% and 2.6% in outpatient consultations due to respiratory diseases in children on the sixth and seventh days following exposure, respectively. Levels of particulate matter from biomass burning in the Amazon were associated with adverse effects on the respiratory health of children (Carmo et al., 2010). A cross-sectional analysis of cardiovascular mortality among people over 65 years in the Brazilian Amazon, where the predominant source of air pollution is from wildfires, found a significant association between the percentage of hours of PM2.5 over 25 $\mu g/m^3$ and cardiovascular mortality (Nunes et al., 2013).

Implications for Public Policies and Research

Considering the fact that there is clear evidence that forest fires are directly correlated to health problems, and that climate change is likely to increase the magnitude of this problem, there is urgency for action.

Deforestation is a result of a complex sum of factors, which vary both in space and time. Therefore, it is necessary to take into account this complexity to understand the dynamics of deforestation. Drivers of deforestation can be grouped into (1) economic factors (market structure, conditions and access to credit, incentives, prize of commodities, international demand, etc.), (2) governance factors (budget and structure of environmental institutions, efficiency of command and control instruments, management efficacy of protected areas and indigenous lands); (3) political factors (power of the lobby of the agricultural sector; changes in legislation at the federal, state and municipal levels), (4) infrastructure development (road construction, hydroelectric dams, etc.) and development of extractive industries (mining, oil, and gas), and (5) structural factors (population growth and migration, poverty and inequality, etc.). Reducing deforestation and forests fires, therefore, is no simple task (Viana, 2017).

A roadmap for action should include a mix of policy instruments to: (1) strengthen governance in the frontier, (2) increase funding mechanisms for sustainable development in the Amazon, (3) change the paradigm of financing for infrastructure, (4) reduce corruption and increase transparency, (5) improve corporate behavior, (6) strengthen civil society organizations, (7) education and communication, (8) ecosystem services and poverty, (9) promote community-based sustainable development pathways, especially in protected areas and indigenous lands, and (10) build political support (Viana, 2017).

Environmental policies have had a setback in recent elections in Brazil, both at the national and subnational levels. There is a general perception that environmental policies are slowing economic growth and thus need to be weakened (Viana, 2019).

The connection between deforestation, forest fires and health can be an important tool to change environmental policies in the Amazon. Health relates directly to the interests of all citizens and their family members. High air pollution is seen and easily understood by the general public. Health problems are publicized in the media, and this creates an opportunity to drive policy change.

Institutions engaged in the design and implementation of environmental public policies should give greater emphasis to the health dimension of deforestation. The same applies to those engaged in public health public policies: they should emphasize the forest fire–health nexus. There is a need for a massive communication strategy to create greater public awareness and thus influence policy-making on the forest fire–health nexus.

There are some questions that require attention in the design and implementation of sound public policies: How can we communicate effectively and promote dialogue on public policies associated with the forest fires, air pollution, and health? How can we disseminate research results on the nexus between forest fires, air pollution, and health?

Although the body of evidence is quite sufficient for practical action, there are several research questions that deserve further work: How can we address the ethical components of climate injustice towards indigenous peoples and isolated communities and their linkages to health and forest fires? How can we improve public health systems to deliver health care to problems associated with air pollution in the Amazon, especially for indigenous peoples and isolated communities? How can we improve the understanding of the consequences of air pollution to public health in the Amazon? What are the scenarios for forest fires, air pollution, and health in their short term and long term in light of climate change?

Conclusions and Call to Action

Public policies in the Amazon aimed at reducing deforestation and forest fires are traditionally driven by an environmental rationale, without appropriate consideration to the human health dimension. Public health policies do not give due consideration to air pollution derived from forest fires. Climate policies do not give the necessary attention to forest fires. This needs to change if the Amazon is to succeed in meeting the Sustainable Development Goals by 2030.

A better understanding of the nexus between forest fires and health can change societal perceptions and become drivers of necessary change in public policies. This is key to strengthen policies aimed at reducing forest fires in the Amazon.

There is a need for urgent action to avoid the tipping point of the Amazon forest, which would have profound consequences for Brazil, South America, and the planet. The following actions are recommended:

1. **Policy dialogue, design, and implementation of cross-sectoral policies aiming at zero deforestation** in all nine Amazon countries by 2025. The process should be informed by science and supported by massive communication about the health–deforestation nexus, in all Amazon regions. Other regions of South America should be informed about the importance of the Amazon to maintain rainfall regimes. Zero net deforestation policies should include reforestation, such as agroforestry systems to boost resilience and compensate for residual deforestation. Research on the efficacy and efficiency of different policy instruments should be supported.
2. **New financial mechanisms to support implementation of zero net deforestation policies.** These mechanisms should include a carbon tax and payments for ecosystem services to compensate Amazon populations, especially indigenous peoples, for the benefits (carbon, biodiversity, water, others) provided by their forests. Research on the importance and value of the Amazon for rainfall regimes and climate stability in the region and globally should be supported.
3. **Health policies to improve care for air pollution-related diseases.** These policies should have a special focus on indigenous peoples, who suffer from one of the world's greatest climate injustices, because they have not been part of the problem of greenhouse gas emissions. Research on the health impacts and appropriate care for deforestation-borne air pollution should be supported.
4. **Develop new partnerships engaging the business and civil society organizations in the design and implementation of practical solutions for sustainable development of the Amazon.** These solutions should support national and subnational governments to increase efficiency, efficacy and transparency. These partnerships should be broad, including local, national and global actors. The Amazon is too big to fail.

References

Artaxo, P., Oliveira, P. H., Lara, L. L., Pauliquevis, T. M., Rizzo, L. V., Pires Junior, C., et al. (2006). Efeitos climáticos de partículas de aerossóis biogênicos e emitidos em queimadas na Amazônia. *Revista Brasileira de Meteorologia, 21,* 168–122. Retrieved from http://sigma.cptec.inpe.br/aerossois/documentos/LBA040-2006_PArtaxo.pdf

Baccini, A., Walker, W., Carvalho, L., Farina, M., Sulla-Menashe, D., & Houghton, R. A. (2017). Tropical forests are a net carbon source based on aboveground measurements of gain and loss. *Science, 358,* 230–223. Retrieved from http://science.sciencemag.org/content/358/6360/230

Butt, N., de Oliveira, P. F., & Costa, M. H. (2011). Evidence that deforestation affects the onset of the rainy season in Rondonia, Brazil. *Journal of Geophysical Research, 116,* D11120. Retrieved from https://agupubs.onlinelibrary.wiley.com/doi/epdf/10.1029/2010JD015174

Cancedo-González, & Viqueira. (2018). Forest fires and public health. *Anales de la Real Academia Nacional de Farmacia, 84,* 289–300. Retrieved from https://www.analesranf.com/index.php/aranf/article/download/1902/1888

Carmo, C. N., Hacon, S., Longo, K. M., Freitas, S., Ignotti, E., Leon, A. P., et al. (2010). Associação entre material particulado de queimadas e doenças respiratórias na região sul da Amazônia brasileira. *Revista Panamericana de Salud Pública, 27,* 10–16. Retrieved from https://www.scielosp.org/pdf/rpsp/2010.v27n1/10-16/pt

Case M. (2006). Climate Change Impacts in the Amazon: Review of the Scientific Literature (World Wildlife Fund) 8th Conference, of the Parties, to the Convention on Biological Diversity (pp. 20–31). Retrieved November 27, 2019 https://d2ouvy59p0dg6k.cloudfront.net/downloads/amazon_cc_impacts_lit_review_final_2.pdf

Cook, B. I., Smerdon, J. E., Seager, R., & Coats, S. (2014). Global warming and 21st century drying. *Climate Dynamics, 43,* 2607–2627. Retrieved from http://www.readcube.com/articles/10.1007/s00382-014-2075-y

Davidson, E. A., de Araújo, A. C., Artaxo, P., Balch, J. K., Brown, I. F., Bustamante, M. M. C., et al. (2012). The Amazon basin in transition. *Nature, 481,* 321–328. Retrieved from https://www.nature.com/articles/nature10717

FAO (2011). *The State of Forests, in the Amazon Basin, Congo Basin and Southeast Asia.* Retrieved from http://www.fao.org/3/i2247e/i2247e00.pdf

Fowler. (2003). Human health impacts of forest fires in the southern United States: A literature review. *Journal of Ecological Anthropology, 7,* 39–63. Retrieved from https://scholarcommons.usf.edu/jea/vol7/iss1/3/

Faustini, A., Alessandrini, E. R., Pey, J., Perez, N., Samoli, E., Querol, X. et al. (2015). Short-term effects of particulate matter on mortality during forest fires in Southern Europe: results of the MED-PARTICLES Project. *Occupational and Environmental Medicine, 72,* 323–329. https://doi.org/10.1136/oemed-2014-102459

Frankenberg, E., McKee, D., & Thomas, D. (2005). Health consequences of forest fires in Indonesia. *Demography, 42,* 109–112. Retrieved from https://link.springer.com/article/10.1353/dem.2005.0004

Gaffney, O., Crona, B., Dauriach, A., & Galaz, V. (2018). *Sleeping financial giants: Opportunities in financial leadership for climate stability.* Retrieved from https://sleepinggiants.earth/wp-content/uploads/2018/09/Sleepingfinancial-giants-report-24-September-2018.pdf

INPE. (2019). Retrieved from http://www.obt.inpe.br/OBT/assuntos/programas/amazonia/prodes

Johnston, F., Hanigan, I., Henderson, S., Morgan, G., & Bowman, D. (2011). Extreme air pollution events from bushfires and dust storms and their association with mortality in Sydney, Australia 1994–2007. *Environmental Research, 111,* 811–816. Retrieved from https://www.sciencedirect.com/science/article/pii/S0013935111001241?via%3Dihub

Longo, M., Saatchi, S., Keller, M., Ferraz, A., Bowman, K. W., Bloom A. A., et al. (2018). Tropical forest structure modulates flammability and response to extreme droughts in the Amazon. *American Geophysical Union.* Retrieved from http://adsabs.harvard.edu/abs/2018AGUFM.B31K2630L

Lovejoy, & Nobre. (2018). Amazon tipping point. *Science Advances, 4*(2), eaat2340. Retrieved from http://advances.sciencemag.org/content/advances/4/2/eaat2340.full.pdf

Machin, A. B., Nascimento, L. F., Mantovani, K., & Machin, E. B. (2019). Effects of exposure to fine particulate matter in elderly hospitalizations due to respiratory diseases in the South of the Brazilian Amazon. *Brazilian Journal of Medical and Biological Research, 52*, e8130. https://doi.org/10.1590/1414-431X20188130

Marengo, J. A., Souza, C. M., Thonicke, K., Burton, C., Halladay, K., Betts, R. A., et al. (2018). Changes in climate and land use over the Amazon region: Current and future variability and trends. *Frontiers Earth Science, 6*, Article 228. Retrieved from https://www.frontiersin.org/articles/10.3389/feart.2018.00228/full

Morgan, G., Sheppeard, V., Khalaj, B., Ayyar, A., Lincoln, D., Jalaludin, B., et al. (2010). Effects of bushfire smoke on daily mortality and hospital admissions in Sydney, Australia. *Epidemiology, 21*, 47–55. Retrieved from https://insights.ovid.com/pubmed?pmid=19907335

Nobre. (2014). O Futuro Climático da Amazônia: Relatório de Avaliação Científica. *Articulación Regional Amazónica, 1*, 42. Retrieved from http://www.pbmc.coppe.ufrj.br/documentos/futuro-climatico-da-amazonia.pdf

Nunes, K. V. R., Ignotti, E., & de Souza Hacon, S. (2013). Circulatory disease mortality rates in the elderly and exposure to PM2.5 generated by biomass burning in the Brazilian Amazon in 2005. *Cadernos de Saúde Pública, 29*, 589–598. Retrieved from http://www.scielo.br/pdf/csp/v29n3/a16v29n3.pdf

RAISG (2015). *Deforestación en la Amazonia (1970–2013)*. Retrieved from https://www.amazoniasocioambiental.org/es/publicacion/deforestacion-en-la-amazonia-1970-2013-atlas/

Rittmaster, R., Adamowicz, W. L., Amiro, B., & Pelletier, R. T. (2006). Economic analysis of health effects from forest fires. *Canadian Journal of Forest Research, 36*, 868–877. Retrieved from https://www.nrcresearchpress.com/doi/abs/10.1139/x05-293#.XJ9Xk5hKjIV

Salati, E., Dall'Olio, A., Matsui, E., & Gat, J. R. (1979). Recycling of water in the Amazon, Brazil: An isotopic study. *Water Resources Research, 15*, 1250–1258. Retrieved from https://agupubs.onlinelibrary.wiley.com/doi/abs/10.1029/WR015i005p01250

Sankaran, & Sheldon. (2018). Transboundary pollution in Southeast Asia: Welfare and avoidance costs in Singapore from the forest burning in Indonesia. "Boston College Working Papers in Economics 960, Boston College Department of Economics." American Geophysical Union, Fall Meeting 2018, abstract #GH11C-0923. Retrieved from http://adsabs.harvard.edu/abs/2018AGUFMGH11C0923S

Tan-Soo, & Pattanayak. (2019). Seeking natural capital projects: Forest fires, haze, and early-life exposure in Indonesia. *Proceedings of the National Academy of Sciences of the United States of America, 116*, 5239–5245. Retrieved from https://www.pnas.org/content/pnas/116/12/5239.full.pdf

UNEP. (2018). *Emissions Gap Report*. Retrieved from http://wedocs.unep.org/bitstream/handle/20.500.11822/26895/EGR2018_FullReport_EN.pdf

Viana, (2017). Too big to fail: A roadmap for sustainable development of the Amazon. In: *II International Conference on Sustainable Development, SDSN*, New York.

Viana, (2019). Não há desenvolvimento sem proteção ambiental. *El Pais Brasil*. Retrieved from https://brasil.elpais.com/brasil/2019/03/15/opinion/1552674544_685747.html

WHO (2019). *Ambient air pollution—A major threat to health and climate*. Retrieved from https://www.who.int/airpollution/ambient/en/

WWF (2018). *Wildlife in a warming world*. Retrieved from https://c402277.ssl.cf1.rackcdn.com/publications/1149/files/original/WWF_-_Wildlife_in_a_Warming_World_-_2018_FINAL.pdf?1520886759

Part IV
Climate Change and Health: Perspectives from Physicians

CHAPTER 14
Psychological Impacts of Climate Change and Recommendations

Lise Van Susteren and Wael K. Al-Delaimy

Summary Although the harm to our physical health from the climate crisis is increasingly reported, still underrecognized is the harm the climate crisis is having on us *psychologically*. Yet it is the psychological impacts of *global overheating*—and the consequent cascading destabilization of ecosystems—that will carry the biggest burden and be the most difficult to remedy. It requires our utmost attention. Understanding the gravity of the mounting psychological harm underscores the urgent need for all those concerned, especially public officials, to take action. The devastating psychological and physical impacts of the tragedy of lead-tainted water in Flint, Michigan, in microcosm, serves as a recent example to those who would downplay harm to our health in the face of ongoing warnings and pleas for action. The psychological aspects of the climate crisis are increasingly drawing the attention of mental health professionals. They are uniquely qualified and urgently needed to address the denial, discounting, or distancing that feed inaction, and to point out the deepening injustice of putting vulnerable populations at risk, especially our children and future generations. They are also needed to advocate for programs that address climate anxiety and trauma, and to design programs that build resilience. The mounting risk of an epidemic of fear, outrage and despair calls for mental health professionals to play a pivotal role in what has become a do-or-die effort to restore humanity and the rest of the natural world to safety. All of the losses associated with climate change—from extreme weather events and chronic climate conditions to the devastating physical injuries, illnesses, and deaths and the attendant displacement, disruptions and downstream indirect ripple effects—carry with them an emotional toll. The magnitude and relentlessness of the destruction, as well as the insinuation into every aspect of our lives—economic, personal, political—must be recognized as we consider the mental health and psychosocial impacts of the deepening crisis.

L. Van Susteren (✉)
Physicians for Social Responsibility, Washington, DC, USA

W. K. Al-Delaimy
University of California San Diego, La Jolla, CA, USA

© The Author(s) 2020   177
W. K. Al-Delaimy, V. Ramanathan, M. Sánchez Sorondo (eds.), *Health of People, Health of Planet and Our Responsibility*, https://doi.org/10.1007/978-3-030-31125-4_14

Extreme Weather Events

The global rise in temperature is driving local spikes of searing heat waves and increasing the frequency and intensity of storms, wildfires, and floods. In the last two decades, extreme weather has wounded, displaced or required emergency assistance for four billion people; half a million have died (United Nations Office for Disaster Risk Reduction, 2015).

Extreme weather events drive emotional turmoil in every phase. Anticipatory fear is experienced in the early stages, followed by the trauma of the event itself, then sorrow and grief at the losses. Feelings of outrage may emerge—slowing down the working through—when it is believed that government and institutions are not providing enough help or didn't undertake preventive measures. Despair and feelings of helplessness deepen as reports surface that conditions will get worse.

Those in disaster-prone areas are the most affected. Anticipatory anxiety sometimes rises to the level of "pre-traumatic stress disorder" among those who fear that another disaster may be just around the corner. In anecdotal reports, residents of New Orleans report, for example, that sometimes just seeing a storm cloud triggers deep anxiety (Hauser, 2015; Reckdahl, 2017). Communities under stress can fray and decompensate—increasing violence and other psychosocial ills. Mental health professionals are seeing a full range of psychiatric disorders and conditions emerge as a result of these stressors: major depression, anxiety, PTSD, adjustment disorders, a rise in drug and alcohol abuse as people attempt to cope, and domestic violence— including child abuse (American Psychiatric Association [APA], 2017; World Health Organization, 2005). In addition, the more destructive the event, the higher the incidence of PTSD and the greater the risk of suicide (Edwards & Wiseman, 2011).

Processing disaster After a *natural* disaster, an identifiable low point is seen, followed by the feeling that the worst is over, and the recovery process can begin. How we process the event is determined in part by how we answer critical questions: *why* did this happen, *who* or *what* is responsible, and could the disaster have been prevented? Disasters experienced as natural are easier to reconcile because they are experienced as "fate"—beyond our control. When disasters are no longer experienced solely as natural, as "acts of God or nature," but instead are experienced as having arisen or been made worse because of the behavior of humans—due to a mistake, carelessness, or worse, as *the result of deliberate disregard for consequences*—then it is much harder for us to recover and the psychological harm is more serious (Folkman, Lazarus, Dunkel-Schetter, DeLongis, & Gruen, 1986).

How we "carry on" is influenced by our background, current mental state, personality, and life experiences. The nature of the event plays a role: the intensity of the feeling of powerlessness, the "merciless" character of incidents, the pace, suddenness, degree of damage, loss of life and injury and the extent to which people personalize these incidents will be factors influencing the extent of emotional injury. Though studies quantifying it are limited, human empathy tells us of the emotional toll: When the place you call home is burned down by wildfires, blown away by tornados, flooded or overrun by hurricanes, when you lose your possessions, maybe your pets, your livelihood, the comfort of a familiar community, and witness the injuries, illnesses, and deaths, the resulting mix of fear, anger, sorrow, and trauma can easily send a person to a breaking point. Stress drives up the secretion of cortisol; high levels of this hormone are not damaging when they are short-lived, but *persistently* elevated levels can be exceedingly damaging. Immune function can be compromised, sleep patterns disturbed, digestion disrupted, memory impaired, and the cardiovascular system harmed—all with an attendant emotional toll (Shaw et al., 2019). Everyday life is full of burdens and anxieties; confronted by the additional stress of the climate crisis, some individuals will decompensate.

Exposed to repeated or ongoing disasters, victims may not have a chance to recover emotionally before the next disaster hits, compounding the harm because each incident deepens emotional vulnerability. Victims also contend often with the realization that "no lessons have been learned". The healing process is helped along when we can look back, identify where we went wrong and take action to legitimately say we are doing all that we can to prevent further injury. But on the contrary, even as the dangers continue to unspool at an ever-higher rate, the extraction and use of fossil fuels continues, with recent US government policies *promoting* their use—exacerbating the mounting psychological toll.

Wildfires Increasing temperatures from climate change dry out land and vegetation, causing bigger, more frequent and intense wildfires. Homes and communities are increasingly finding themselves in fire-prone areas. While the physical impacts of fires can be measured as lives lost, structures burned, and communities disrupted, the psychological damage is immeasurable. The acute phase is dominated by fear—or *terror*. The mercilessness of flames, incinerating everything in their path, has a particularly devastating effect on those who witness what has happened—and they may be unable to shake loose from intrusive and horrifying thoughts of what it is like to burn to death. When the flames are finally spent or defeated, seeing what was once a comforting refuge but is now nothing or little more than a moonscape brings overwhelming grief.

Various phases of disasters have been studied, along with many common emotional responses to them (SAMHSA Phases of Disaster, 2018). Based on the clinical experience of the first author with victims of trauma, including of natural disasters, many emotional stressors have not yet been measured but warrant being pointed out. In an initial phase, vows to rebuild and faith in outside support do help in the first steps towards emotional recovery, but fears of "next time" often plague survivors. Wrenching, conflicting feelings about remaining in the vulnerable area, financial strain, and friction from different coping styles may erode family stability. Lingering feelings of helplessness, despair, and foreboding make restoring a sense of security more challenging. Emotionally charged triggers can reawaken traumatic memories—in the case of wildfires, by the simple smell of smoke, the sound of sirens, the once-pleasant crackling of an ordinary fire. Acute displacement may become permanent—rupturing the comfort and stabilizing effect of familiar relationships, places, and daily routines. Some scars may be buried or disguised—expressed in ways that make them not only hard to "count" but worse, harder to confront and work through. The stress on individuals and communities is compounded by mental health systems that are underprepared to deal with the trauma.

Violent storms and floods Violent storms and floods are responsible for half the disasters from extreme weather events (UNISDR, 2015). Higher temperatures increase the moisture content in the atmosphere, increasing the severity and frequency of storms. As with other extreme weather events, violent storms and floods unleash catastrophic physical and emotional harm—bringing about the disorders and conditions cited earlier. Of all extreme weather events, flooding affects the greatest number of people. The financially stretched are most vulnerable, with the physical difficulty of getting to safety too great or the cost too high—which the financially privileged may sometimes forget when they ask why people did not get out of harm's way. Even temporary housing is frequently beyond the reach of those who live paycheck to paycheck.

The far-reaching psychological impacts of violent storms saw their prototype in Hurricane Katrina in 2005, now "updated" by other storms, in particular the devastating Hurricane Maria that leveled Puerto Rico in 2018. Grief, anxiety, violence, outbursts of outrage, and blame at the government's slow response are among the acute emotional stressors of many victims. Many report feeling that the government "doesn't care" about them (Morin & Rein, 2005).

A flood-ravaged home brings a particular type of anguish. Based on experience, violently being run out of one's refuge from the outside world underscores the fear that no place is safe. Seeing belongings a soggy ruin, often turning black and moldy as they undergo the slow but inexorable process of rotting, brings unconscious associations to dying and death. The sense of powerlessness over "nature unleashed" can weigh heavily on victims.

The widespread presence of mold in homes and other buildings is common in the aftermath of floods; the physical symptoms associated with exposure to mold are well documented (Potera, 2007). Exposure to mold is also linked to psychiatric disorders, especially depression, and reduced cognitive functioning (Shenassa, Daskalakis, Liebhaber, Braubach, & Brown, 2007). Other impacts include anxiety, irritability, and fatigue (Hope, 2013). Toxic mold-induced mental illness and cognitive decline are rarely correctly diagnosed or treated.

Research confirms the association between the chronic stress of climate disaster emotional trauma and cardiovascular health (Peters et al., 2014). Residents of New Orleans suffered heart attacks at a rate three times higher than the rate reported before the storm (Peters et al., 2014). The absence of social support when communities are torn apart is one of the strongest predictors of posttraumatic stress (Brewin, Andrews, & Valentine, 2000; Ozer, Best, Lipsey, & Weiss, 2003).

Summer Heat Waves

As global temperatures rise, the incidence of extreme heat waves is increasing. Nearly one-third of the world's population is now exposed to deadly heat waves (Mora, Counsell, Bielecki, & Louis, 2017). In the last decade, with less than 1% increase in warming, the loss of human life has risen 2300% (Mora, Counsell, Bielecki, & Louis, 2017). Significant psychological stress is also associated with heat waves. With rising temperatures comes rising rates of aggression (Bulbena, Sperry, & Cunillera, 2006; Raj, 2014). It is linked to increased catecholamine release in response to the stress of the heat (Al-Hadramy & Ali, 1989). For each standard deviation of increased temperature or more extreme rainfall, studies show a 4% increase in conflict between individuals, and a 14% increase in conflict between groups (Hsiang et al., 2013). These findings are valid across all regions and among all ethnic groups. Increased acts of aggression include assaults, murders, and suicides, especially violent suicides. As temperatures continue to rise, a global increase in unrest should be anticipated.

Temperatures *modestly* above the human comfort zone are associated with a decline in cognitive functioning: slowed response time, diminished accuracy, and less sophisticated patterns of decision-making. Declining cognitive function could become increasingly problematic for those unable to escape the heat (Goodman, Hurwitz, Park, & Smith, 2018). The mental health toll from heat waves is hard to measure. Like blizzards and storms, the disruptions of heat waves range from irritating inconveniences to profoundly stressful losses. Declining business and productivity—which drops at high temperatures—are consequences of extreme heat (Somanathan, Somanathan, Sudarshan, & Tewari, 2018). Those who must work outdoors or where AC is not available may not be able to work at all. With little financial cushion, it may not be possible to pay bills or even buy food. The emotional toll on families and the communities they live in can be horrendous. With the persistent high temperatures and the ever-expanding desert that they create, scientists predict that the Middle East may well become uninhabitable by the end of the century (Funkhouser, 2016).

Particularly vulnerable are the poor, the elderly, the sick, and the very young. Individuals with preexisting mental disorders are also especially susceptible. During hotter-than-average periods, they appear to get sicker than expected, show greater aggressiveness towards others, and require more frequent use of restraints (Bulbena et al., 2006). The good judgment to stay cool and hydrated may be compromised; some individuals in this population live outdoors and lack access to air conditioning. Many take medication (psychotropics) that impede the ability to perspire—the body's chief means of reducing internal temperature.

Chronic Climate Conditions

Drought Conditions

Drought conditions are worsening with rising temperatures, particularly in the western United States. Drought has its own attendant, often chronic, mental health impacts (Stanke, Kerac, Prudhomme, Medlock, & Murray, 2013). An unrelenting day-after-day anxiety and despair can descend on those who depend on nature's rains for their livelihood—farmers, day workers, owners of related businesses—who watch and wait for water that does not come. A serious and protracted decline in available water may force many to make life-altering changes, with stressful impacts on individuals, families and communities (Vins, Bell, Saha, & Hess, 2015).

Suicide is the result of complex dynamics, but prolonged drought is considered (in most studies) a major contributing factor among at-risk populations for suicide among farmers in India, rural Australia, and South Africa (Carleton, 2017; Hanigan, Butlera, Kokicc, & Hutchinson, 2012; Vins et al., 2015).The psychological toll on families who have lost their fathers, sons, uncles and brothers, in addition to what is often their sole source of income, is devastating. Entire countries now compete for water from the once great rivers of the world—the Mekong, Jordan, the Brahmaputra, the Tigris and Euphrates, the Nile—bringing rationing and water wars as tension mounts (Bhalla, 2012). The decline in available water brings food insecurity—unleashing a flood of its own woes—with individual impacts ranging from the personal stress of hunger and rationing to malnutrition, starvation and death when conditions persist. In the crops the world's poor are most dependent on, malnutrition from higher concentrations of atmospheric CO_2 *alone* causes malnutrition from micronutrient deficiencies responsible for cognitive deficits and other physical harms, carrying with them an emotional toll that can be lifelong (Smith & Myers, 2018).

As climate change worsens and causes more droughts, millions will continue to see their access to drinking water threatened. Access to water grips us emotionally in ways other needs do not because we cannot survive beyond a few days without water. When a community has a dwindling supply of clean water—from diminished snowpack, accelerating and preseason melting of glaciers, drying up of rivers, streams, lakes, and aquifers, or from saltwater intrusion or other climate-induced contamination, a deepening sense of anxiety and foreboding follows.

In regions dependent on agriculture and already rife with ethnic tensions, strained by poverty and political divisiveness, climate related disasters are "threat multipliers" for outbreaks of violence (Schleussner, Donges, Donner, & Schellnhuber, 2016). Ongoing periods of water scarcity, spiked by periods of drought, particularly in the growing season of these vulnerable regions, trigger not only spasms of violence but sustained cycles of conflict (Von Uexkull, Croicu, Fjelde, & Buhaug, 2016).

News reports have covered how new waves of migrations created by regional conflicts often drive people to relocate in places that do not want them and cannot help them (Baker, 2015). Those with the fewest resources are often unable to relocate—"trapping them in high risk areas" (Randall, 2013). Although sometimes controversial—and complex to assess, peer-reviewed reports specifically tie the recent scarcity of water in Syria to increased migration and the geopolitical tensions it spawned (Müller, Yoon, Gorelick, Avisse, & Tilmant, 2016).

An influx of refugees can stir fear, uncertainty, and anger—due to pressures that mount from scare resources, inadequate infrastructure, problems adapting to changing circumstances and cultural differences, limited or absent health and mental health services, along with pervasive and profound physical and psychological trauma among refugees (Langlois, Haines, Tomson, & Ghaffar, 2016). Data from

157 countries demonstrated that drought severity and likelihood of armed conflict played a role in explaining increased asylum seeking between 2011 and 2015 (Abel, Brottrager, Crespo Cuaresma, & Muttarak, 2019).

Confrontations in southwestern and southeastern states over water are leading to legal battles. In California, drought conditions have forced rationing, with news reports that neighbors engage in "drought shaming," fighting, and tattling on each other over their water use, foreshadowing what may lie ahead (Huffaker, 2015).

Sea Level Rise

Global sea level is likely to increase 10–13 feet in as few as 50 years unless there is a major rapid reduction in greenhouse gas emissions. "Parts of our coastal cities would still be sticking above the water, but you couldn't live there" (Hansen et al., 2016). Already, many living in low-lying coastlines and small island states are being displaced by sea level rise, but these numbers are dwarfed by predictions that the accelerating pace of rising seas will flood some of the world's most populous cities—creating a tidal wave of psychosocial trauma as a bottleneck of climate refugees compete for higher ground (Hearty, 2012; Parker, 2017). In the United States alone, tens of millions live on coastlines and depend on coastline stability for their livelihoods, well-being, and survival; worldwide hundreds of millions of people live within 30 ft of sea level (McGranahan, Balk, & Anderson, 2007).

With some *acute* climate events, victims can "arm themselves" by taking actions that help protect them should they face a similar event. Sea level rise is different. Although governments can work to adopt measures to mitigate harm in the short term, future sea level rise will continue, inundating low-lying cities, countries, and islands. Fears of permanent displacement to unfamiliar places gnaw at victims, provoking additional stress—financial, social, and personal. Current and future climate refugees, who have been displaced from their homes, search for safety and security, but chaos and violence often follow them, and can spread to others along the way. Massive relocation will become necessary; it is already underway in places such as Shishmaref in Alaska, St Charles Island in LA, and now Charleston NC. Parts of Miami, FL are flooded nearly daily.

Believing that we can take action, that we are empowered to protect ourselves, is critical to maintaining our mental health. Nurturing the hope of returning home, restoring, and rebuilding after cataclysmic losses builds resilience. But no such action or hope can undo the ravages of locked in sea level rise. Layered on are the crushing financial losses of walking away from what is typically our biggest investment—our homes—now with little or no value. The plight of climate refugees is increasingly the focus of mental health professionals. The psychological stress of refugees is high. The refugees are often blamed for psychosocial ills. Katrina refugees were initially accused of precipitating a crime wave in Houston, where many were taken in, before a closer analysis showed this was a misreading of statistics (Hamilton, 2010). A flare-up of anti-refugee sentiment has rocked countries and governments across Europe and on the American border with Mexico where the divisive battle over building a wall has further polarized the USA and precipitated a record-breaking government shutdown.

Other Sources of Climate-Related Psychological Impacts

New Disease Threats

As ecosystems are altered by climate change, old diseases are expanding their reach and new diseases are emerging. While physical health data is examined and reported, the complex psychosocial toll—difficult to quantify—often is not addressed as a result.

Warnings that new diseases and old diseases are emerging are heard frequently. Deeply unnerving, they precipitate a pervasive if not always fully conscious sense of vulnerability that adds to existing stressors in life. Millions worried, for example, about exposure to the Zika virus—with widespread repercussions: local economies took a hit as fears drove tourists away, pregnancies were put off; travel plans unraveled; and decisions about family life and individual personal choices were driven by fears of infection. Families suffered the anguish of having babies born with deformities and with the knowledge that a baby who was exposed might appear normal but have disabilities down the road.

Lyme disease is well known in some cases to leave its victims with long-lasting and painful emotional wounds. Cognitive deficits and mood disorders with depression and anxiety are not uncommon. Individuals struggle with the torment of treatments that are often ineffective, especially when the diagnosis—sometimes difficult to make—comes after years of raging infection and unexplained symptoms. Other climate-linked viruses, parasites and bacteria storm on and off stage depending on local weather conditions, or are waiting in the wings for their chance, or continue their inexorable expansion: Dengue fever, also known as "bone-break fever" because it is so painful; Chikungunya; Hantavirus; West Nile; Chagas disease; and of course, Malaria. All of them carry an attendant psychosocial toll that must be addressed.

Air Pollution

The World Health Organization reports that nine out of ten people breathe unhealthy air. Air pollution is primarily the result of our use of fossil fuels: when coal, oil, and gas are burned, they create pollutants that undergo chemical reactions to form particulate matter. Warmer temperatures speed up these reactions, causing pollution to become more concentrated. The presence of particulate matter in our bodies triggers inflammation. Many diseases and conditions are linked to exposure to polluted air. Ultrafine particulate matter is especially worrisome because it can cross directly from lung tissue into the bloodstream and be carried throughout the entire body. Already confirmed in nonhuman primates, initial studies suggest that ultra-fine particulate matter also crosses directly from the nasal mucosa via nerve endings that transport it directly into the human brain (Obederdöster, Elder, & Rinderknecht, 2009).

Though more research is needed to address some inconsistencies, evidence is mounting that inflammation of brain tissue from air pollution is linked to dementia, including Alzheimer's type; is linked as a comorbid factor in the development of Parkinson's disease and exacerbation of symptoms; and is linked to amyotrophic lateral sclerosis (ALS) (Calderón-Garcidueñas & Villareal-Ríos, 2017; Chen et al., 2017; Lee et al., 2017; Seelen et al., 2017).

The incidence of these diseases is expected to rise and already brings a crushing economic burden in the hundreds of billions of dollars. The cost in human suffering alone should be argument enough to make eliminating the burning of fossil fuels a number one public health priority (Gladman & Zinman, 2015; Kirson et al., 2016; Kowal, Dall, Chakrabarti, Storm, & Jain, 2013).

Evidence supports a link between neuro-inflammation and classic psychiatric illness: major depressive disorders, bipolar disorder, schizophrenia, and obsessive-compulsive disorders (Souhel, Pearlman, Alper, Najjar, & Devinsky, 2013). Air pollution is also associated with increased psychosis in adolescents (Newbury, Arseneault, Beevers, et al., 2019). Even *low levels of pollution* in Sweden—primarily from traffic—are associated with an increased risk of mental illness in children (Oudin, Bråbäck, Åström, Strömgren, & Forsberg, 2016).

The American Psychological Association reported children exposed to particulate matter in the air were more likely to have symptoms of anxiety or depression (Weir, 2012). Emergency room visits for anxiety—including panic attacks and threats to commit suicide—are significantly higher on days with poor air quality (Szyszkowicz, Willey, Grafstein, Rowe, & Colman, 2010).

Vulnerable Populations

The elderly, the sick, the disabled, the poor, and those whose jobs are tied to the natural environment (farming, fishing, forestry, etc.) are particularly susceptible to the psychological impacts of climate disruption. Residents in areas destroyed by climate disasters or where fossil fuel production has destroyed the natural landscape are especially likely to suffer from "Solastalgia": a gripping existential pain that comes from seeing places that once gave the treasured feeling of "home" now lost or irreparably damaged (Albrecht, 2005).

The mentally ill are especially vulnerable: they are less resilient, have fewer resources, face disrupted services that put them at risk of decompensation, and take psychotropic drugs that impede the ability to perspire—making them more prone to overheating. Often overlooked groups include: first responders, who see the injuries, deaths and devastation in its rawest state; climate experts, whose professional lives are spent cataloguing the growing evidence of decline and with a growing sense of doom see their life's work damaged, destroyed or disappearing (Guest, 2014; Nicholls et al., 2014); and activists, the "Climate Cassandras", who, desperately issuing anguished warnings, and in the grip of images of future disasters they cannot get out of their minds, may suffer from "pre-traumatic stress disorder" (Van Susteren, 2013).

The poor and rural populations as well as those who live a day-to-day existence face falling income, food shortages, and rising prices. Decisions must be made about who eats and who does not. Families and communities fray when competition erodes the sharing support system that was characteristically relied upon. The world's poor suffer the anguish of climate injustice far from the consciousness of those whose overconsumption and disproportionate use of resources are most responsible. Ten percent of the world's richest are responsible for the emissions of 50% of the world's greenhouse gases (Gore, 2015).

Ocean acidification, warming, and ocean deoxygenation are impacts from climate disruption that are responsible for dramatic declines in marine life. Experts warn that without immediate dramatic action to improve the health of our oceans, within a few decades oceans will no longer have the capacity to feed millions of people who depend on them to survive. As catches dwindle from depleted sources, food production falls, jobs decline, incomes fall, and hunger rises. Cascading economic, health, social, and psychological ills follow (IPSO, 2013).

Children and young people are at special risk. The message that the planet is dying and that the survival of future generations is on the line *registers in* climate-aware young people. Though findings are sometimes inconsistent and more work is called for, many are *less likely* to distance or discount the warnings of scientists, educators and government agencies that the window on our ability to "act in time" is closing. They know that conditions will worsen and the cumulative destructive toll, emotional and physical, will fall on their shoulders. Children are frightened, report they are sad and angry; a bleak and pessimistic view of the future may be more prevalent in adolescents and young adults (Ojala, 2012; Strife, 2012). Chronic childhood stress, persistently driving up cortisol levels, can alter brain development, notably in the hippocampus and prefrontal cortex, with the potential to cause impaired cognition and a life-long predisposition to anxiety, depression, and susceptibility to additional stress (Carrion & Wong, 2012; Simpson, Weissbecker, & Sephton, 2011). Susceptibility to anxiety can be passed onto succeeding generations through epigenetic inheritance (Skinner, 2014). Children who experience multiple early life stressors are at greater risk for suicide (Björkenstam, Kosidou, & Björkenstam, 2017).

The first reported case of climate crisis-induced psychosis occurred in an Australian boy in 2009, who, in his drought-stricken country, refused to drink water because he feared millions would die as a result. The doctor who treated him at Melbourne's Children Hospital said he has since seen many other children suffering from climate anxiety in varying degrees of severity (Wolf & Salo, 2008).

Summary of Psychosocial Impacts

"Not everything that counts can be counted..." (William Bruce Cameron).

Personal anecdotal clinical expertise suggests that it is the inchoate, insidious, complex, and unconscious psychological states driven by climate trauma—not lending themselves to studies and precise numbers—that can be the most profoundly damaging and drive systemic emotional conditions that society will find difficult to treat and surmount. Just as the functioning of families breaks down in the face of a chaotic home environment, in the face of the turbulence from declining national security, without trust that our institutions can protect us, the fabric of society can break down. The waves of injury from climate trauma and frustration from inaction are reverberating across our communities, the workplace, and our families, cumulatively taking a toll on the national mood. Repercussions from a stressed-out national mood drive our economy, our politics, and our relations with other countries. It shapes our culture and increasingly affects how we treat each other. How will we deal with a growing state of fear from diminished productivity, conflict between haves and have-nots, callousness, and numbness in our response to suffering—"compassion fatigue"—an alienated citizenry, distrust of each other?

We can be anxious and not know why, and we can be anxious and not know it. Sometimes anxiety is manifested as anger or another emotion or state that doesn't even suggest the root cause. We can be anxious and give it the wrong name—often if the real source is especially unsettling. We don't always recognize the psychological toll anxiety is having on us. Much of the violence in the world can be explained by unaddressed anxiety emanating from fears of impotence and vulnerability. Climate change evokes a profound sense of both. Whether we know it or not, whether we like it or not, whether we accept it or not, we believe the climate crisis is causing varying degrees of anxiety among most of the population.

Special Message to Mental Health Professionals: Recommendations and a Call to Action

Trying to avoid thinking about climate-related issues, or other conditions that cause anxiety, is an attempt to try to control our fears, but fear and avoidance are at the root of denial and inaction.

Years of experience have shown mental health professionals how to move people beyond this state, replacing resistance with the empowering realization that the solution is to take action. As mental health professionals we know the power of words—we know how to break down defenses, put together messages that help people finally begin to understand. We get science, we get suffering, we get urgency, we get lifelong consequences. We see the value of hope, we know how to engender it even in dark times. We are good at encouraging people to grow, to be conscious of the needs of others.

As we consider where the public needs to be on climate change and what we have to offer, we can see that we have largely failed to realize the *power* of our expertise.

It no longer makes sense, because of the challenges presented by the climate crisis and the unique skills we have, to confine our professional lives only to traditional roles and services in our offices, academic settings and clinics. We need activism now to *initiate* services in this time of urgency—helping people out of their denial, tending to their wounds, working to create resilience. Just as we have qualifications in other specialties—we need "Climate Mental Health Specialists" (CMHSs), who are specifically trained to offer "trauma informed care" for the many levels and kinds of distress generated by climate instability— both direct and indirect, addressing the plight of refugees and the needs of newly blended populations, helping to organize communities and collaborating with primary care and public health experts, fanning out into settings to educate and advocate.

More research is needed to identify best practices, determining societal and infrastructural resources that are truly helpful, and building upon this evidence base to train the next generation of CMHSs.

Our efforts in every sector send the message—that the fight for sane climate policy is not the exception, but the legitimate expected behavior in this time of crisis. With our understanding of the social sciences, we know that people can be persuaded to take action *just because* other people are, and promote and harness the power of social norms to accelerate the pace of change.

Our canons of ethics tell us it is our duty to protect the health of the public and to participate in activities that contribute to it.

Committed professionals are joining others to call upon our elected officials, policy makers, and other leaders to show why the public's health is in jeopardy, what steps must be taken, and how we can use our expertise to speed up the process. Many professional organizations and academic institutions are setting up curricula for training and making collaborative declarations describing the dangers of climate change and the urgent need for action.

But at this epic moment, our unique contributions and the urgent need for global commitment have gone missing. But it does not have to stay this way—because we can change that, too.

References

Abel, G. J., Brottrager, M., Crespo Cuaresma, J., & Muttarak, R. (2019). Climate, conflict and forced migration. *Global Environmental Change, 54*, 239–249.

Albrecht, G. (2005). Solastalgia: A new concept in health and identity. *PAN: Philosophy Activism Nature, 3*, 41–55.

Al-Hadramy, M.S., & Ali, F. (1989). Catecholamines in heat stroke. *Military Medicine*. Oxford: Oxford University Press.

American Psychiatric Association (2017). *How climate-related natural disasters affect mental health*. Philadelphia: American Psychiatric Association. Retrieved February 9, 2020 from https://www.psychiatry.org/patients-families/climate-change-and-mental-health-connections/affects-on-mental-health

Baker, A. (2015). How climate change is behind the surge of migrants to Europe. *Time Magazine*. Time USA, LLC.

Bhalla, N. (2012). Thirsty South Asia's river rifts threaten "water wars". *Discover Thomson Reuters*.

Björkenstam, C., Kosidou, K., & Björkenstam, E. (2017). Childhood adversity and risk of suicide: Cohort study of 548 721 adolescents and young adults in Sweden. *British Medical Journal, 357*, j1334.

Brewin, C. R., Andrews, B., & Valentine, J. D. (2000). Meta-analysis of risk factors for posttraumatic stress disorder in trauma-exposed adults. *Journal of Consulting and Clinical Psychology, 68*, 748–766.

Bulbena, A., Sperry, L., & Cunillera, J. (2006). Psychiatric effects of heat waves. *Psychiatric Services, 57*, 1519.

Calderón-Garcidueñas, L., & Villareal-Ríos, R. (2017). Living close to heavy traffic roads, air pollution, and dementia. *The Lancet, 389*, 675–677.

Carleton, T. (2017). Crop-damaging temperatures increase suicide rates in India. *Proceedings of the National Academy of Sciences of the United States of America, 114*, 8746–8751.

Carrion, V. G., & Wong, S. S. (2012). Can traumatic stress alter the brain? Understanding the implications of early trauma on brain development and learning. *Journal of Adolescent Health, 51*, S23–S28.

Chen, C.-Y., Hung, H.-J., Chang, K.-H., Chung, Y. H., Muo, C.-H., Tsai, C.-H., et al. (2017). Long-term exposure to air pollution and the incidence of Parkinson's disease: A nested case-control study. *PLoS One, 12*(8), e0182834.

Edwards, T., & Wiseman, J. (2011). Climate change, resilience and transformation: Challenges and opportunities for local communities. In I. Weissbecker (Ed.), *Climate change and human well-being: Global challenges and opportunities* (pp. 185–200). New York: Springer.

Folkman, S., Lazarus, R. S., Dunkel-Schetter, C., DeLongis, A., & Gruen, R. J. (1986). Dynamics of a stressful encounter: Cognitive appraisal, coping, and encounter outcomes. *Journal of Personality and Social Psychology, 50*, 992–1003.

Funkhouser, D. (2016) Climate may make some regions 'uninhabitable' by end of century. *State of the Planet Climate May Make Some Regions Uninhabitable by End of Century Comments.* Earth Institute, Columbia University.

Gladman, M., & Zinman, L. (2015). *The economic impact of amyotrophic lateral sclerosis: A systematic review*. National Center for Biotechnology Information, U.S. National Library of Medicine.

Goodman, J., Hurwitz, M., Park, J., & Smith, J. (2018). *Heating and learning* (NBER Working Paper Series No. 24639). National Bureau of Economic Research.

Gore, T. (2015). *Extreme carbon inequality: Why the Paris climate deal must put the poorest, lowest emitting and most vulnerable people first*. Oxfam Media Briefing 02-12-2015. Retrieved on February 9, 2020 from https://policy-practice.oxfam.org.uk/publications/extreme-carbon-inequality-why-the-paris-climate-deal-must-put-the-poorest-lowes-582545

Guest, R. (2014). Portraits of scared scientists reveal truth about climate change. *Lost at E Minor.*

Hamilton, R. (2010). The huddled masses. *The Texas Tribune.*

Hanigan, I. C., Butlera, C. D., Kokicc, C. N., & Hutchinson, M. F. (2012). Suicide and drought in New South Wales, Australia, 1970–2007. *Proceedings of the National Academy of Sciences of the United States of America, 109*, 13950–13955.

Hansen, J., Sato, M., Hearty, P., Ruedy, R., Kelley, M., Masson-Delmotte, V., et al. (2016). Ice melt, sea level rise and superstorms: Evidence from paleoclimate data, climate modeling, and modern observations that 2 °C global warming could be dangerous. *Atmospheric Chemistry and Physics: Journal of the European Geosciences Union, 16*, 3761–3812.

Hauser, A. (2015). We are the Walking Dead. *The Weather Channel (features).*

Hearty, P. (2012). Rising seas threaten low-lying coastlines of the world: Low coastlines under watery siege. *Ecology Global Network.* Ecology Communications Group, Inc. (ECG).

Hope, J. (2013). A review of the mechanism of injury and treatment approaches for illness resulting from exposure to water-damaged buildings, mold, and mycotoxins. *The Scientific World Journal, 2013,* 767482.

Hsiang, S., Burke, M., & Miguel, E. (2013). Quantifying the influence of climate on human conflict. *American Association for the Advancement of Science, 341,* 1235367.

Huffaker, S. (2015). The water wars begin in parched California as the rich fight drought restrictions. *The National Post, June 15, 2015.*

IPSO. (2013). State of the ocean 2013 report: Interactions between stresses, impacts and some potential solutions. *Marine Pollution Bulletin, 74,* 491–552.

Kirson, N., Desai, U., Ristovska, L., Cummings, A. K., Birnbaum, H., Ye, W., et al. (2016). Assessing the economic burden of Alzheimer's disease patients first diagnosed by specialists. *BMC Geriatrics, 16,* 138.

Kowal, S.L., Dall, T. M., Chakrabarti, R., Storm, M.V., & Jain, A. (2013). *The current and projected economic burden of Parkinson's disease in the United States.* Movement Disorders Society. US National Library of Medicine National Institutes of Health.

Langlois, E., Haines, A., Tomson, G., & Ghaffar, A. (2016). Refugees: Towards better access to health-care services. *The Lancet, 387,* 319–321.

Lee, H., Myung, W., Kim, D. K., Kim, S. E., Kim, C. T., & Kim, H. (2017). *Short-term air pollution exposure aggravates Parkinson's disease in a population-based cohort.* National Center for Biotechnology Information. U.S. National Library of Medicine.

McGranahan, G., Balk, D., & Anderson, B. (2007). The rising tide: Assessing the risks of climate change and human settlements in low elevation coastal zones. *Sage Journals: International Institute for Environment and Development, 19,* 17–37.

Mora, C., Counsell, C., Bielecki, C., & Louis, L. (2017). Twenty-seven ways a heat wave can kill you: Deadly heat in the era of climate change. *Circulation: Cardiovascular Quality and Outcomes, 10,* e004233.

Morin, R., & Rein, L. (2005). Some of the uprooted won't go home again. *The Washington Post Company.*

Müller, M. F., Yoon, J., Gorelick, S. M., Avisse, N., & Tilmant, A. (2016). Impact of the Syrian refugee crisis on land use and transboundary freshwater resources. *Proceedings of the National Academy of Sciences of the United States of America, 113,* 14932–14937.

Newbury, J. B., Arseneault, L., Beevers, S., et al. (2019). Association of air pollution exposure with psychotic experiences during adolescence. *JAMA Psychiatry, 76,* 614–623.

Nicholls, N., Harper, A., Rahmstorf, S., Carilli, J., Buontempo, C. Santoso, A., et al. (2014). *The scientists: Is this how you feel?* Creative Commons Attribution. Retrieved February 9, 2020 from https://www.isthishowyoufeel.com/this-is-how-scientists-feel.html

Oberdörster, G., Elder, A., & Rinderknecht, A. (2009). Nanoparticles and the brain: Cause for concern? *Journal of Nanoscience and Nanotechnology, 9,* 4996–5007.

Ojala, M. (2012). Regulating worry, promoting hope: How do children, adolescents, and young adults cope with climate change? *International Journal of Environmental Science and Education, 7,* 537–561.

Oudin, A., Bråbäck, L., Åström, D. O., Strömgren, M., & Forsberg, B. (2016). Association between neighbourhood air pollution concentrations and dispensed medication for psychiatric disorders in a large longitudinal cohort of Swedish children and adolescents. *BMJ Open 6:*e010004.

Ozer, E. J., Best, S. R., Lipsey, T. L., & Weiss, D. S. (2003). *Predictors of posttraumatic stress disorder and symptoms in adults: A meta-analysis. Psychological Bulletin, 129,* 52–73.

Parker, L. (2017). *Sea level rise will flood hundreds of cities in the near future: Many shore communities in the U.S. face inundation in the coming decades.* National Geographic Society. National Geographic Partners, LLC. Retrieved February 9, 2020 from https://www.national-geographic.com/news/2017/07/sea-level-rise-flood-global-warming-science/

Peters, M., Moscona, J., Katz, M., Deandrade, K., Quevedo, H., Iwari, S., et al. (2014). National disaster and myocardial infarction: The six years after Hurricane Katrina. *Mayo Clinic Proceedings, 89*, 472–477.

Potera, C. (2007). Molding a link to depression. *Environmental Health Perspectives, 115*, A536.

Raj, A. (2014). Heat-fueled rage. *Scientific American Mind, 25*, 16–17. https://doi.org/10.1038/scientificamericanmind0114-16

Randall, A. (2013). Moving stories: The Sahel. *Climate and Migration Coalition*. Climate Outreach.

Reckdahl, K. (2017). New Orleans scrambles to repair drainage system after severe flooding. *The New York Times*. The New York Times Company.

SAMHSA. *Phases of Disaster*. U.S. Department of Health and Human Services 1-10-2018. Retrieved April 16, 2019 from https://www.samhsa.gov/dtac/recovering-disasters/phases-disaster

Schleussner, C.-F., Donges, J., Donner, R., & Schellnhuber, H. (2016). Armed conflict risks enhanced by climate-related disasters in ethnically fractionalized countries. *Proceedings of the National Academy of Sciences of the United States of America, 113*, 9216–9221.

Seelen, M., Toro Campos, R. A., Veldink, J. H., Visser, A. E., Hoek, G., Brunekreef, B., van der Kooi, A. J., de Visser, M., Raaphorst, J., van den Berg, L. H., & Vermeulen, R. C. (2017). Long-term air pollution exposure and amyotrophic lateral sclerosis in Netherlands: A population-based case-control study. *Environmental Health Perspectives, 125*, 097023.

Shaw, W., Labott-Smith, S., Burg, M., Hostinar, C., Alen, N., van Tilburg, M., et al. (2019). *Stress effects on the body: Musculoskeletal system*. Washington, DC: American Psychological Association. Retrieved February 9, 2020 from https://www.apa.org/helpcenter/stress/effects-musculoskeletal

Shenassa, E. D., Daskalakis, C., Liebhaber, A., Braubach, M., & Brown, M. (2007). Dampness and mold in the home and depression: An examination of mold-related illness and perceived control of one's home as possible depression pathways. *American Journal of Public Health, 97*, 1893–1899.

Simpson, D. M., Weissbecker, I., & Sephton, S. E. (2011). Extreme weather-related events: Implications for mental health and well-being. In I. Weissbecker (Ed.), *Climate change and human well-being. International and cultural psychology*. New York: Springer.

Skinner, M. (2014). Environmental stress and epigenetic transgenerational inheritance. *BMC Medicine, 12*, 153.

Smith, M., & Myers, S. (2018). Impact of anthropogenic CO_2 emissions on global human nutrition. *Nature Climate Change, 8*, 834–839.

Somanathan, E., Somanathan, R., Sudarshan, A., & Tewari, M. (2018). *The impact of temperature on productivity and labor supply: Evidence from Indian manufacturing*. Energy Policy Institute at the University of Chicago. WORKING PAPER. NO. 2018–69.

Souhel, N., Pearlman, D., Alper, K., Najjar, A., & Devinsky, O. (2013). Neuroinflammation and psychiatric illness. *Journal of Neuroinflammation, 10*, 43.

Stanke, C., Kerac, M., Prudhomme, C., Medlock, J., & Murray, V. (2013). *Health effects of drought: A systematic review of the evidence*. PLOS Currents Disasters. Jun 5 . Edition 1. https://doi.org/10.1371/currents.dis.7a2cee9e980f91ad7697b570bcc4b004

Strife, J. (2012). Children's environmental concerns: Expressing ecophobia. *The Journal of Environmental Education, 43*, 37–54.

Szyszkowicz, M., Willey, J. B., Grafstein, E., Rowe, B., & Colman, I. (2010). Air pollution and emergency department visits for suicide attempts in Vancouver, Canada. *Environmental Health Insights, 4*, 79–86.

United Nations Office for Disaster Risk Reduction (2015). *United Nations Office for Disaster Risk Reduction (UNISDR)*. The Human cost of weather-related disasters 1995–2015. Retrieved April 14, 2019, from https://www.unisdr.org/2015/docs/climatechange/COP21_WeatherDisastersReport_2015_FINAL.pdf

Van Susteren, L. (2013). *The Bureau of Linguistic Reality*. Retrieved from: https://bureauoflinguis-ticalreality.com/portfolio/pre-traumatic-stress-disorder-2

Vins, H., Bell, J., Saha, S., & Hess, J. (2015). The mental health outcomes of drought: A systematic review and casual process diagram. *International Journal of Environmental Research and Public Health, 12*, 13251–13275.

Von Uexkull, N., Croicu, M., Fjelde, H., & Buhaug, H. (2016). Civil conflict sensitivity to growing-season drought. *Proceedings of the National Academy of Sciences of the United States of America, 113*, 12391–12396.

Weir, K. (2012). Smog in our brains: Researchers are identifying startling connections between air pollution and decreased cognition and well-being. *American Psychological Association, 43*, 32.

Wolf, J., & Salo, R. (2008). Water, water, everywhere, nor any drop to drink: Climate change delusion. *Australia & New Zealand Journal of Psychiatry, 42*, 350.

World Health Organization. (2005). *Violence and Disasters. World Health Organization.* Department of Injuries and Violence Prevention. Retrieved February 9, 2020 from https://www.who.int/violence_injury_prevention/publications/violence/violence_disasters.pdf?ua=1

CHAPTER 15
Air Pollution and Cardiovascular Disease: A Proven Causality

Conrado J. Estol

"It was the best of times, it was the worst of times, it was the age of wisdom, it was the age of foolishness, it was the epoch of belief, it was the epoch of incredulity…"
Charles Dickens, A Tale of Two Cities

Summary Epidemiological data has shown that air pollution accounts for at least six million deaths per year worldwide. Cardiovascular disease ranks first among the causes of death from pollution, yet it is largely neglected. There is compelling evidence from studies in different world regions showing a causal relationship between air pollution and cardiovascular disease, but this information has not reached the scientific community at large and much less the general public. Gaseous and particle pollutants access the circulation through the lungs and accelerate atherosclerosis by altering blood pressure, lipids, oxidative processes, and insulin resistance. They affect all individuals from the maternal uterus to old age. Almost the entire population in Asian countries is exposed to severely elevated levels of pollutants and most other countries have vascular-damaging levels. More than 80% of the affected individuals belong to emerging regions. The use of N95 masks reduces the exposure to PM2.5, the most damaging particle. Physicians should consider air pollution as a recognized risk factor for vascular disease and act accordingly with patients exposed to different levels of pollutants. The only way to limit the worsening levels of pollution worldwide is to increase the joint efforts among science, governments and other leadership to generate a climate change conscious society and increase efforts to reach a decarbonized world by mid-century.

C. J. Estol (✉)
Stroke Unit, Sanatorio Guemes, Buenos Aires, Argentina

Heart and Brain Medicine, Breyna, Buenos Aires, Argentina
e-mail: conrado.estol@stat-research.com

© The Author(s) 2020
W. K. Al-Delaimy, V. Ramanathan, M. Sánchez Sorondo (eds.), *Health of People, Health of Planet and Our Responsibility*, https://doi.org/10.1007/978-3-030-31125-4_15

Background

Climate change and air pollution pose a direct threat to human life on earth. The amount of data supporting this notion is compelling and encompasses a broad spectrum of disciplines, including energy, hydrology, physics, chemistry, anthropology, medicine, biology, economics, and philosophy, among others. Yet this is likely the most neglected menace to humanity. In the popular quote that opened the chapter, Dickens eloquently expresses the opposing forces of love and odium, seemingly in a rather equal balance. Can we humans working as a group tilt these apparently matched forces?

This is the era of communications and of the ubiquity of knowledge. So, why are most of the data and messages about climate change and diseases caused by air pollution missing their targets? Probably because we mostly preach to those who already know and want to listen. Most of the world is simply not aware of this self-caused health catastrophe. Air pollution is an epidemic and the medical community is looking the other way. Breathing during our daily personal, working, and leisure time accelerates the occurrence of heart and brain infarction, cognitive decline and serious lung diseases, ranging from infections to cancer. The result is the occurrence of millions of preventable deaths. Despite this appalling statistic, air pollution is a significant risk factor for death that every human faces with indifference. Is this the result of some form of misinformation? We are now celebrating 30 years since evidence-based medicine radically changed "eminence"-based anecdotes and opinions into information based on well-designed research. Those of us with a specific knowledge in topics that most people fail to fully grasp and comprehend ought to disseminate this information in a scientific, reliable, complete, and clear message. In this chapter, I describe the data that show how air pollution damages the vascular system and thus becomes a leading risk factor for cardiovascular disease and death.

The Impact of Cardiovascular Disease Worldwide

Cardiovascular disease leads to 50,000 deaths every day, representing a third of all deaths worldwide. Such deaths include coronary heart disease (myocardial infarction), cerebrovascular disease (stroke), aortic aneurysm, peripheral vascular disease of the legs, vascular kidney failure, and others. Coronary heart disease is the first cause of death worldwide (Mozaffarian, Benjamin, Go, et al., 2015). In 2013, stroke was the second cause of death globally, with 25.7 million people who survived a stroke and 6.5 million people who died from stroke (Feigin, Norrving, & Mensah, 2017). Vascular disease is also a leading cause of cognitive decline and dementia, affecting up to a third of patients with cerebrovascular disease (Mijajlović, Pavlović, Brainin, et al., 2017).

More than 85% of strokes occur in developing regions, where prevalence of the disease is lower compared to developed countries because case fatality rates are higher due to suboptimal health systems. Cardiovascular disease incidence has increased 100% in the developing world over the last decade (Feigin, 2005). A study in 44 nations revealed that vascular risk factors are similarly distributed and largely undertreated (Bhatt, Steg, Ohman, et al., 2006).

Stroke occurs one to two decades earlier in individuals in emerging countries compared to developed countries with up to 50% of events affecting people under 70 years of age (20% in developed regions) (Strong, Mathers, Leeder, & Beaglehole, 2005).

The aforementioned data suggest that cardiovascular disease could be a straight-forward problem to tackle (Yusuf, Hawken, Ounpuu, et al., 2004). Supporting this, various studies have shown that by following vascular disease management guide-lines, adhering to treatment, avoiding cigarette smoking, having an appropriate nutritional intake and exercising, up to 80% of vascular events could be avoided (Chiuve, McCullough, Sacks, & Rimm, 2006; Hackam & Spence, 2007).

However, between 2005 and 2016, incidence of ischemic heart disease and stroke increased by 53% and 25%, respectively (Global Burden of Disease 2015 Risk Factors Collaborators, 2016). Unfortunately, most patients do not receive adequate treatment (as an example, 80% of treated hypertensive patients are not controlled!).

Air Pollution: A Poorly Recognized Vascular Disease Risk Factor

Although the deleterious effects of air pollution on human health were reported almost three decades ago, this major health risk factor has only recently come to the attention of the general medical community (Dockery, Pope, Xu, et al., 1993). In a recent analysis, Feigin et al. studied all modifiable vascular risk factors included in the Global Burden of Disease Study in 188 countries carried out from 1990 to 2013 (Feigin, Roth, Naghavi, et al., 2016). They analyzed the population attributable fraction (PAF) of all risk factors to define their individual contribution to stroke. More than three-quarters of the stroke burden could be avoided by controlling behavioral and metabolic risk factors. Smoking, poor nutrition and sedentary habits had a PAF of 74%. Hypertension, diabetes, cholesterol, and a high body mass index accounted for 72% of the burden. However, the authors found an unexpected 29% of the stroke burden attributed to air pollution, which was most significant in emerg-ing regions (and highest in sub-Saharan Africa and South Asia) (Feigin et al., 2016).

Interestingly, when considering the health consequences of air pollution, cogni-tive decline and dementia are rarely mentioned. However, population-based data suggest that seven modifiable vascular risk factors accelerate cerebrovascular dis-ease, which account for 50% of dementia cases. Because disease of the cerebral microcirculation has been associated with the risk of Alzheimer disease, those same modifiable risk factors have been shown to account for one-third of Alzheimer dementia (Ashby-Mitchell, Burns, Shaw, & Anstey, 2017; Norton, Matthews, Barnes, Yaffe, & Brayne, 2014). Cognitive disease should be included when mea-suring the total disease burden associated with air pollution.

The World Health Organization (WHO), the American Heart Association (AHA) and the European Cardiology Society have all formally declared air pollution as a risk factor for vascular disease (Brook, Rajagopalan, Pope, et al., 2010; Newby, Mannucci, Tell, et al., 2015).

Particulate Matter

Air pollution consists of a gaseous phase composed of carbon monoxide, ozone, sulfur dioxide, and nitrogen dioxide, and a particulate phase that encompasses 10, 2.5, and 0.1 µm particle matter (PM). Fuel combustion accounts for 85% of PM and 100% of the gaseous toxins.

Different estimates reveal that ten million people die worldwide from all causes of pollution, including air, water, soil, and occupational. Of them, between five and six million died from air pollution in 2015; for approximately 60%, the cause of death was cardiovascular disease (the remainder died from lung cancer, chronic obstructive pulmonary disease, and respiratory infections). By 2050, the number of deaths will have increased by 50% (Lelieveld, Evans, Fnais, et al., 2015). Air pollution currently represents the fifth leading cause of death, accounting for 7.6% of all deaths; 92% of the affected population is located in low- and middle-income countries (Cohen, Brauer, Burnett, et al., 2017). It is also the fifth most common vascular disease risk factor following hypertension, smoking, diabetes, and obesity (Global Burden of Disease 2015 Risk Factors Collaborators, 2016). Available data show an association between acute and chronic exposure to air pollution and different forms of cardiovascular disease, such as congestive heart failure, myocardial infarction, stroke, arrhythmias and sudden death (Brook et al., 2010; Ensor, Raun, & Persse, 2013; Link, Luttmann-Gibson, Schwartz, et al., 2013; Newby et al., 2015).

PM Toxic Levels: How Low is Harmless?

The WHO air quality guidelines suggest that a $PM_{2.5}$ concentration of 10 µg/m^3 is "safe" to avoid a causal role with vascular and lung disease. However, if these particles are pathogenic, then only a zero level can be considered safe, just as any number of cigarettes above zero is associated with disease caused by smoking. However, beyond the fact that reaching a zero PM level is probably practically impossible, the most recent estimation for the lowest mean $PM_{2.5}$ concentrations that could be considered a minimum risk exposure level is 4.2 ± 1.8 µg/m^3 (Lelieveld et al., 2019). This threshold value is significantly lower than that accepted by WHO and the 8.8 µg/m^3 advised by the Global Burden of Disease study, which explains the significantly larger proportion of $PM_{2.5}$-related deaths with current calculations (Lelieveld et al., 2015; Lim, Vos, Flaxman, et al., 2012).

Worldwide Air Pollution

During the period from 1990 to 2015, the overall population-weighted exposure to PM increased from 39.7 to 44.2 $\mu g/m^3$. Air pollution's toxic effects are most significant in Southeast Asia, where 99% of the population is exposed to levels that are beyond WHO air quality guidelines (van Donkelaar, Martin, Brauer, & Boys, 2015). New Delhi and Beijing show averages of 100 $\mu g/m^3$ with peaks of 1000 $\mu g/m^3$. India (with a population-weighted exposure of 74 $\mu g/m^3$) and China (58 $\mu g/m^3$) alone account for at least 50% of all PM-related cardiovascular deaths.

In the USA, improvements have been achieved with an average exposure of 8.4 $\mu g/m^3$ reached in 2015 from an average 14 $\mu g/m^3$ in 2000, reflecting a 40% decrease in $PM_{2.5}$ concentration (Cohen et al., 2017). If this improvement were to be replicated worldwide, deaths from air pollution would decrease by approximately 46% (Morishita, Thompson, & Brook, 2015). A 10 $\mu g/m^3$ reduction results in a half-year increase in life expectancy. A recent study in the USA that evaluated 61 million individuals showed that increases of 10 $\mu g/m^3$ were associated with a 7.3% increase in mortality. Racial minorities and people with low incomes were most affected (Di, Wang, Zanobetti, et al., 2017). When Cohen et al. in the Global Burden of Disease study evaluated worldwide trends of average $PM_{2.5}$ concentrations from 1990 to 2015, all countries had values above the WHO air quality guidelines of 10 $\mu g/m^3$. Many (mostly in Asia) reached five to almost ten times the average reported in the USA, with evolution curves that have become steeper since 2010 in the most populated countries (Bangladesh, China and India); only Nigeria showed a trend above the standard but with a decreasing projection (Cohen et al., 2017). Multiple efforts with varying results have been made to fight and control air pollution in the USA. In some areas, these efforts are counterbalanced; for example, in California, according to the strength and direction of winds, 10–20% of air pollution may come from China across the Pacific (Lin, Pan, Davis, et al., 2014).

Is $PM_{2.5}$ Inhalation Currently Preventable?

The use of N95 breathing masks is the only method that has proven effective to decrease particle inhalation by more than 95% (Münzel et al., 2017). One study showed a significant reduction in mean blood pressure and electrocardiogram changes in individuals wearing N95 masks (Langrish, Wang, Lee, et al., 2012). A study to measure the stress hormonal response to air pollution was performed in Shanghai in 55 healthy college students (Li, Cai, Chen, et al., 2017). A double-blind, crossover analysis was performed using indoor air filter purifiers and sham purifiers for 9 days with a 12-day wash-out period. The filters resulted in a 50% decrease in PM, from 53 to 24 $\mu g/m^3$. The students underwent measurements of glucocorticoids, catecholamines, free fatty acids, blood pressure, and insulin resistance to convey data on systemic inflammation and blood circulation status.

The comparison between students revealed that those not breathing purified air had changes in most biomarkers indicative of increased oxidative stress, inflammation, thrombosis and increased blood pressure. A relevant concern of this study is whether the 50% reduction noted is enough, because the resulting PM levels are still significantly above a safe threshold.

How Does Air Pollution Reach the Vascular System?

Fossil fuel emissions contribute to most of the PM associated with cardiovascular disease. The gaseous contributors ozone and nitrogen oxide double the pathogenic effects of PM. These toxins cause endothelium dysfunction due to increased reactive oxygen species and decreased nitric oxide, increased blood pressure and insulin resistance, abnormal lipid concentrations and arrhythmias that lead to increased thrombosis (Brauer, Freedman, et al., 2016; Ostro, Hu, Goldberg, et al., 2015; Shah, Lee, McAllister, et al., 2015). Arrhythmias may also lead to cardiac embolism to the brain, and an increased sympathetic tone correlates with an abnormal cerebrovascular resistance that impairs cerebral blood flow. $PM_{2.5}$ is one-third the size of red blood cells and enters the body's circulation via the lung alveoli; by directly translocating across the lung–vessel barrier or by entering macrophage cells, they penetrate this barrier into the circulating blood to cause the aforementioned endothelial changes (Nemmar, Vanbilloen, Hoylaerts, et al., 2001).

What Is the Evidence Linking Air Pollution and Cardiovascular Disease?

The effect of $PM_{2.5}$ on atherosclerosis has been shown in animal models. In one study, 28 mice were fed with normal or high-fat chow and exposed to $PM_{2.5}$ or filtered air for 6 h/day, 5 days a week for 6 months. A significant increase in the amount of atherosclerosis, vasoconstrictor response and vessel inflammation was measured in the mice aorta exposed to $PM_{2.5}$ (Sung, Wang, Jin, et al., 2005). The Multi Ethnic-Study of Atherosclerosis and Air Pollution (MESA) was conducted over 10 years in nine locations from six USA states. A greater degree of coronary calcification in multi-slice CT was determined in subjects exposed to $PM_{2.5}$ and nitrogen dioxide. Hypertensive and older subjects were the most vulnerable (Kaufman, Adar, Allen, et al., 2012; Kaufman, Adar, Barr, et al., 2016). Significant changes in forearm blood flow have been shown with smoking, which are exacerbated when another variable such as high cholesterol is also present (Heitzer, Yla-Herttuala, Luoma, et al., 1996). In another study, 31 subjects were exposed to $PM_{2.5}$, ozone, their combination, or filtered air. Diastolic blood pressure was lowest in the filtered air group and most significantly elevated in the PM plus ozone group

(Brook, Urch, Dvonch, et al., 2009). In a double-blinded, crossover study, Kristen et al. exposed 45 nonsmoking subjects (age 18–49) to diesel exhaust PM for 2 h or to filtered air (exposure times were separated by 2 weeks). Blood pressure was measured every 30 min during exposure and four times during the first 24 h. Systolic blood pressure was significantly and rapidly increased in the diesel group compared to controls despite the short exposure time (Kristen, Cosselman, Krishnan, Oron, et al., 2012). Short-term exposure to air pollution outside the laboratory has also been shown to cause acute vascular events. A French study of the gaseous phase of air pollution showed that people aged 55–64 years without prior clinically evident vascular disease, exposed for 1–2 days to increased concentrations of ozone, had a significantly higher risk of developing an acute coronary event (Ruidavets, Cournot, Ferrières, et al., 2005). Other studies provide extensive evidence of the association between air pollution and hospital admissions and emergency room consults for vascular events, myocardial infarction, stroke, arrhythmias, and congestive heart failure (Brook, Franklin, Cascio, et al., 2004; Claeys, Rajagopalan, Nawrot, & Brook, 2017; Forastiere & Agabiti, 2013; Gold & Samet, 2013; Link & Dockery, 2010; Mustafić, Jabre, Caussin, et al., 2012; Stafoggia, Cesaroni, Peters, et al., 2014).

Epidemiological data also show that when deaths are compared for each month of the year before and after smoke-free legislation was passed in the USA, a significant reduction in the number of deaths was noted (Pell, Haw, Cobbe, Newby, et al., 2008). Shah et al. performed a meta-analysis to assess exposure to air pollutants and acute stroke (Shah et al., 2015). The authors reviewed approximately 100 articles reporting 6.2 million vascular events in 28 countries. A robust association was found between admissions/death from stroke with both PM and gaseous air pollutants (except for ozone); the strongest association occurred on the day of exposure to $PM_{2.5}$. The most significant findings were seen in low- and middle-income countries. Previous studies had shown an association between long-term PM exposure and stroke (Dockery et al., 1993; Miller, Siscovick, Sheppard, et al., 2007; Pope, Burnett, Krewski, et al., 2009). A recent review analyzed the association between air pollution and stroke focusing on the specific subtypes of stroke (hemorrhagic, small and large vessel, cardioembolic), patient's age, effect of temperature in the cold and warm season, types of pollutants, duration of exposure, and the presence of vascular risk factors or previous vascular events (Estol, 2019).

The Global Burden of Stroke study by Feigin et al. revealed a significant loss of stroke disability-adjusted life-years (DALYs) from air pollution among metabolic and behavioral risk factors (Feigin et al., 2016). The importance of the authors' findings was showing that 90% of stroke burden is due to modifiable risk factors. Air pollution was the third largest contributor to stroke burden, with 33 million stroke-related DALYs in 2013. Although the world average of stroke burden attributed to air pollution was 30%, it ranged from 34% in low- to middle-income countries to 10% in high-income countries. Cohen et al. showed that stroke and coronary heart disease represent a large proportion of deaths due to $PM_{2.5}$ and have shown a steady and significant increase since 1990 (Cohen et al., 2017).

The 2017 report by the Lancet Commission on pollution and health provides data in support of pollution being the most important environmental cause of disease and premature death worldwide (Landrigan, Fuller, Acosta, et al., 2017). Air pollution causes more deaths than water, occupational and soil pollution combined; it results in three times more deaths than HIV, tuberculosis and malaria combined and 15 times more deaths than those caused by war and other violence. In 2015, air pollution accounted for 26% of coronary heart disease deaths and for 23% of stroke deaths, reflecting 21% of cardiovascular deaths overall (Landrigan et al., 2017). These deaths have an age peak in children under the age of 5 years, although the largest absolute number of deaths occurs over 55 years of age. Because of the significant number of years in age expectancy, the largest number of DALYs lost affect children.

Air Pollution and Congenital Heart Disease

The fetus offers the most vulnerable state for pathogenic effects of toxin exposure and is exposed to a higher relative toxin dose compared to the mother. The most robust correlation was found between the air pollutant nitrogen oxide and coarctation of the aorta, Fallot tetralogy, and ventricular and atrial septal defects (Jerrett, McConnell, Wolch, et al., 2014; Mao, Sun, Zhang, et al., 2017). The causative mechanism appears to be DNA hypomethylation from the interaction of nitrogen oxide and methionine. It is also feasible that due to cardiac structural changes and a greater incidence of metabolic syndrome, exposed fetuses remain predisposed to have a higher risk of cardiovascular events (Chen, Zmirou-Navier, Padilla, & Deguen, 2014).

Conclusions

The health costs of air pollution are hidden behind a myriad of heart attack and stroke hospital admissions. A vast proportion of affected patients admitted for a first episode have no known common vascular risk factors, so they are labeled as heart or brain infarctions "of unknown etiology." I have provided ample evidence to support the causality between acute and chronic exposure to air $PM_{2.5}$ (and other pollutants) and the occurrence of acute myocardial infarction and stroke. However, this is rarely considered causative by physicians providing hospital care to patients and much less by clinic administrators and politicians who are responsible for generating and implementing legislation for air pollution control. Air pollution accounts for at least 6% of global health care expenses (Landrigan et al., 2017). However, the increasing costs of health care, and especially those related to the growing epidemic of noncommunicable diseases (led by heart disease and stroke) are easily identified in hospital financial statements.

Investments in air pollution control are largely cost effective. Following the 1970 Clean Air Act in the USA, air pollution decreased by 70% with a greater than 210% related GDP increase (U.S. EPA, n.d.). In contrast, in California for every 10 µg/m³ increase in PM, there is an associated 6% decrease in productivity. The myth promoting the idea that pollution is an unavoidable consequence in the transition to economic development, the so-called Kuznets hypothesis, has long been proven wrong (Kuznets, 1955). The problem is large and complex enough that different government areas, including financial, health, and environmental agencies, must coordinate efforts to effectively tackle this enormous challenge. Efforts to educate the general public and the medical community need to evolve in parallel with government, multinational, and corporate decisions to decrease the causes of air pollution and climate change as different components of the same problem that directly endanger human life. In the encyclical letter *"Laudato Si,"* Pope Francis makes a thorough analysis of the threat and danger posed to our planet by current human behavior (Francis, 2015). The Holy Father reminds us that several popes before him expressed concern about an ongoing destructive attitude of humanity towards our "common home." Pope Francis made an "urgent appeal" for dialogue about the future of our planet dominated by a "throwaway culture" while being converted in an "immense pile of filth."

By thoroughly covering the broad spectrum of climate change, air pollution and health, this book ought to be the tipping point to help reverse the present course of this avoidable disaster facing both the current and future human generations.

Acknowledgements The author is grateful to Sarah MacKenzie, Ph.D., for editorial assistance.

References

Ashby-Mitchell, K., Burns, R., Shaw, J., & Anstey, K. J. (2017). Proportion of dementia in Australia explained by common modifiable risk factors. *Alzheimer's Research Therapy, 9*, 11.

Bhatt, D. L., Steg, P. G., Ohman, E. M., et al. (2006). International prevalence, recognition, and treatment of cardiovascular risk factors in outpatients with atherothrombosis. *Journal of the American Medical Association, 295*, 180–189.

Brauer, M., Freedman, G., & Frostad, J., et al. (2016). Ambient air pollution exposure estimation for the global burden of disease 2013. *Environmental Science & Technology, 50*, 79–88.

Brook RD, Franklin B, Cascio W, et al., for the Expert Panel on Population and Prevention Science of the American Heart Association (2004). Air pollution and cardiovascular disease: A statement for healthcare professionals from the expert panel on population and prevention science of the American Heart Association. *Circulation, 109*, 2655–2671.

Brook, R. D., Rajagopalan, S., Pope, C. A., et al. (2010). Particulate matter air pollution and cardiovascular disease: An update to the scientific statement from the American Heart Association. *Circulation, 121*, 2331–2378.

Brook, R. D., Urch, B., Dvonch, J. T., et al. (2009). Insights into the mechanisms and mediators of the effects of air pollution exposure on blood pressure and vascular function in healthy humans. *Hypertension, 54*, 659–667.

Chen, E. K., Zmirou-Navier, D., Padilla, C., & Deguen, S. (2014). Effects of air pollution on the risk of congenital anomalies: A systematic review and meta-analysis. *International Journal of Environmental Research and Public Health, 11*, 7642–7668.

Chiuve, S. E., McCullough, M. L., Sacks, F. M., & Rimm, E. B. (2006). Healthy lifestyle factors in the Primary prevention of coronary heart disease among men: Benefits among users and nonusers of lipid lowering and antihypertensive medications. *Circulation, 114*, 160–167.

Claeys, M. J., Rajagopalan, S., Nawrot, T. S., & Brook, R. D. (2017). Climate and environmental triggers of acute myocardial infarction. *European Heart Journal, 38*, 955–960.

Cohen, A. J., Brauer, M., Burnett, R., et al. (2017). Estimates and 25-year trends of the global burden of disease attributable to ambient air pollution: An analysis of data from the Global Burden of Diseases Study 2015. *Lancet, 389*, 1907–1918.

Di, Q., Wang, Y., Zanobetti, A., et al. (2017). Air pollution and mortality in the Medicare population. *New England Journal of Medicine, 376*, 2513–2522.

Dockery, D. W., Pope, C. A., Xu, X., et al. (1993). An association between air pollution and mortality in six U.S. cities. *New England Journal of Medicine, 329*, 1753–1759.

Ensor, K. B., Raun, L. H., & Persse, D. (2013). A case-crossover analysis of out-of-hospital cardiac arrest and air pollution. *Circulation, 127*, 1192–1199.

Estol, C. J. (2019). Is breathing our polluted air a risk factor for stroke? *International Journal of Stroke, 14*, 340–350.

Feigin, V. L. (2005). Stroke epidemiology in the developing world. *Lancet, 365*, 2160–2161.

Feigin, V. L., Norrving, B., & Mensah, G. A. (2017). Global burden of stroke. *Circulation Research, 120*, 439–448.

Feigin, V. L., Roth, G. A., Naghavi, M., et al. for the Global Burden of Diseases, Injuries, and Risk Factors (2016). Global burden of stroke and risk factors in 188 countries, during 1990–2013: A systematic analysis for the Global Burden of Disease Study 2013 and Stroke Experts Writing Group. *Lancet Neurology, 15*, 913–924.

Forastiere, F., & Agabiti, N. (2013). Assessing the link between air pollution and heart failure. *Lancet, 382*, 1008–1010.

Pope Francis. (2015). *Laudato Si': On Care for Our Common Home* [Encyclical].

Global Burden of Disease 2015 Risk Factors Collaborators (2016). Global, regional, and national comparative risk assessment of 79 behavioral, environmental and occupational, and metabolic risks or clusters of risks, 1990–2015: A systematic analysis for the Global Burden of Disease Study 2015. *Lancet, 388*, 1659–1724.

Gold, D. R., & Samet, J. M. (2013). Air pollution, climate, and heart disease. *Circulation, 128*, e411–e414.

Hackam, D. G., & Spence, J. D. (2007). Combining multiple approaches for the secondary prevention of vascular events after stroke: A quantitative modeling study. *Stroke, 38*, 1881–1885.

Heitzer, T., Yla-Herttuala, S., Luoma, J., et al. (1996). Cigarette smoking potentiates endothelial dysfunction of forearm resistance vessels in patients with hypercholesterolemia. *Circulation, 93*, 1346–1353.

Jerrett, M., McConnell, R., Wolch, J., et al. (2014). Traffic-related air pollution and obesity formation in children: A longitudinal, multilevel analysis. *Environmental Health, 13*, 49.

Kaufman, J. D., Adar, S. D., Allen, R. W., et al. (2012). Prospective study of particulate air pollution exposures, subclinical atherosclerosis and clinical cardiovascular disease: The Multi-Ethnic Study of Atherosclerosis and Air Pollution (MESA Air). *American Journal of Epidemiology, 176*, 825–837.

Kaufman, J. D., Adar, S. D., Barr, R. G., et al. (2016). Association between air pollution and coronary artery calcification within six metropolitan areas in the USA (the Multi-Ethnic Study of Atherosclerosis and Air Pollution): A longitudinal cohort study. *Lancet, 388*, 696–704.

Kristen, E., Cosselman, K. E., Krishnan, R. M., Oron, A. P., et al. (2012). Blood pressure response to controlled diesel exhaust exposure in human subjects. *Hypertension, 59*, 943–948.

Kuznets, S. (1955). Economic growth and income equality. *American Economic Review, 45*, 1–28.

Landrigan, P. J., Fuller, R., Acosta, N. J. R., et al. (2017). The Lancet Commission on pollution and health. *Lancet, 391*, 462–512. https://doi.org/10.1016/S0140-6736(17)32345-0

Langrish, L. X., Wang, S., Lee, M. M., et al. (2012). Reducing personal exposure to particulate air pollution improves cardiovascular health in patients with coronary heart disease. *Environmental Health Perspectives, 120*, 367–372.

Lelieveld, J., Evans, J. S., Fnais, M., et al. (2015). The contribution of outdoor air pollution sources to premature mortality on a global scale. *Nature, 525*, 367–371.

Lelieveld, J., Klingmuller, K., Pozzer, A., et al. (2019). Cardiovascular disease burden from ambient air pollution in Europe reassessed using novel hazard ratio functions. *European Heart Journal, 40*, 1590–1596. https://doi.org/10.1093/eurheartj/ehz135

Li, H., Cai, J., Chen, R., et al. (2017). Particulate matter exposure and stress hormone levels: A randomized, double-blind, crossover trial of air purification. *Circulation, 136*, 618–627.

Lim, S. S., Vos, T., Flaxman, A. D., et al. (2012). A comparative risk assessment of burden of disease and injury attributable to 67 risk factors and risk factor clusters in 21 regions, 1990–2010: A systematic analysis for the Global Burden of Disease Study 2010. *Lancet, 380*, 2224–2260.

Lin, J., Pan, D., Davis, S. J., et al. (2014). China's international trade and air pollution in the United States. *Proceedings of the National Academy of Sciences of the United States of America, 111*, 1736–1741.

Link, M. S., & Dockery, D. W. (2010). Air pollution and the triggering of cardiac arrhythmias. *Current Opinion in Cardiology, 25*, 16–22.

Link, M. S., Luttmann-Gibson, H., Schwartz, J., et al. (2013). Acute exposure to air pollution triggers atrial fibrillation. *Journal of the American College of Cardiology, 62*, 816–825.

Mao, G., Sun, Q., Zhang, X., et al. (2017). Individual and joint effects of early-life ambient $PM_{2.5}$ exposure and maternal prepregnancy obesity on childhood overweight or obesity. *Environmental Health Perspectives, 125*, 067005. https://doi.org/10.1289/EHP261

Mijajlović, M. D., Pavlović, A., Brainin, M., et al. (2017). Post-stroke dementia—A comprehensive review. *BMC Medicine, 15*, 11.

Miller, K. A., Siscovick, D. S., Sheppard, L., et al. (2007). Long-term exposure to air pollution and incidence of cardiovascular events in women. *New England Journal of Medicine, 356*, 447–458.

Morishita, M., Thompson, K. C., & Brook, R. D. (2015). Understanding air pollution and cardiovascular diseases: Is it preventable? *Current Cardiovascular Risk Reports, 9*, 30.

Mozaffarian, D., Benjamin, E. J., Go, A. S., et al. (2015). Heart disease and stroke statistics. 2016 update. A report from the AHA. *Circulation, 138*, e38–e360.

Münzel, T., Daiber, A., Steven, S., Tran, L. P., Ullmann, E., et al. (2017). Effects of noise on vascular function, oxidative stress, and inflammation: Mechanistic insight from studies in mice. *European Heart Journal, 38*, 2838–2849.

Mustafić, H., Jabre, P., Caussin, C., et al. (2012). Main air pollutants and myocardial infarction: A systematic review and meta-analysis. *Journal of the American Medical Association, 307*, 713–721.

Nemmar, A., Vanbilloen, H., Hoylaerts, M. F., et al. (2001). Passage of intratracheally instilled ultrafine particles from the lung into the systemic circulation in hamster. *American Journal of Respiratory and Critical Care Medicine, 164*, 1665–1668.

Newby, D. E., Mannucci, P. M., Tell, G. S., et al.. for the ESC Working Group on Thrombosis, European Association for Cardiovascular Prevention and Rehabilitation, ESC Heart Failure Association (2015). Expert position paper on air pollution and cardiovascular disease. *European Heart Journal, 36*, 83–93.

Norton, S., Matthews, F. E., Barnes, D. E., Yaffe, K., & Brayne, C. (2014). Potential for primary prevention of Alzheimer's disease: An analysis of population-based data. *Lancet Neurology, 13*, 788–794.

Ostro, B., Hu, J., Goldberg, D., et al. (2015). Associations of mortality with long-term exposures to fine and ultrafine particles, species and sources: Results from the California Teachers Study Cohort. *Environmental Health Perspectives, 123*, 549–556.

Pell, J. P., Haw, S., Cobbe, S., Newby, D. E., et al. (2008). Smoke-free legislation and hospitalizations for acute coronary syndrome. *New England Journal of Medicine, 359*, 482–491.

Pope, C. A., Burnett, R. T., Krewski, D., et al. (2009). Cardiovascular mortality and exposure to airborne fine particulate matter and cigarette smoke: Shape of the exposure-response relationship. *Circulation, 120*, 941–948.

Ruidavets, J. B., Cournot, M., Ferrières, J., et al. (2005). Ozone air pollution is associated with acute myocardial infarction. *Circulation, 111*, 563–569.

Shah, A. S. V., Lee, K. K., McAllister, D. A., et al. (2015). Short term exposure to air pollution and stroke: Systematic review and meta-analysis. *British Medical Journal, 350*, 1295.

Stafoggia, M., Cesaroni, G., Peters, A., et al. (2014). Long-term exposure to ambient air pollution and incidence of cerebrovascular events: Results from 11 European cohorts within the ESCAPE project. *Environmental Health Perspectives, 122*, 919–925.

Strong, K., Mathers, C., Leeder, S., & Beaglehole, R. (2005). Preventing chronic diseases: How many lives can we save? *Lancet, 366*, 1578–1582.

Sung, Q., Wang, A., Jin, X., et al. (2005). Long term air pollution exposure and acceleration of atherosclerosis and vascular inflammation in an animal model. *Journal of the American Medical Association, 294*, 3003–3010.

U.S. EPA. (n.d.). Comparison of Growth Areas and Emissions, 1980–2013.

van Donkelaar, A., Martin, R. V., Brauer, M., & Boys, B. L. (2015). Use of satellite observations for long-term exposure assessment of global concentrations of fine particulate matter. *Environmental Health Perspectives, 123*, 135–143.

Yusuf, S., Hawken, S., Ounpuu, S., et al. (2004). Effect of potentially modifiable risk factors associated with myocardial infarction in 52 countries (the INTERHEART study). *Lancet, 364*, 937–952.

CHAPTER 16
Healthy People, Healthy Planet: Holistic Thinking

Jamie Harvie and Erminia Guarneri

Summary In the context of our broad ecological calamity, we highlight the linkages between climate change and human health. Holism is offered as a unifying model and process to align and foster individual, community, and planetary health and thereby address our existential crisis. We explain healthcare's significant climate footprint and the obligation and opportunity for the healthcare sector to think and act holistically. Our linear economy is used as another example to explain how mechanistic thinking fuels ecological degradation, inequality, and adverse health outcomes. Action steps that support a healthy planet for all now and for all future generations are offered along with guidance for healthcare systems and health professionals.

Introduction

In the context of our planetary ecological calamity—climate change, stratospheric ozone depletion and ultraviolet (UV) radiation, biodiversity loss, land degradation and desertification, water pollution and water scarcity, and persistent toxic contamination—(World Health Organization, 2014; EPA, 2017b) the global human community is facing an existential crisis. Indigenous communities have long understood the indivisible relationship between individual, community, and ecological health. In the Anthropocene, we are relearning this ancient wisdom by necessity as we confront the stark reality that the health and survival of human populations is inti-

J. Harvie (✉)
Institute for a Sustainable Future, Duluth, MN, USA
e-mail: harvie@isfusa.org

E. Guarneri
Pres. Academy Integrative Health and Medicine, Medical Director Guarneri
Integrative Health, La Jolla, CA, USA

© The Author(s) 2020
W. K. Al-Delaimy, V. Ramanathan, M. Sánchez Sorondo (eds.), *Health of People, Health of Planet and Our Responsibility*, https://doi.org/10.1007/978-3-030-31125-4_16

mately connected to the health and resilience of global ecosystems, both of which are in crisis together. We cannot be healthy on an unhealthy planet.

Medical wisdom helps explain these crises as symptoms of an underlying problem, a root cause. What we discover is that we must think differently. With urgency, we are called to relearn and adopt a new worldview. This call to action is central to the emerging Great Transition, a global paradigm shift from a mechanistic anthropocentric model to an ecological living systems model. Rather than focusing on discreet parts, we must look at the world for what it is, a complex interrelated holistic system of relationships. If we are successful, we will discover a host of strategies that can release the emergence of powerful, compassionate, equitable approaches to the multitude of problems confronting our common home, fostering health and creating livable communities for all and future generations.

Climate and Health

Climate change alone has been called the largest human health threat by public health, medical, and healthcare organization across the globe (Watts, 2018; World Health Organization, 2016). Direct and indirect health effects of climate change include temperature-related death and illness, air quality impacts, mental health effects, extreme weather events, water-related illness, and conflict. Food safety, malnutrition, undernutrition, and food security are additional concerns, particularly in relation to floods and drought. Rising temperatures increase ground-level ozone, which in turn creates difficulty breathing, the aggravation of asthma and the development of new cases of asthma, emphysema, chronic bronchitis, and pneumonia. Ozone can also affect the heart, leading to cardiac arrhythmia and heart attacks as well as increasing the number of low-birthweight babies (EPA, 2017a). Aeroallergens such as pollen are higher in extreme heat. Higher levels of aeroallergens trigger asthma, affecting close to 300 million people (Baxi & Phipatanakul, 2010). Heat stress can make working conditions unbearable and increase the risk of cardiovascular, respiratory, and renal diseases (U.S. Global Change Research Program, 2016). Vulnerable populations include children, pregnant women, the elderly, and socially marginalized groups, all of whom experience greater susceptibility to climate-sensitive health impacts. Even animals are at risk; pollinators such as bees can develop heat exhaustion and decreased immune response (McKinstry et al., 2017).

Climate change lengthens the transmission season and expands the geographical range of many diseases like yellow fever, malaria, chikungunya, Lyme, West Nile, and dengue.

Emotional, spiritual and mental health impacts are also recognized effects associated with climate change and can include trauma, fear, fatalism and loss of loved ones, livelihoods, social support, identity, and a sense of control (Clayton, Manning, Krygsman, & Speiser, 2017).

It is estimated that up to 90% of health is accounted for by social and environmental factors outside of healthcare (Minnesota Department of Health, 2014). In the context of the global burden of chronic disease, the health sector is awakening to the limitations of the biomedical approach, which focuses on disease treatment (Abelson, Rupel, & Pincus, 2008; Bendelow, 2013; Huber et al., 2011). Industrial agriculture, a sector which contributes approximately 30% of global climate emissions (Food and Agriculture Organization, 2011)—far greater than transportation—employs a similar mechanistic model with an overwhelming array of unintended impacts to individual, community, and planetary health. Our industrial economy is another linear model that inaccurately assumes finite resources and unlimited growth, ultimately contributing to resource depletion, ecological degradation, and concentrated wealth (Ellen McArthur Foundation, 2013; Piketty & Goldhammer, 2017; World Economic Forum, 2014). If we can find solace in our climate emergency, it is that the situation is forcing us to rapidly reevaluate the design of human society and the necessity for the application of a holistic, or living systems, model.

Holistic Thinking

It is not surprising that we are experiencing profound challenges. Many of our institutions and approaches were built on a mechanistic, cause-and-effect scientific model that does not fully explain the complexity of humans, our relationships and the interconnectedness of all life (Rosas, 2015). Acknowledging this truth helps us understand why, as Pope Francis states, "*It cannot be maintained that empirical science provides a complete explanation of life, the interplay of all creatures and the whole of reality*" (Francis, 2015). Moreover, a linear model precludes full systemic awareness.

A systems worldview shifts our perception of critical matters in new and profound ways. As this model requires us to change our focus from parts to the whole, it is often described as a holistic approach. Some of its novel attributes include the following shifts (Capra, 2012):

- From parts to the whole
- From objects to relationships
- From isolated knowledge to contextual knowledge
- From quantity to quality
- From structure to process
- From contents to patterns

Systems models support approaches and organizational paradigms that are characterized by collaboration and connection, rather than hierarchy and control. Systems thinking helps rebalance empirical scientific knowledge by re-prioritizing cultural wisdom and knowledge and by embracing diverse expertise and experiential contexts. Through a shift in value from quantity to quality, an entirely new holistic set of qualitative metrics can be developed. If we are going to succeed in addressing the human-made challenges of our common home, it seems obvious that this living systems model enables us to align our ways of operating in accordance with life on a complex planet, reflective of our true nature and thereby fostering health.

Holism: A Systems Model of Health

Efforts to better incorporate this holistic understanding into institutionalized definitions of health are ongoing. International experts now recognize that the once-heralded World Health Organization definition of health, formulated in 1948, as "a state of complete physical, mental, and social well-being and not merely the absence of disease or infirmity" is insufficient, as the term "complete" suggests an unachievable end state, unintentionally perpetuating the medicalization of society (Huber et al., 2011). "Positive health" is one of the most recent definitions offered by international experts to reflect the dynamic nature of health, or "health as the ability to adapt and to self-manage in the face of social, physical and emotional challenges" (Huber et al., 2011). What we are uncovering is ancient learning from indigenous cultures, from which we all are rooted. The health and well-being of individuals is inseparable from nature and inseparable from the health of community (Fig. 16.1) (Harvie, 2016).

The well-known US poet and farmer Wendell Berry made this thoughtful statement about health:

> "Health is not just the sense of completeness in ourselves but also is the sense of belonging to others and to our place; it is an unconscious awareness of community, of having in common. It may be that this double sense of singular integrity and of communal belonging is our personal standard of health for as long as we live."

Holistic Models of Health

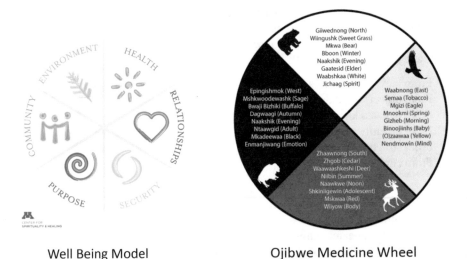

Well Being Model   Ojibwe Medicine Wheel

Fig 16.1 **a** Well-being model. **b** Ojibwe medicine wheel. http://www.curvelakeculturalcentre.ca/culture/medicine-wheel/

Though not a clinician or healthcare expert, Berry's observation is consistent with ancient wisdom, modern science and most recent health concepts which understand the indivisibility of individual, community, and planetary health (NACCHO, 2019; Adelson, 2007; Hermida, 2011). Solastalgia, or distress caused by environmental change, negatively affects health, while views of nature improve health (Albrecht et al., 2007; Ulrich, 1984). Similarly, a sense of connection, empathy, spirituality and love confer intrinsic health properties (Egolf et al., 1992; Puchalski, 2001; Rakel et al., 2011). Studies show that when we experience a sense of awe from viewing nature, we reduce stress, become more generous, and feel a sense of plentitude (Bai et al., 2017; Joye & Bolderdijk, 2015; Rudd et al., 2012; Stellar et al., 2015). We can infer from these studies that nature and relationship have helped us co-evolve as a communal species.

Individuals and communities measure health qualitatively, which is different (though equally valid) from those of public health and healthcare experts (Huber et al., 2016), The respect for inherent individual knowledge and self-efficacy and connection is a fundamental point, increasingly understood as a foundational principle, an essential ingredient required to foster agency and thereby improve health.

Within North America, Europe and elsewhere, holistic medicine and nursing approaches are being increasingly embraced. Functional medicine is a systems biology based approach that focuses on identifying and addressing the root cause of disease and sees health as homeodynamic, a function of dynamic balance between internal and external factors. Integrative medicine (IM) is healing-oriented medicine that takes into account the whole person—mind, body, emotion, and spirit—including all aspects of lifestyle. It emphasizes the therapeutic relationship between practitioner and patient.

Regardless of any specific health definition or clinical approach, these efforts to elevate our resilience and our relationship to planetary health demonstrate the emergence of collective wisdom. A holistic model of health better explains that we are ever changing, social, sentient beings in intimate relationship with all living beings on a dynamic planet. Individual, community, and planetary health are inseparable. The holistic model of health care is fundamental to helping us unlock and galvanize our efforts to support and protect planetary health and ecological integrity. What we do to the planet, we do to ourselves.

Our challenge will be how to amplify our voices into collective action and resist the allure of deeply entrenched mechanistic thinking across society. Across the globe, we are witnessing the emergence of new paradigm models across society. Our opportunity is to deepen our commitment to these systems models, strategies and approaches and bring them to light. By so doing, we will foster healthy communities and better align with the many values that ultimately make us resilient, healthy and whole and allow us to thrive.

Towards a Health Commons

One manifestation of the industrial food system model is the obesity crisis and the global burden of chronic disease, which have aggravated a financial crisis in the healthcare sector. Climate change is anticipated to further exacerbate this financial stress. These stressors are also leading to our increasing awareness of a broader fragility of the healthcare sector itself. Much like the industrial food model, its design and operations have largely been built on a mechanistic model, putting its long-term health and that of the planet in peril.

For example, the way that health care is organized, financed and delivered may contribute to health disparities by limiting access to certain populations, inadvertently incentivizing utilization. In addition, the health care sector has a sizable ecological footprint including discharge of toxic chemicals, pharmaceuticals, air and water pollution, and solid waste—all associated with negative health outcomes. Healthcare is also a substantial contributor of greenhouse gas (GHG) emissions. The U.K. National Health Service generates an estimated 39% of all public sector emissions (Health and Care System (HCS) Carbon Footprint, n.d.). The US healthcare sector represents an estimated 10% of total US GHG emissions. If the US healthcare sector were itself a country, it would rank 13th in the world for GHG emissions, ahead of the entire UK (Eckelman & Sherman, 2016).

The pharmaceutical footprint is an estimated 21% of the UK healthcare footprint, greater than building energy (Fig. 16.2). Reducing the pharmaceutical impact

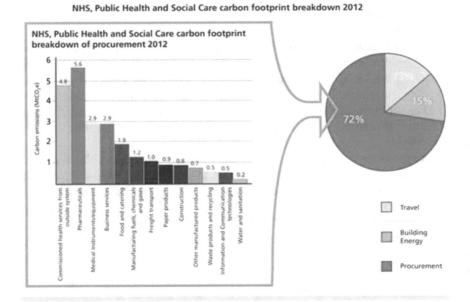

Fig. 16.2 National Health System Carbon Footprint Breakdown 2012 (NHS Sustainable Development Unit, 2013)

by 2.5% is estimated to offer the highest carbon reduction impact (Bouley et al., 2017), which offers strong insights regarding the potential of wellness or resiliency as a powerful climate mitigation strategy.

Viewed systemically, it becomes clear that our focus on disease treatment has caused us to lose sight of a more comprehensive whole, and has inadvertently created a self-magnifying disease treatment sector with deleterious climate and ecological health impacts that imperil its own sustainability.

Studies show that progress toward reducing health disparities and improving well-being will involve support for community-based strategies, enhanced understanding of the social determinants of health and an increased diversity of the healthcare workforce (Jackson & Nadine, 2014), Symposia convening healthcare and community highlight the imperative to adopt a systemic approach to health, emphasizing themes that include the concept that health is place-based, holistic thinking and the need to include a diversity of community expertise. In parallel to the failures of the industrial food model, the healthcare sector itself is called to think holistically. Healthcare itself is but one part of a health commons—the collective resources and relationships within a defined geographic boundary that foster health (Fig. 16.3).

Across the world, medical professionals are among the most trusted occupations. Yet we must remember that health as a biophysical process remains deeply embedded in our collective psyche, including healthcare. This transition will challenge a powerful medical industry with economic interests vested in supporting and maintaining our attention on individualized treatment. Similarly, the idea that health is transactional, something given, or received—rather than dynamic and relational—is a powerful, alluring concept that is deeply entrenched and will remain difficult to resist.

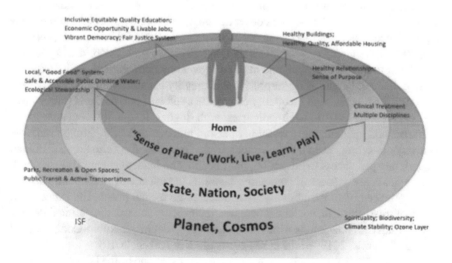

Fig. 16.3 Commons Health Model. http://www.isfusa.org

Nevertheless, in the context of climate change, we can no longer wait. There is an urgency to radically redesign our communities to equitably foster health. To do so, we must shift the locus of health to whole persons and the system of relationships in which they exist.

Being Well Together: Towards a Livable Economy

The last 150 years of industrial evolution have been dominated by a one-way or linear model of production and consumption in which goods are manufactured from raw materials, sold, used, and then discarded or incinerated as waste (World Economic Forum, 2014). This linear model has been supported by an economic system requiring an endless growth of production, which itself is contingent on consumption or consumerism (Movahed, 2016).

The nature of the economic system reinforces the primacy of economic growth, which promotes the externalization of social and environmental costs. At the current rate of consumption, only 58 years of minerals and metals remain available. Additionally, our oceans and marine life are filled with plastics, and the web of life is contaminated with persistent toxic chemicals. Ironically, CO_2, the waste product from the same fossil fuels that feed the global economy, is now predicted to destroy it.

Growth has been uneven, with extremes in inequality across the globe. Income disparities have become so pronounced that America's top 10% now average more than 9 times as much income as the bottom 90%, while the top 1% average over 40 times more income than the bottom 90%. The world's wealthiest individuals, those owning over $100,000 in assets, total only 8.1% of the global population but own 84.6% of global wealth, while the world's ten richest billionaires own $505 billion in combined wealth, a sum greater than the total goods and services most nations produce on an annual basis. The concentration of economic power and influence is highly concentrated with 147 companies, the majority financial institutions, controlling 40% of the global economy (Vitali et al., 2011). In such a dominant system, it is difficult to imagine how anyone might discover any hope in their ability to affect change. It should come as no surprise that inequality negatively affects our health.

At the local level, high inequality can create a sense of personal and public insecurity (Phillips, 2016). Rich countries with higher inequality consume more resources and generate more waste per person, influencing health through multiple pathways (Dorling, 2011). In a vicious cycle, reducing the formation of human capital, unequal access to education, poor health, and inadequate nutrition are both causes and consequences of inequality. Those on the lower end of economic disparity experience higher infant mortality and decreased mental health, life expectancy, levels of trust, altruism, social cooperation, reciprocity, and trust in political institutions (Attanasio, Fitzsimons, Grantham-McGregor, Meghir, & Rubio-Codina, 2012; Bowles & Gintis, 2013; Burns et al., 2014; Elgar & Aitken, 2010).

Viewed holistically, inequality may in fact represent one of the largest influences on the health of individuals, communities, and the planet. Citizens across the planet are experiencing a deepening spiritual emptiness as the economic model pulls us away from a vital sense of connection and relationship with one another and the planet; this connection helps make us resilient and well. The extremes in inequality are rapidly eroding the sense of trust and cooperation necessary for the functioning of civil societies and the global economy itself. In addition, in our singular quest for economic growth, we have lost sight of what makes us healthy and whole. Our hearts and minds tell us we need a new regenerative economic paradigm, designed first and foremost to support the health and resilience of all living beings together on our common home. This model is taking form.

Participatory budgeting is a democratic process in which community members directly decide how to spend part of a public budget. Designed to include those often left out of the democratic process such as low-income individuals and youth, participatory budgeting has now spread to thousands of cities across the world and been promoted as a best practice for democratic governance by the United Nations.

British Columbia, Canada, has adopted a tax (or fee) and dividend to regulate carbon emissions. Tax and dividend improves health by equitably distributing financial benefits to all citizens, allowing for the just transition to a carbon-free economy (Citizens' Climate Lobby Canada, 2019).

Inequity is driving a renewed focus on cooperative and worker ownership models, especially as research demonstrates how employee-ownership offers significant benefits including higher median wage income, job stability and household wealth, and benefits such as flexible work schedules, retirement plans, parental leave, and tuition reimbursement.

By including impacted communities and workers in their strategy, the Global Alliance for Incinerator Alternatives (GAIA) and others have been instrumental in helping shift policies towards zero waste and away from a "design for the dump" or "built to be burned" approach. Across the globe, we have witnessed a necessary shift away from plastic bags, Styrofoam, and other single-use polluting products. The concept of a circular economy, an industrial system that is restorative or regenerative by intention and design, is now rapidly advancing. Studies show that a shift to a circular economy, in which waste is designed out, can reduce climate emissions by up to 70%.

Global movements such as Transition Towns and Sharing Cities and nationwide networks such as Businesses Alliance for Local Living Economies (BALLE) within the USA are cultivating community-rooted enterprises, reclaiming the commons, democratizing and reorienting finance and building an economy that works for all. For this work to grow, it will require us to move beyond the singular economic measure of progress and the gross domestic product, and adopt holistic metrics such as the Genuine Progress Indicator, Gross National Happiness Index, Inclusive Wealth Index or co-create a new one that truly reflects the needs and aspirations of humans for a healthy meaningful life.

Despite the debilitating climate change predictions, punishing inequality and all the other crises affecting our planet, we are witnessing creativity, self-organization and the worldwide emergence of a new economy that can work for all (Fig. 16.4). Collectively, these are helping to demonstrate that we can change our current trajectory and there are viable economic alternatives that put the health and resilience of people and the planet at the center.

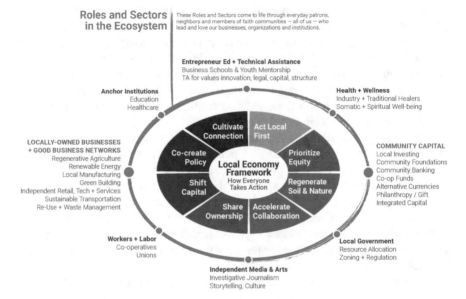

Fig. 16.4 Local economy ecosystem to build a healthy, equitable local economy (From BALLE, https://bealocalist.org/local-economy-framework/)

Action Steps

Climate Justice

A holistic worldview informs us that climate change is not strictly an environmental issue but one that also connects and links all of humanity. Strategies and approaches must be grounded in a human rights framework that guards the rights of the most vulnerable people and shares burdens and benefits of climate change equitably. Organizations and approaches must be grounded in climate justice principles, such as the Mary Robinson Foundation Climate Justice Principles.

Wealth Creation

We all do better when we all do better. Citizens, institutions and other purchasing entities can use their purchasing power and investments to align their values by preferentially supporting businesses with employee stock ownership programs (ESOPs) or worker, consumer and other cooperative models that are designed to feature principles of human well-being such as justice, equity, ecological steward-ship, and community. Prioritize equity in hiring and job creation.

Dietary Wellness

Industrial meat production has an outsized global carbon footprint. Cities, hospitals, churches, schools, and universities can initiate support for policies, practices, and programs such as the global Meatless Monday's Initiative and catalyze a global shift towards plant-based diets. High-caloric beverages are significant contributors to obesity with a direct and indirect climate footprint. Institutions can adopt policies to phase-out sales. Examples from Mexico and the USA and other countries demonstrate how sugary beverage taxes can increase consumption of healthy beverages and indirectly mitigate climate change, while revenue can be directed to support publicly owned water systems or be allocated through participatory budgeting.

Good Food Purchasing and Food System Models

Cities can encourage anchor institutions such as hospitals, schools, and universities to adopt purchasing policies similar to the US-based Center for Good Food Purchasing, which provides steps and metrics that catalyze good food production and demand. Cities and economic development agencies can invest and support community-owned holistic food systems models, such as The Food Commons.

Health Commons and Fostering Health

Governments, communities, healthcare, nursing, and medical schools can locally adopt recommendations similar to the NHS Manifesto for Health Creation (NHS Alliance, 2017), Manifesto for Sustainable Global Health (Manifesto for Sustainable Global Health, 2017), Creating Health Collaborative or the Commons Health. These suggestions encourage place-based community decision making, agency and shift away from a disease model.

Integrative Health and Medicine

Governments should support and adopt the Stuttgart Declaration, which calls on governments to recognize integrative health and medicine as a whole society approach that will help to reach the Sustainable Development Goals. They should encourage more holistic training in medical and nursing schools, including multidisciplinary fellowships such as the Academy of Integrative Health and Medicine Fellowship.

Climate Neutrality and Climate Smart Healthcare

Citizens and governments can encourage hospitals across the globe to adopt climate neutrality goals and climate-smart healthcare strategies such as those outlined in the World Bank Group document, Climate Smart Healthcare (Bouley et al., 2017), or adopt the 2020 Health Care Climate Challenge designed to mobilize health care institutions around the globe to protect public health from climate change. Clinicians can participate in the Medical Society Consortium on Climate and Health.

The Commons and the Healing Power of Nature

In the context of scarcity, it will be vital to develop strategies to embrace, preserve and equitably share our commons. The commons include but are not limited to forests, fisheries, parks, public-owned water treatment, renewable energy, libraries, medicinal plants, and so on. It is vital that governments and citizens develop legislation, legal structures and self-governance strategies that maintain these assets for our commonwealth and future generations. Clinicians can prescribe the healing power of nature and advocate for equitable access to green space.

Active Living and Public Transportation

Communities and clinicians can advocate for accessible and affordable public transportation and prioritize pedestrian and bicycle policies and practices, such as bicycle lockers, showers and connectivity to public transportation.

Zero Waste

Together, citizens can adopt and mobilize their homes, communities, schools, universities, hospitals, and businesses to adopt zero waste policies and eliminate the landfilling and incineration of waste. Food waste reduction strategies are a high priority. Prioritizing reduction and reuse other strategies, such as those proposed by GAIA, include municipal composting, phase-outs on Styrofoam, plastic straws, and other single-use items such as carryout bags.

Being Well Together

On a planet in crisis, we will need to improve and hone lost skills for working in community and in collaboration. This transition will require practices to open hearts and open minds such as mindfulness, meditation and deep immersion in skills and methods to harness the collective wisdom and self-organizing capacity of groups. Take action on something together. Our health is a function of our relationships.

Skills for Health Professionals

Healthcare professionals should build belief in their own resilience and cultivate active coping and self-regulation skills for themselves and their patients. They should adopt and guide patients to maintain practices that help to provide a sense of meaning and that foster optimism, as well as highlight and promote connectedness to family, place, culture, and community. Clinicians should become climate-literate professionals and engage with fellow health professionals. They should be vocal, model leaders within their communities, and support national and international climate solutions (Clayton et al., 2017).

Conclusion

For the last half century or more, we have ignored the growing signs that our planetary life support system is in peril. Through floods, drought, and storms, climate change has cracked a dam of hubris and denial and is stirring an awareness about the inter-relationship of all life. Now, we are awakening to the realization that humanity's challenges are our own doing, a function of linear models and associated hierarchical structures and operating systems incongruous with the complexity of life.

One challenge that health professionals understand is that, in times of crisis, the human body and mind may "freeze." Immobilization is a natural defense mechanism, yet it results in an incapacity to act. So while one inclination is to create a heightened sense of fear and urgency about the climate emergency, it may, in fact, create inaction.

Perhaps one of our biggest challenges is to understand the pervasive power and influence of our dominant narrative. We are challenged by patterns of thought that have been reinforced daily through our culture since birth. It makes us want to focus on climate change or carbon as a singular issue, when climate change is a complex product of multiple factors. Many of us have lost the skills to work in groups; without these necessary skills, the notion of collaborating with individuals having diverse perspectives can be paralyzing. We can be intimidated by a cultural belief in the primacy of science and question our internal wisdom when something does not feel right, even though deep-rooted human values will almost always offer the best path forward hand-in-hand with scientific knowledge.

In fact, leading with one's heart and an awareness of the connection to all life is the essence of planetary health and resilience. Though we are challenged with little time to spare, it is this awareness that offers us necessary direction. By adopting a new regenerative operating system with health and well-being at the core, we have the opportunity to elevate a new narrative together, and we can co-create a healthy and resilient world for all.

References

Abelson, B., Rupel, A., & Pincus, T. (2008). Limitations of a biomedical model to explain socio-economic disparities in mortality of rheumatic and cardiovascular diseases. *Clinical and Experimental Rheumatology, 26*, S25–S34.

Adelson, N. (2007). *Being alive well: Health and the politics of Cree well-being*. Toronto: University of Toronto Press.

Albrecht, G., Sartore, G.-M., Connor, L., Higginbotham, N., Freeman, S., Kelly, B., et al. (2007). Solastalgia: The distress caused by environmental change. *Australasian Psychiatry, 15*, S95-S98. https://doi.org/10.1080/10398560701701288

Attanasio, O., Fitzsimons, E., Grantham-McGregor, S., Meghir, C., & Rubio-Codina, M. (2012). *Early childhood development: Identifying successful interventions and the mechanisms behind them* (pp. 8). London: International Growth Centre (IGC).

Bai Y., Maruskin L. A., Chen S., Gordon A. M., Stellar J. E., McNeil G. D., et al (2017). Awe, the diminished self, and collective engagement: Universals and cultural variations in the small self. *Journal of Personality and Social Psychology, 113*, 185–209. https://doi.org/10.1037/pspa0000087

Baxi, S. N., & Phipatanakul, W. (2010). The role of allergen exposure and avoidance in asthma. *Adolescent Medicine: State of the Art Reviews, 21*, 57–71.

Bendelow, G. (2013). Chronic pain patients and the biomedical model of pain. *AMA Journal of Ethics*. Retrieved on February 16, 2020 from journalofethics.ama-assn.org/2013/05/msoc1-1305.html

Bouley, T., Roschnik, S., Karliner, J., Wilburn, S., Slotterback, S., Guenther, R., et al. (2017). *Climate-smart healthcare: Low-carbon and resilience strategies for the health sector*. The World Bank. Retrieved on February 16, 2020 from documents.worldbank.org/curated/en/322251495434571418/Climate-smart-healthcare-low-carbon-and-resilience-strategies-for-the-health-sector

Bowles, S., & Gintis, H. (2013). *A cooperative species: Human reciprocity and its evolution*. Princeton: Princeton University Press.

Burns, J. K., Tomita, A., & Kapadia, A. S. (2014). Income inequality and schizophrenia: increased schizophrenia incidence in countries with high levels of income inequality. *International Journal of Social Psychiatry, 60*, 185–196. https://doi.org/10.1177/0020764013481426

Capra, F. (2012) Systems thinking and the state of the world: Knowing how to connect the dots. In *Grow small, think beautiful: Ideas for a sustainable world from Schumacher College*. Edinburgh: Floris Books. doi: 0863158358

Citizens' Climate Lobby Canada (2019). *Carbon fee and dividend*. Retrieved on February 16, 2020 from canada.citizensclimatelobby.org/carbon-fee-and-dividend/

Clayton, S., Manning, C. M., Krygsman, K., & Speiser, M. (2017). *Mental health and our changing climate: Impacts, implications, and guidance*. Washington, DC: American Psychological Association, and ecoAmerica.

Dorling, D. (2011). Inequality: A real risk to our planet. *Resurgence & Ecologist, 297*, 20–22.

Eckelman, M. J., & Sherman, J. (2016). Environmental impacts of the U.S. health care system and effects on public health. *Plos One, 11*, e0157014. https://doi.org/10.1371/journal.pone.0157014

Egolf, B., et al. (1992). The Roseto effect: A 50-year comparison of mortality rates. *American Journal of Public Health, 82*, 1089–1092. https://doi.org/10.2105/ajph.82.8.1089

Elgar, F. J., & Aitken, N. (2010). Income inequality, trust and homicide in 33 countries. *European Journal of Public Health, 21*, 241–246. https://doi.org/10.1093/eurpub/ckq068

Ellen McArthur Foundation (2013). *Towards a circular economy—Economic and business rationale for an accelerated transition*. Retrieved on February 16, 2020 from www.ellenma-carthurfoundation.org/assets/downloads/publications/Ellen-MacArthur-Foundation-Towards-the-Circular-Economy-vol.1.pdf

EPA (2017a). *Climate adaptation—Ground-level ozone and health*. Environmental Protection Agency. Retrieved on February 16, 2020 from www.epa.gov/arc-x/climate-adaptation-ground-level-ozone-and-health.

EPA (2017b). *Persistent organic pollutants: A global issue, a global response*. Environmental Protection Agency. Retrieved on February 16, 2020 from www.epa.gov/international-cooperation/persistent-organic-pollutants-global-issue-global-response

Food and Agriculture Organization (2011). *Energy smart food for people and climate—An Issue Paper*. Rome. Retrieved on February 16, 2020 from www.fao.org/3/i2454e/i2454e00.pdf.

Pope Francis (2015). Laudato Si. *The Vatican*. Retrieved from w2.vatican.va/content/francesco/en/encyclicals/documents/papa-francesco_20150524_enciclica-laudato-si.html

Harvie, J. (2016). *The Next Health System*. Retrieved on February 16, 2020 from TheNextSystem.org. thenextsystem.org/next-health-system.

Health and Care System (HCS) (2020). Carbon Footprint. *HCS Carbon Footprint*. Retrieved on February 16, 2020 from www.sduhealth.org.uk/policy-strategy/reporting/hcs-carbon-footprint.aspx.

Hermida, C. (2011). *Sumak Kawsay: Ecuador builds a new health paradigm*. MEDICC Review, Medical Education Cooperation with Cuba. Retrieved on February 16, 2020 from www.scielosp.org/scielo.php?script=sci_arttext&pid=S1555-79602011000300015&lng=en&nrm=iso%3E.

Huber, M., Bolk, L., Knottnerus, J. A., Green, L., van der Horst, H., Jadad, A. R., et al. (2011). How should we define health? *British Medical Journal, d4163*, 343. Retrieved on February 16, 2020 from http://www.bmj.com/bmj/section-pdf/187291?path=/bmj/343/7817/Analysis.full.pdf

Huber, M., Huber, M., van Vliet, M., Giezenberg, M., Winkens, B., Heerkens, Y., Dagnelie, P. C., & Knottnerus, J. A. (2016). Towards a 'patient-centred' operationalisation of the new dynamic concept of health: A mixed methods study. *BMJ Open, 5*e010091. https://doi.org/10.1136/bmjopen-2015-010091

Jackson, C. S., & Nadine Gracia, J. (2014). Addressing health and health-care disparities: The role of a diverse workforce and the social determinants of health. *Public Health Reports, 129*, 57–61. https://doi.org/10.1177/00333549141291s211

Joye, Y., & Bolderdijk, J. W. (2015). An exploratory study into the effects of extraordinary nature on emotions, mood, and prosociality. *Frontiers in Psychology, 5*, 1577. https://doi.org/10.3389/fpsyg.2014.01577

Manifesto for Global Sustainable Health (2017). *Sustainable Health Symposium Global Perspectives*. Retrieved on February 16, 2020 from https://aru.ac.uk/-/media/Files/global-sustainability-institute/Sustainable-Health-Symposium/SHS_Manifesto.pdf

McKinstry, M., Chung, C., Truong, H., Johnston, B. A., & Snow, J. W. (2017). The heat shock response and humoral immune response are mutually antagonistic in honey bees. *Scientific Reports, 7*, 8850. https://doi.org/10.1038/s41598-017-09159-4

Minnesota Department of Health. (2014). *Advancing health equity in Minnesota. In Minnesota Department of Health*. Retrieved on February 16, 2020 from https://www.health.state.mn.us/communities/equity/reports/ahe_leg_report_020114.pdf

Movahed, M (2016, February 15). *Does capitalism have to be bad for the environment?* Retrieved on February 16, 2020 from https://www.weforum.org/agenda/2016/02/does-capitalism-have-to-be-bad-for-the-environment/

National Aboriginal Community Controlled Health Organisation (NACCHO) (2019) *Definitions*. Retrieved on February 16, 2020 from http://www.naccho.org.au/about/aboriginal-health/definitions/

NHS Alliance (2017). *A manifesto for health creation.* New NHS Alliance. Retrieved on February 16, 2020 from nhsalliance.org/wp-content/uploads/2018/11/A-Manifesto-For-Health-Creation.pdf

NHS Sustainable Development Unit (2013). *Carbon Footprint Update for NHS in England 2012.* Retrieved on February 16, 2020 from www.sduhealth.org.uk/documents/carbon_footprint_summary_nhs_update_2013.pdf.

Phillips, B. J. (2016). Inequality and the emergence of vigilante organizations: The case of Mexican Autodefensas. *Comparative Political Studies, 50,* 1358–1389. https://doi.org/10.1177/0010414016666863

Piketty, T., & Goldhammer, A. (2017). *Capital in the twenty-first century.* Cambridge: The Belknap Press of Harvard University Press.

Puchalski, C. M. (2001). *The role of spirituality in health care.* Proceedings (Baylor University. Medical Center), Baylor Health Care System. Retrieved on February 16, 2020 from www.ncbi.nlm.nih.gov/pmc/articles/PMC1305900/

Rakel, D., Barrett, B., Zhang, Z., Hoeft, T., Chewning, B., Marchand, L., & Scheder, J. (2011). Perception of empathy in the therapeutic encounter: Effects on the common cold. *Patient Education and Counseling, 85,* 390–397. https://doi.org/10.1016/j.pec.2011.01.009

Rosas, S. R. (2015). Systems thinking and complexity: Considerations for health promoting schools. *Health Promotion International, 32,* 301–311. https://doi.org/10.1093/heapro/dav109

Rudd, M., Vohs, K. D., & Aaker, J. (2012). Awe expands people's perception of time, alters decision making, and enhances well-being. *Psychological Science, 23,* 1130–1136. https://doi.org/10.1177/0956797612438731

Stellar, J.E., John-Henderson, N., Anderson, C.L., Gordon, A.M., McNeil, G.D., & Keltner, D. (2015). Positive affect and markers of inflammation: Discrete positive emotions predict lower levels of inflammatory cytokines. *Emotion, 15,* 129–133. https://doi.org/10.1037/emo0000033

U.S. Global Change Research Program (2016). *The Impacts of Climate Change on Human Health in the United States: A Scientific Assessment.* Retrieved on February 16, 2020 from health2016.globalchange.gov/

Ulrich, R. (1984). View through a window may influence recovery from surgery. *Science, 224,* 420–421. https://doi.org/10.1126/science.6143402

Vitali, S., Glattfelder, J.B., & Battiston, S. (2011). The network of global corporate control. *PLoS One, 6,* e25995. https://doi.org/10.1371/journal.pone.0025995

Watts, N. (2018). The 2018 report of the Lancet Countdown on health and climate change: Shaping the health of nations for centuries to come. *Lancet, 392,* 2479–2514. https://doi.org/10.1016/S0140-6736(18)32594-7

World Economic Forum. (2014) *Towards the circular economy: Accelerating the scale-up across global supply chains.* Retrieved on February 16, 2020 from reports.weforum.org/toward-the-circular-economy-accelerating-the-scale-up-across-global-supply-chains/the-limits-of-linear-consumption/

World Health Organization (2014). *Global environmental change.* World Health Organization. Retrieved on February 16, 2020 from www.who.int/globalchange/environment/en/

World Health Organization (2016). *Preventing disease through healthy environments: A global assessment of the burden of disease from environmental risks.* Geneva: World Health Organization. Retrieved on February 16, 2020 from www.who.int/quantifying_ehimpacts/publications/preventing-disease/en/

Part V
Climate Change and Health: Social Impacts

CHAPTER 17
Climate Change, Public Health, Social Peace

Hans Joachim Schellnhuber and Maria A. Martin

Summary Human well-being results from multiple interactions between highly diverse issues and actors. Crucial factors range from basic hygiene to social cohesion and environmental integrity. In this century, the development of a critical triangle of challenges—consisting of climate change, public health, and social peace—will determine whether a good life for all people on Earth is possible, at least in principle. Those three challenges may seem rather unrelated at first glance, but they are closely tied together and cannot be properly understood in isolation. In this chapter, we highlight essential aspects and relationships according to the current state of the respective arts.

Climate Change

With respect to the global mean surface temperature, the year 2017 was the second-warmest on record, according to the European Centre for Medium-Range Weather Forecasts (ECMWF) (Anon., n.d.-b). That is a remarkable event, since it happened in the absence of El Niño, a recurring phenomenon in the tropical Pacific that transfers large amounts of upwelling heat from the ocean to the atmosphere every 3–5 years, thus boosting global tropospheric temperatures. Moreover, 2017 turned out to be an "*annus horribilis*," with a terrifying routine of breaking disaster records: unprecedented storms (Irma, etc.), downpours (in Houston, etc.), floods (in Nepal, etc.), heat waves (Lucifer, etc.), and wildfires (in California, etc.) shook societies around the planet, giving us a foretaste of what the *new environmental normal* might look like in times of climate change.

In particular, the hurricanes of 2017 received worldwide attention. One after another wreaked havoc with exceptionally high wind speeds (Hurricane Irma), slow movement and record rainfalls (Hurricane Harvey), and dramatic economic impacts

H. J. Schellnhuber (✉) · M. A. Martin
Potsdam Institute for Climate Impact Research, Potsdam, Germany
e-mail: john@pik-potsdam.de

© The Author(s) 2020
W. K. Al-Delaimy, V. Ramanathan, M. Sánchez Sorondo (eds.), *Health of People, Health of Planet and Our Responsibility*, https://doi.org/10.1007/978-3-030-31125-4_17

(Hurricane Maria). The basic mechanism of how the surface temperature rise affects tropical cyclones is clear and undisputed: warmer ocean waters provide more energy to the storms, making them increasingly powerful. But the amount of rainfall they bring along will also increase: fundamental thermodynamics as expressed in the notorious Clausius–Clapeyron relation implies that the atmosphere can hold about 7% more water vapor with every degree of warming. Thus, the physics sets the stage for extreme rainfall events in the future, events that will be even more devastating than the one that shook East Houston in the wake of Hurricane Harvey. The down-pours were already so exceptional that the US National Weather Service had to introduce a new color into its graphs in order to properly represent what happened. Finally, coastal storm surges become far more dangerous with rising seas due to global warming. One example that remains pertinent because it caused extensive subway flooding was the impact of Superstorm Sandy in the year 2012: without the roughly 30 cm of sea level rise that the New York area had already experienced by 2012, the flooding would have done significantly less harm.

And sea levels might rise even faster than anticipated by the most recent report from the Intergovernmental Panel on Climate Change (IPCC, 2013). Newer research findings, based on conceptual models that include the entire range of known contributions to the ocean rise, suggest that in a business-as-usual scenario we might have to expect an average rise of about 1.3 m worldwide by the end of this century, which would be twice that previously estimated (Nauels et al., 2017). Note, however, that there is no academic consensus yet on the specific processes of ice cliff disintegration that an earlier study (DeConto & Pollard, 2016) suggested could drive an accelerating Antarctic contribution to the ocean surge. Nevertheless, exploring the consequences of the assumption that those processes are real is quite illuminating: Nauels and colleagues connect the proposed ice physics with a new generation of emission scenarios, the *shared socioeconomic pathways*, and conclude that the sea level rise could still be limited to around 0.5 m in 2100—if an almost complete coal phaseout was achieved by midcentury. In other words, the truly big inundation can still be avoided, but the escape path becomes steeper every year without ambitious climate action.

One question, which is routinely raised in the media following extreme weather events, is whether a specific event has been caused by climate change. Given the complexity of the climate system and the nonlinearity of its dominant dynamics, this question is impossible to answer (and almost meaningless) in the strict sense. However, no-nonsense statistical statements can be made more and more often. The field that courageously tackles this task is called attribution studies and usually employs huge numbers of computationally expensive ensemble simulations (see, for example, van Oldenborgh et al. (2017) for an analysis with respect to Hurricane Harvey, and Mann et al. (2017) regarding planetary waves and extreme weather events). Although great steps forward are being taken, one might remain skeptical as to the ability of the models to capture the specific mechanisms involved in a single extreme event's formation. On the other hand, the scientific community may waste its time anyway by trying to answer the unanswerable, so it should, rather, focus on the impacts of events that are individually unaccountable but statistically unavoidable on the basis of the first principles of physics. One tricky aspect in this context has recently been highlighted in the *Bulletin of the American Meteorological Society*: in a warmer climate, not only is the damage arising from hurricanes likely to become more severe but also the forecasting might become more difficult too, as the storms intensify more rapidly just before landfall (Emanuel, 2017). The bulletin also issues a popular and quite useful series titled *Explaining Extreme Events from a Climate Perspective* (Anon., n.d.-c).

Not only American coasts are affected by hurricanes: Atlantic Storm Ophelia, for example, struck Ireland after forming unusually far to the east in October 2017. Actually, this might become more common in the future (Haarsma et al., 2013). The impacts of that specific hurricane were spectacular in two ways. Firstly, desert particles from Africa were swirled up in the air and carried far north, such that for an entire day the sunlight over England was transformed into a reddish glow, typical for dawn or dusk only. Secondly, and far more dramatically, the winds fanned the flames of devastating Iberian wildfires very late in the season. Wildfires are globally on the rise and are becoming more destructive, also because new settlements keep on intruding into formerly untouched landscapes. The wildfires in California in 2018, for example, brought about a huge loss of life, and this was probably only the beginning on a warming planet. High temperatures and absence of rain, often in combination with high winds, are basic ingredients for severe fire conditions, and more and more land is affected by monthly heat extremes now (Coumou & Robinson, 2013; Coumou, Robinson, & Rahmstorf, 2013). In fact, the number of record-breaking temperature extremes is already about five times larger than it would be in an unperturbed climate.

This development enhances the risks of fire and drought, but the bitter irony of anthropogenic climate change is that the contemporary rise in extremes also includes rainfall (as mentioned above). Recently, *Nature Climate Change* published a study applying interesting new diagnostics to computer simulations, involving both the humidity in the air and its vertical motion. It demonstrated that the basic thermodynamic relationships between temperature and precipitation intensity, which often capture the local reality well, need to be complemented by atmospheric fluid dynamics to get the big picture right. It turns out that in the midlatitudes (i.e., regions with low to intermediate temperatures) hourly precipitation extremes might increase by up to 15% per degree of warming (Fig. 17.1), roughly doubling the classical Clausius–Clapeyron relation (see Pfahl, O'Gorman, & Fischer (2017); see also Lenderink & Fowler (2017) for a "News & Views" contribution). This is in line with the more general analysis that a weakening of midlatitude circulation leads to more hot, cold, dry, and wet extremes (Lehmann & Coumou, 2015).

Returning briefly to the question about the anthropogenic causation of an individual meteorological extreme event, we wish to highlight a recent heat wave with an especially suitable name, Lucifer, which haunted the Mediterranean regions in the summer of 2017: this specific disaster was found to have been made four times more likely by anthropogenic climate change (World Weather Attribution Group, WWA; Anon., n.d.-e). During that heat wave, temperatures stayed above 30 °C around the clock for three consecutive days. Yet another record may have been reached in the Iranian city of Ahvaz on an afternoon in June 2017, when the temperature climbed to almost 54 °C. If that measurement was correct, it would practically

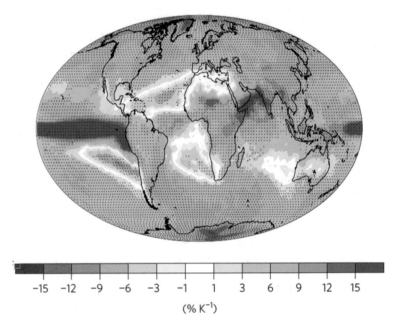

Fig. 17.1 The most extreme rainfall intensity will increase by up to 15% per degree of warming (Pfahl et al., 2017)

tie with the hottest temperature recorded on Earth since the beginning of modern measurements. The Ahvaz episode was made much worse by high concomitant air humidity, as we will explain below.

Public Health and Social Peace

When one thinks about public health in the context of climate change, one of the first questions that come to mind is if and how the human body will be able to adapt to the altered conditions. Steven Sherwood and Matthew Huber have suggested that unabated climate change might occasionally enforce the transgression of a rather critical threshold for human physiology (Sherwood & Huber, 2010). A combined measure of humidity and heat, called the *wet-bulb temperature*, sets limits to our body's acclimatization capacity. In a business-as-usual scenario, this limit might be approached in some world regions by the end of this century and could even be exceeded in a few locations, where unfavorable geographic conditions reign (Im, Pal, & Eltahir, 2017; Pal & Eltahir, 2015). Two of these literal hot spots happen to be in superdensely populated areas—namely, in the agricultural regions of the Indus and the Ganges river basin. The wet-bulb temperature sets an upper limit to adaptability; however, heat extremes with temperatures far below this threshold can cost lives, especially those of the sick, old, and very young. A recent metastudy on

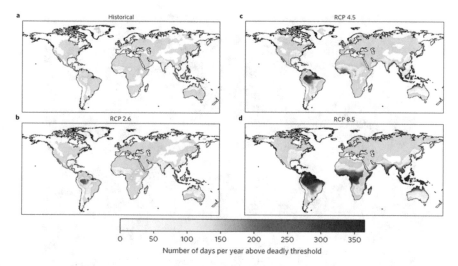

Fig. 17.2 Geographical distribution of deadly climatic conditions in different emission scenarios (Mora et al., 2017). *RCP* representative concentration pathway

observed excess heat mortality (Mora et al., 2017) found that around 30% of the world's population is already suffering from potentially fatal climatic conditions on more than 20 days of the year. In a business-as-usual scenario, this percentage is estimated to rise to almost three quarters of the human population by 2100 (Fig. 17.2).

Beyond extreme heat, climate change affects and threatens public health in various other ways (see, e.g., McMichael, Woodruff, & Hales (2006))—that is, through storms, floods, and wildfires such as the 2017 events in North America, Europe, and Asia, highlighted above. The impacts of climate change on public health also have a crucial socioeconomic dimension: shifting environmental conditions will be felt most strongly in poor and already risk-exposed communities. This applies both to the socially disadvantaged within industrialized countries and to poor societies in developing countries—a matter of social justice across scales. Within affluent countries, physical weakness and lack of monetary freedom reduce resilience against extreme events such as heat waves, fires, or flooding. In developing countries, health challenges related to diarrhea, vector-borne diseases, and childhood stunting may become much more salient.

"The Uninhabitable Village," a recent interactive report by the *New York Times*, gives a vivid and shocking impression of quintessential humanitarian aspects of global warming (Anand & Singh, 2017). It tells the story of families left behind by their husbands and fathers—Indian farmers—who killed themselves because they felt personally responsible for crop failures that climate researchers believe to result mainly from rising temperatures and the associated extreme events (Fig. 17.3). Suicides by farmers are part of an escalating tragedy on that subcontinent, with hundreds of thousands of cases occurring over the past three decades. This sad phenomenon is partially rooted in local social and economic challenges that are tied to global

Fig. 17.3 Interactive reportage from the *New York Times* (Anand & Singh, 2017)

inequalities in intricate ways. However, the problem is not merely one of status, wealth, and allocation; it is also one of fierce environmental reality. Hot days and the number of suicides correlate, and researchers have estimated that nearly 60,000 cases during the last 30 years might have been caused by the warming trend (Carleton, 2017). The social ripple effects of the phenomenon propagate through space and time: with women and children left behind and deprived of basic income, trauma is passed on to the next generations and families are left with daunting choices. They have to decide between staying and trying to survive under worsening conditions in already nearly uninhabitable regions or, alternatively, migrating to the cities, where their points of arrival are generally slums with dire living conditions and prospects.

Another major example of a climate change–related health threat is the rising sea. Salinization and other associated impacts force communities to give up their homesteads, which no longer provide them with essential nutrients. Some small island developing states (SIDS), such as certain Pacific islands, are already becoming uninhabitable as a result of the degradation of formerly fertile ground. Since these secluded islands are difficult to reach, supply from the outside is not a reliable alternative. Remoteness and a dire lack of adaptive capacity force islanders to consider moving away decades before they actually lose their lands to the sea.

Environmental migration has actually become a global phenomenon now, which is both embedded in and transforming the traditional movement patterns. This type of migration is difficult to quantify, mainly because of a lack of data. Seasonal or permanent movements following single extreme events or long-term environmental degradation are poorly documented so far, not least because those movements happen predominantly in developing countries without pertinent scientific communities.

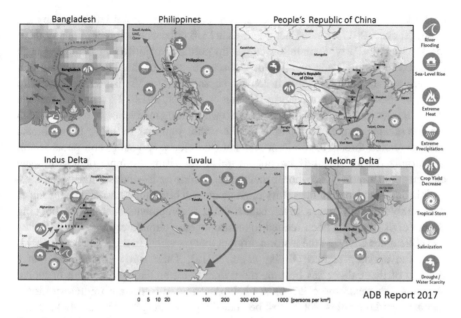

Fig. 17.4 Climate impacts and possible migration routes in the Asia-Pacific region (Vinke et al., 2017). *ADB* Asian Development Bank, *UAE* United Arab Emirates

Moreover, each migrfromion decision is based on a complex mix of very individual motivations. There is, however, new impetus in the scientific community to gain a better understanding of the overall phenomenon. Some of the most recent insights have been compiled in the *Atlas of Environmental Migration* (Ionesco, Mokhnacheva, & Gemenne, 2016): more than ten million people are forcibly displaced by weather-related sudden-onset hazards (such as floods, storms, wildfires, or heat waves) each year. Many others migrate in response to slow-onset hazards. Some households even use migration to prepare for future crises looming because of natural disaster risks, demonstrating that at least in certain cases, adaptation options can be realized by mobility decisions—which does not necessarily mean, however, that the decision to move is taken truly voluntarily (Melde, Laczko, & Gemenne, 2017). In 2015, 85% of all displacement due to sudden-onset events happened in South and East Asia. A recent report on the human dimensions of climate change in that region, issued by the Asian Development Bank with major contributions by the Potsdam Institute for Climate Impact Research (Vinke et al., 2017), sheds light on relevant aspects such as human health, urban development, security, migration, and trade networks. Among others, climate impacts and possible migration routes are explored (Fig. 17.4).

The potential indirect consequences of a single extreme event can sometimes be enormous. The devastations wrought by Hurricane Maria, for instance, which hit Puerto Rico exceptionally hard on September 20, 2017, and destroyed large parts of the electrical infrastructure there, eventually may have significant political implications at a distance. The *New York Times* reported on the huge number of islanders (estimated to have amounted to more than 200,000 in the first 2 months after the event) flowing into the Orlando area as a response to the havoc left behind by the storm (Tackett, 2017). Since Puerto Ricans are US citizens who predominantly favor the Democratic Party, their inflow might tip political majorities in the swing state Florida. Solomon Hsiang and Trevor Houser, directors of the Climate Impact Lab (Anon., n.d.-a), calculated that it could take 26 years for the island's economy to return to its state before the storm (Hsiang & Houser, 2017). This implies that the Puerto Ricans moving to Florida might not have a strong motivation to return home in the foreseeable future. So, apart from the social injustice resulting from climate change–related threats to the least resilient (the old, very young, and poor) *within* industrialized countries, this provides a rather surprising example of how climate change might impact those who are mainly responsible for it—industrialized countries.

It is quite evident that migratory movements can be a threat to social peace under unfavorable circumstances. Even catastrophic developments such as the ongoing Syrian civil war have been related to anthropogenic climate change. Colin Kelley and colleagues analyzed the historic drought that prevailed in the years before 2011, when the uprising started (Kelley et al., 2015). Their study emphasized that a drought of that caliber has become more than twice as likely as a result of global warming. This extreme event most likely contributed to certain migratory movements toward the cities, yet it must be placed within a tangle of drivers that include unsustainable water management, rising food prices, corruption, and political instability, as well as the specific motives of the different social and cultural groups involved. Recently, a number of studies have shed light on pertinent aspects. One study revealed that groups depending on agriculture and politically marginalized people are particularly vulnerable to droughts and may resort to violence because of their lack of coping capacity (von Uexkull et al., 2016). Another study found that the risk of armed conflict in the aftermath of climate-related disasters is significantly higher in ethnically fractionalized countries (Schleussner et al., 2016). Of course, almost nowhere is climate change the single cause of conflicts, which generally result from complex socioeconomic dynamics and depend heavily on local contexts. This is reflected in a lively debate on the subject, especially with respect to the Syrian case (Selby et al., 2017). However, in the sense of *threat multipliers*, climate impacts have the potential to exacerbate certain existing conflicts or to push other ones out of latency.

A Radically Logical Proposal

How can we deal with all of these challenges brought about by global warming, even if the temperature rise is confined to "only" 2 °C? We do think that multilaterally arranged climate migration—not compulsory eviction—will be a key element of an appropriate adaptation strategy. Defending all areas on Earth against sea level rise, desertification, and increasingly intense "natural" disasters will not be possible, so a relocation process of unprecedented dimensions needs to be considered (WBGU, 2016). This cannot be organized rigorously top-down, however, and the whole enterprise must be guided by humanitarian values. There are already ambi-

Fig. 17.5 Fridtjof Nansen
(1897)

tious projects such as the *Nansen Initiative* (Anon., n.d.-d), which was launched in 2012 by Switzerland and Norway, and is aimed at establishing an international consultative process in order to raise awareness and build consensus among states on a protection agenda for persons displaced by climate change.

But the challenges are enormous: identifying climate refugees among other migrants is highly difficult, especially since voluntary and forced migration are not easily discernible. There is also a legal gap in dealing with this group of people, particularly when it comes to cross-border movements. So far, the very term *climate refugees* (which suggests that it describes refugees in the sense of the 1951 Refugee Convention and the obligation to grant asylum) has not been established in international law, and the necessary debate is only beginning now.

However, climate migration is already a reality (in the South Pacific, for instance) and the human calamities resulting from it may become unspeakable in the future unless a novel institutional set of measures comes to the rescue. Here we mention just one, which could become a true game-changer. Our suggestion departs from a famous historical analogue and a name already mentioned above: Fridtjof Nansen (1861–1930; see Fig. 17.5)—a top diplomat for the League of Nations, a Nobel Peace Prize laureate, and arguably the most important pioneer of polar research worldwide—created a certificate, called the *Nansen Passport*, for stateless persons who were deprived of their citizenship in the cruel vagaries of the First World War and its aftermath. The document was launched on July 5, 1922, and eventually accepted by 53 countries, which opened their borders to the holders (such as Igor Stravinsky, Rudolf Nureyev, and Aristotle Onassis) of that supranational passport.

Today, a similar certificate—a *planet passport* or *climate passport*—could become a key instrument for dealing, in a liberal and humane way, with migration triggered by climate change. The passport should give people who lose their home—and possibly even their national territory—to the impacts of anthropogenic global warming access to all countries bearing a significant responsibility for that loss. The fundamental idea underlying this proposal can be epitomized by the following: *If someone destroys the house of another person, he should be obliged to open his own house to that person.*

There are many evident and also more subtle aspects of our proposal that need to be carefully discussed, not least by experts in international law and moral philosophy. Here we confine ourselves to highlighting just two critical dimensions of such a certificate—namely (a) *eligibility* and (b) *accessibility*. The first one is fairly straightforward in the case of inhabitants of island states threatened in their entirety by sea level rise or citizens of countries (such as Bangladesh) that will lose large portions of their territory to the sea. However, things become more complicated when one considers displacement due to heat, drought, or erosion, for instance, as the previous discussion about the difficulties in attribution studies has shown. One could think of setting up an *international climate court* that would decide on eligibility on the basis of the best scientific evidence available. Secondly, a *principle of the perpetrator* would have to lay the basis for identifying the states bound to accept climate passport holders. For example, a 2% rule (in the vein of the 2 °C guardrail for climate mitigation) could objectively name the few (and mostly economically prosperous) countries (or blocs) that have been individually responsible for a significant share of the global cumulative emissions since the industrial revolution (see Fig. 17.6).

Cumulative CO$_2$ Emissions 1850–2011 (% of World Total)

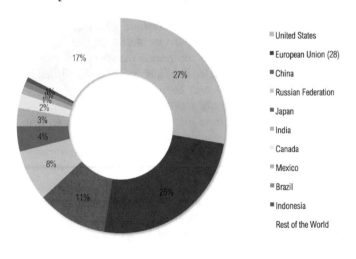

WORLD RESOURCES INSTITUTE

Fig. 17.6 Share of total cumulative emissions by country or political union. *Source*: World Resources Institute

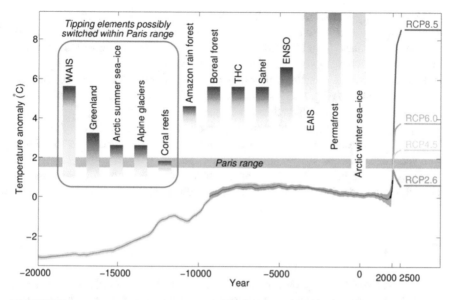

Fig. 17.7 Tipping elements of the Earth System and their potential disruption ranges in the context of the global mean temperature development over the last 20,000 years (Schellnhuber et al., 2016). *EAIS* East Antarctic Ice Sheet, *ENSO* El Nino southern oscillation, *RCP* representative concentration pathway, *THC* thermohaline circulation, *WAIS* West Antarctic Ice Sheet

A climate passport (or the already serious discussion of such a format) could help set a countertrend against the currently observable *race to the bottom* of moral standards in the treatment of refugees, manifesting itself not least in the rise of reactionary, xenophobic, and nationalistic political parties in the developed world. Also, if a given nation comes to the conclusion that the burden of giving shelter to climate-displaced persons is too heavy, then it might at least raise its mitigation ambitions and help reduce the very causes of the displacement abroad. Rapid decarbonization is clearly the choice to be made then.

Conclusion and Outlook

What are the implications of the most recent climate research findings for our civilization? If we step back a little and let our eyes roam over the possible future states of the Earth System, we have to consider the fact that the changes humans are causing will not happen, as a rule, in a gradual and reversible manner. In particular, there are tipping points in the global climate system (Lenton et al., 2008), where the operation and interaction of crucial components of the planetary machinery, called "tipping elements"—such as the large ice sheets, the Amazon and Congo rainforests, the major ocean currents, the monsoon patterns, and the jet streams—can change their character in a highly nonlinear and possibly irreversible way (for more information please consult our contribution to the Pontifical Academy of Sciences

(PAS) and Pontifical Academy of Social Sciences (PASS) publication *Sustainable Humanity, Sustainable Nature: Our Responsibility* (Schellnhuber & Martin, 2014)).

Putting the latest insights on tipping elements into a relationship with the bygone, stable environmental regime of the Holocene and with the new, rapidly changing one of the Anthropocene, it becomes clear that the world might look very different in only a couple of centuries (Fig. 17.7). A drastic change of course is essential to prevent massive climatic disruptions (Schellnhuber, Rahmstorf, & Winkelmann, 2016), which, through tipping cascades, might eventually result in the Earth System locking into a hothouse state (Steffen et al., 2018), with conditions similar to the climate 15–17 million years ago.

Therefore, staying within the temperature range agreed upon in Paris will be crucial for limiting the associated damage to public health and social peace.

References

Anand, G. & Singh, V. (2017). The uninhabitable village. *New York Times*. Retrieved February 9, 2020 from https://www.nytimes.com/interactive/2017/10/26/world/middleeast/india-farmers-drought.html

Anon. (n.d.-a). *Climate impact lab*. Retrieved February 9, 2020 from http://www.impactlab.org/

Anon. (n.d.-b). *European Centre for Medium-Range Weather Forecasts*. Retrieved February 9, 2020 from https://www.ecmwf.int/

Anon. (n.d.-c). Explaining extreme events from a climate perspective. *BAMS*. Retrieved February 9, 2020 from https://www.ametsoc.org/ams/index.cfm/publications/bulletin-of-the-american-meteorological-society-bams/explaining-extreme-events-from-a-climate-perspective/

Anon. (n.d.-d). *Nansen initiative*. Retrieved February 9, 2020 from https://www.nanseninitiative.org/secretariat/

Anon. (n.d.-e). *World weather attribution*. Retrieved February 9, 2020 from https://www.worldweatherattribution.org/

Carleton, T. A. (2017). Crop-damaging temperatures increase suicide rates in India. *Proceedings of the National Academy of Sciences of the United States of America, 114*, 201701354.

Coumou, D., & Robinson, A. (2013). Historic and future increase in the global land area affected by monthly heat extremes. *Environmental Research Letters, 8*, 034018.

Coumou, D., Robinson, A., & Rahmstorf, S. (2013). Global increase in record-breaking monthly-mean temperatures. *Climatic Change, 118*, 771–782.

DeConto, R. M., & Pollard, D. (2016). Contribution of Antarctica to past and future sea-level rise. *Nature, 531*, 591–597.

Emanuel, K. (2017). Will global warming make hurricane forecasting more difficult? *Bulletin of the American Meteorological Society, 98*, 495–501.

Haarsma, R. J., et al. (2013). More hurricanes to hit western Europe due to global warming. *Geophysical Research Letters, 40*, 1783–1788.

Hsiang, S. & Houser, T. (2017). Don't let Puerto Rico fall into an economic abyss. *New York Times*. Retrieved February 9, 2020 from https://www.nytimes.com/2017/09/29/opinion/puerto-rico-hurricane-maria.html

Im, E.-S., Pal, J. S., & Eltahir, E. A. B. (2017). Deadly heat waves projected in the densely populated agricultural regions of South Asia. *Science Advances, 3*, 1–8.

Ionesco, D., Mokhnacheva, D., & Gemenne, F. (2016). *The atlas of environmental migration*. London: Routledge Taylor & Francis.

IPCC, W. G. I (2013). Climate change 2013: The physical science basis. In T. F. Stocker et al. (Eds.), *Contribution of Working Group I to the Fifth Assessment Report of the Intergovernmental Panel on Climate Change*. Cambridge, UK: Cambridge University Press.

Kelley, C. P., et al. (2015). Climate change in the fertile crescent and implications of the recent Syrian drought. *Proceedings of the National Academy of Sciences of the United States of America, 112*, 3241–3246.

Lehmann, J., & Coumou, D. (2015). The influence of mid-latitude storm tracks on hot, cold, dry and wet extremes. *Scientific Reports, 5*, 1–9.

Lenderink, G., & Fowler, H. J. (2017). Hydroclimate: Understanding rainfall extremes. *Nature Climate Change, 7*, 391–393.

Lenton, T. M., et al. (2008). Tipping elements in the Earth's climate system. *Proceedings of the National Academy of Sciences of the United States of America, 105*, 1786–1793.

Mann, M. E., et al. (2017). Influence of anthropogenic climate change on planetary wave resonance and extreme weather events. *Scientific Reports, 7*, 45242.

McMichael, A. J., Woodruff, R. E., & Hales, S. (2006). Climate change and human health: Present and future risks. *Lancet, 367*, 859–869.

Melde, S., Laczko, F., & Gemenne, F. (2017). *Making mobility work for adaptation to environmental changes: Results from the MECLEP global research*. Geneva: International Organization for Migration.

Mora, C., et al. (2017). Global risk of deadly heat. *Nature Climate Change, 7*, 501–506.

Nauels, A., et al. (2017). Linking sea level rise and socioeconomic indicators under the shared socioeconomic pathways. *Environmental Research Letters, 12*, 11.

Pal, J. S., & Eltahir, E. A. B. (2015). Future temperature in southwest Asia projected to exceed a threshold for human adaptability. *Nature Climate Change, 6*, 197.

Pfahl, S., O'Gorman, P. A., & Fischer, E. M. (2017). Understanding the regional pattern of projected future changes in extreme precipitation. *Nature Climate Change, 7*, 423–427.

Schellnhuber, H. J., & Martin, M. A. (2014). *Climate-system tipping points and extreme weather events* (Sustainable humanity, sustainable nature: Our responsibility). Vatican City: Pontifical Academy of Sciences, Extra Series 41.

Schellnhuber, H. J., Rahmstorf, S., & Winkelmann, R. (2016). Why the right climate target was agreed in Paris. *Nature Climate Change, 6*, 649–653.

Schleussner, C.-F., et al. (2016). Armed-conflict risks enhanced by climate-related disasters in ethnically fractionalized countries. *Proceedings of the National Academy of Sciences of the United States of America, 113*, 9216–9221.

Selby, J., et al. (2017). Climate change and the Syrian civil war revisited. *Political Geography, 60*, 232–244.

Sherwood, S. C., & Huber, M. (2010). An adaptability limit to climate change due to heat stress. *Proceedings of the National Academy of Sciences of the United States of America, 107*, 9552–9555.

Steffen, W., et al. (2018). Trajectories of the Earth System in the Anthropocene. *Proceedings of the National Academy of Sciences of the United States of America, 115*, 8252–8259.

Tackett, M. (2017). An exodus from Puerto Rico could remake Florida politics. *New York Times*. Retrieved February 9, 2020 from https://www.nytimes.com/2017/10/06/us/politics/puerto-rico-florida-voters.html

van Oldenborgh, G. J., et al. (2017). Attribution of extreme rainfall from Hurricane Harvey, 2017. *Environmental Research Letters, 12*, 124009.

Vinke, K., et al. (2017). *A region at risk: the human dimensions of climate change in Asia and the Pacific*. Mandaluyong City, Metro Manila: Asian Development Bank.

von Uexkull, N., et al. (2016). Civil conflict sensitivity to growing-season drought. *Proceedings of the National Academy of Sciences of the United States of America, 113*, 12391–12396.

WBGU (Wissenschaftlicher Beirat der Bundesregierung Globale Umweltveränderungen) (2016). *Der Umzug der Menschheit: Die transformative Kraft der Städte*.

CHAPTER 18
Climate Justice and Public Health: Practical Ethics in Urgent Times

Fonna Forman

> We are faced not with two separate crises, one environmental
> and the other social, but rather with one complex crisis, which
> is both social and environmental. Strategies for a solution
> demand an integrated approach to combating poverty, restoring
> dignity to the excluded, and at the same time protecting nature.
>
> —Pope Francis, *Laudato Si'*, 4, 139

Summary This chapter reflects on the disproportionate impact of climate change on the most vulnerable people and argues that the language of "climate justice" may not be the most effective language for communicating the urgency to the general public and for changing attitudes and behaviors around climate change. Drawing on theories of human motivation in the social sciences, the chapter argues that most people, most often, tend to be motivated by what most affects them, their families, their neighborhoods, and so forth. This suggests that focusing on the likely health impacts of climate change might be a more effective way of communicating with those who are not motivated by the narrative of social justice to change their attitudes and behaviors. Because the health impacts of climate change can be generalized to all of us, this is likelier to motivate a shift in attitudes in behavior among larger publics. The chapter concludes with reflections on how to integrate public health more effectively into our communications about climate change.

F. Forman (✉)
Center on Global Justice, University of California, San Diego, La Jolla, CA, USA
e-mail: fonna@ucsd.edu

Climate Change and the Bottom Three Billion: Injustice, Not Misfortune

There is a distinction in moral philosophy between injustice and misfortunate. A misfortune is something that happens through the fault of no one—an accident, a serendipitous catastrophe, the "way of the world"—a calamity that causes destruction and human suffering. But when does misfortunate become injustice? What if the calamity, for example, disproportionately harms those who are poor? And what if these already vulnerable victims of harm have little capacity to adapt? Moreover, let us imagine that there is a humanitarian response to the calamity but that it arrives more slowly and sporadically in the precarious zones than in the less precarious ones. This is a relevant theme following the hurricanes that tore through the Caribbean and southeastern portions of the USA early in the autumn of 2017. Let us imagine further that the harm caused by the calamity is not unpredictable or serendipitous at all; it is instead the result of human decision or neglect—a poorly or inadequately designed levy that breaks, for example, or crumbling infrastructure that finally fails, or building practices that ignore the risk of placing human settlements in vulnerable places. And let us imagine finally that the calamity is a long-term environmental impact of the behaviors and habits of a wealthy minority that make this type of calamity more frequent and the harms it inflicts on the poor more severe.

Where to draw the line between injustice and misfortune must be understood as a political act. As the late political theorist Judith Shklar (1990) noted,

> We must recognize that the line of separation between injustice and misfortune is a political choice, not a simple rule that can be taken as a given. The question is, thus, not whether to draw a line between them at all, but where to do so ..." (Shklar, 1990, p. 5)

It is not uncommon in environmental politics that those responsible for causing harm and those asked to remediate it will try to characterize the harm as misfortune, with arguments like "calamities like this have always happened," "they have nothing to do with my actions," "they are the way of the world," "an act of nature," "an act of God," and so forth. And as power goes, those who cause environmental harm typically have more capacity and political agility than the victims of harm, who often are already vulnerable and marginalized from a society's circle of moral concern.

But let us be clear: the impacts of climate change on the most vulnerable people are not a misfortune; they are an injustice. The idea of climate justice embodies all of the disparities and gaps mentioned above, and many more. Climate justice is about an "adaptation gap" between rich and poor. It is about a "humanitarian response gap" between rich and poor. It is about a "planning gap." It is also about a "transparency gap," an "education gap," a "health access" gap," a "development gap," and an "energy access gap." And, most fundamentally, climate justice is about a "responsibility gap." Climate change is caused disproportionately by the production and consumption habits of the world's richest populations. The richest one billion people on the planet are responsible for about 50% of greenhouse gas emissions, while the poorest three billion, without access to affordable fossil fuels, are responsible for about 5%. In contrast, the "bottom 3 billion" suffer the greatest harms associated with climate change (Dasgupta & Ramanathan, 2014).

Climate change is projected to cause widespread and serious harm to human settlements on the planet, threatening to unravel many of the development gains of the last century (UNICEF, 2015). The effects of climate change cluster and bear down hard on the global poor, those who are both least responsible for the causes and who are least capable of adapting. The health impacts of climate change, particularly on the poor, are predicted to become catastrophic by midcentury if significant reductions in greenhouse gas emissions do not occur (Lancet Commissions, 2015).

The adaptation gap is perhaps most evident in the acceleration of climate-related human displacement in recent years. While metrics are still being refined to causally isolate climate change as a driver of migration, the United Nations Commission for Refugees (UNHCR) already predicted in 2009 that climate change would become the largest driver of population displacements both inside and across national borders (UNHCR, 2009). The global rate of displacement has more than doubled since 1970, from fewer than 2000 persons per million to more than 4000 persons per million in 2014 (NRC/IDMC, 2015, p. 22). The most commonly cited estimate is that 200 million people will be displaced by 2050 (IOM, 2008, pp. 11–12; Myers, 2005). These displacements, even if temporary, have a profound impact on individuals' lives, often involving the loss of a home or crops, and disproportionately harming individuals at the very bottom, who lack the resources to adapt and who become susceptible to human trafficking and forced labor.

In a recent paper on climate migration, Forman and Ramanathan (2018) paid special attention to the disproportionate impacts of slower progressive effects like droughts, soil erosion, forest loss, and the sea level rise (Forman & Ramanathan, 2018). While extreme weather events often cause sudden mass displacements and are increasing in frequency, slower processes seem to have a stronger predictive effect on the likelihood of climate migration (McLeman, 2014). Those living in rural or low-lying coastal areas, whose livelihoods are linked with climate-sensitive sectors like agriculture and fishing, are the most vulnerable and at highest risk, as they are typically the least capable of either adapting in situ or migrating out, since the capacity to leave one's home entails certain financial and social capital such as education, language skills, and support networks (Biermann & Boas, 2008).

In 1991, the Intergovernmental Panel on Climate Change (IPCC) predicted that climate change would accelerate urbanization in developing countries, with populations migrating from coastal lowlands—in particular, densely inhabited delta areas—to inland areas (IPCC, 1990). The world is now urbanizing at a rate of 32 million people each year, exerting unmanageable "demands on urban services" and "increasing political pressure on the state" (Barnett & Adger, 2007, p. 642). In the developing world, rapid urbanization in recent decades has produced dramatic "asymmetrical" growth patterns as the poorest populations have amassed by the millions in precarious informal settlements, often periurban and along rivers and lagoons, uniquely exposing them to the effects of climate change—floods, drought, food and water shortages, and disease (Gadanho, 2014). The explosion of slums at the peripheries of cities across the planet is a humanitarian crisis of gargantuan proportion that cities in the developing world today are unprepared to confront (Davis, 2007). As we have witnessed in Syria, these climate-induced stressors can exacerbate civil unrest and even revolution (Kelly et al., 2015).

Climate Justice: Redistributing Harms and Responsibilities

Viewed through the lens of human suffering, climate change is not only an environmental crisis but also an ethical and political one. Climate justice demands that those who cause harm, and especially those who benefit from that harm, bear primary responsibility for remediating it and for preventing further harm in the future. In other words, climate justice has a temporal dimension that is both backward-looking and forward-looking. The wealthy, polluting population has an ethical responsibility to remediate human suffering in the present, accelerating aid to populations disproportionately struggling to adapt. The polluting population also has an ethical responsibility to mitigate future harm by doing everything necessary to mobilize a low-carbon global economy. This imperative includes helping developing populations leapfrog a carbon-based economy and tempering the promotion of our wasteful, petroleum-based lifestyle as the epitome of human happiness. It could be said that climate justice redistributes responsibilities and harms for the common good. The Mary Robinson Foundation—Climate Justice (MRFCJ), the first international organization committed to climate justice as a human right, frames this ethical mandate as "sharing benefits and burdens equitably" (MRFCJ, n.d.).

There has been perhaps no greater advocate in our generation for the cause of climate justice than Pope Francis in his encyclical letter, *Laudato Si'*, of May 24, 2015. He has become a great ally in the integral fight against poverty, the fight for tolerance and human dignity, and the fight against climate change. In this now well-known passage, he described the two-headed beast we are confronting:

> Today we have to realize that a true ecological approach always becomes a social approach;
> it must integrate questions of justice in debates on the environment, so as to hear both the
> cry of the earth and the cry of the poor. (Pope Francis, *Laudato Si'*, 1, 49)

After the release of *Laudato Si'*, Pope Francis called for urgent coordinated action. It is not surprising that he began by convening mayors. Across the world, municipalities have led the way, demonstrating deeper commitment to coordinated climate action and social justice than nations, with agile, environmentally progressive mayors who know how to mobilize cross-sector partnerships and get things done (Barber, 2013). As the late political theorist and founder of the Global Parliament of Mayors, Benjamin Barber, put it, "Cities are the coolest political institutions on Earth"[1] (Barber, 2017). The mayors who convened at the Vatican in July 2015 ended their meeting with a pledge, which urged world leaders to pass a

> bold climate agreement that confines global warming to a limit safe for humanity, while protecting the poor and the vulnerable from ongoing climate change that gravely endangers their lives. (Pianigiani, 2015)

Climate Justice: Latin American Living Laboratories

I suspect there was also something distinctively Latin American about why Pope Francis began with mayors, a history with which he was quite familiar as a young Argentine. Latin American cities in recent decades have been particularly success-ful at transforming attitudes and behavior around climate change and environmental health while producing more equitable outcomes in the city—advancing social jus-tice and climate justice together.

Some Latin American cities have become almost mythical as living laboratories of equitable green urbanization. The philosopher Antanas Mockus became the mayor of Bogota, Colombia, during its most intense period of violence in the late 1990s and early 2000s. It was a scene of social chaos and urban breakdown, unem-ployment, poverty, and choking air quality, the worst anywhere on the continent. People referred to Bogota as the most dangerous city on the planet. Mockus declared that urban transformation must begin with civic strategies designed to transform social behavior, to change hearts and minds. He became legendary for using arts, culture, and performance to dramatically reduce violence and lawlessness, recon-nect citizens with their government and with each other, increase tax collection, reduce water consumption, and ultimately improve the quality of life for the poor. Shifting social norms and renewing public trust paved the way for Mayor Enrique Peñalosa's renowned TransMilenio bus rapid transit system, at the time the most advanced multinodal transportation system in the world, linking retrofitted bus lanes with bicycle hubs, *cyclovia* (bikeways), and dedicated walking paths that liter-ally stitched that troubled city together. Bogota revolutionized public transportation in Latin America. What is notable here, however, is that shifting social norms came first; the infrastructure and environmental interventions followed (Forman & Cruz, 2017; Forman, 2018). For Peñalosa, like Mockus before him, changing beliefs and social norms was essential to launching a successful egalitarian transportation agenda. He observed,

> A developed country is not a place where the poor have cars; it is a place where the rich ride public transportation. (Peñalosa, 2014)

Mockus and Peñalosa emerged from a long tradition of participatory green urbanization across Latin America, stewarded by mayors who were committed to robust agendas of civic participation to ignite a sense of collective agency and dig-

---

[1] On the Global Parliament of Mayors, see https://globalparliamentofmayors.org/.

nity among the poor, and ultimately to produce greener and more equitable cities: from the Worker's Party mayors in Porto Alegre, Brazil, who experimented with participatory budgeting in the 1980s; to Jaime Lerner in Curitiba, Brazil, who pioneered bus rapid transit and dozens of green interventions across the city; to the "social urbanism" of Mayor Sergio Fajardo in Medellín, Colombia, in the early 2000s, which transformed public spaces and green infrastructure into sites of education and citizenship building, and transformed Medellín into a global model of urban social justice (Forman & Cruz, 2017, 2018; McGuirk, 2014). This tradition still thrives in cities across the continent, from La Paz to Quito to Mexico City, and carries important lessons for equitable green urbanization in cities across the world today.

The Next Step: Public Health as a Social "Lever"

The *Bending the Curve* report produced by the University of California in 2015 devoted a full chapter to the public health impacts of climate change, giving special emphasis to the idea of "climate justice" and the disproportionate public health impacts of global warming on the world's most vulnerable demographics. Titled "Bending the Curve and Closing the Gap," the chapter argued that climate change and global poverty are intricately intertwined and that they must be tackled simultaneously (Forman et al., 2016). We understood public health as an impact requiring urgent intervention. But we did not yet appreciate that public health predictions *themselves* could be deployed as a lever for social change, to stimulate changes in attitudes and behaviors about global warming and the urgencies we are all facing as a planet.

This section reflects further on the ethical implications of integrating public health more robustly as a mechanism for changing public opinion and expediting policy to keep global warming below 2 °C, particularly among those demographics in wealthy, polluting societies that have proven themselves resistant to scientific or ethical arguments. Among the like-minded, we do not need to deploy arguments about the hardships of poverty or the fact that climate change complicates and deepens them. Those working on the front lines in the global health arena tend to be committed already, on ethical grounds, to remediating disproportionate impacts and urgently redistributing resources to assist vulnerable people. But when we engage publics and policy makers who disagree, passively or surreptitiously, we must become realists. Our ethics must become practical and strategic. We must be willing to travel alternative paths—paths that are ethically incomplete, partial, incremental, and circuitous. When the harms we seek to address are severe and urgent, and when tipping points are in sight, ideals can become a hindrance. As development economist Amartya Sen has long argued, we need to welcome "partial realizations" in times of urgency, since ethical compromises can save lives and reduce human suffering. This was the central theme of his 2009 book, *The Idea of Justice*, which sought to chasten the more idealistic and transcendental theories of global justice associated with John Rawls and his followers:

> I would like to wish good luck to the builders of a transcendentally just set of institutions for the whole world, but for those who are ready to concentrate on reducing manifest injustices that so severely plague the world, the relevance of a merely partial ranking [of distributional priorities] for a theory of justice can actually be rather momentous. (Sen, 2009, p. 263)

When we speak to publics and politicians today who are not motivated by altruistic or collective ends, who do not share our distributional priorities (to use Sen's language), and who through habit or agenda are inured to the status quo, we must find the right "lever" to turn them as they are, in very practical ways, toward the ends we seek. We can motivate change at the margins, along ethical lines, when the opportunity costs are low enough. But more fundamental societal changes in attitudes and behavior require a shift in how we communicate.

We know, for example, that people become more receptive to climate-friendly public policy when they better understand the specific impacts of climate change on their own lives and communities (Furth & Gantwerk, 2013; Pincetl, 2010). A study of attitudes and behaviors among residents of low-lying coastal communities in South Florida, commissioned by the Union of Concerned Scientists, found that climate-friendly attitudes and behaviors are likelier when the negative effects of climate change are made concrete and relevant for people, rather than something far-off like melting ice caps and polar bears. Researchers have found that when people understand precisely how the sea level rise will affect their own city and their own neighborhood, their attitudes change (Cosgrove, 2012). They are likelier to be receptive to the concept of global warming and supportive of local climate-friendly public policy. This was even true for a majority of individuals self-described as politically "conservative." Second, the same study found that people are likelier to change their behaviors as well—not just beliefs but behaviors too—when concrete opportunities for local climate action are made available to them. Knowing the risks without having opportunities to act can produce paralysis, so the conclusion from public opinion research is Go Local: Start with Impact, End with Action.

Given this, it seems we will be more effective if we reframe our message about climate change from a language of duty to a language of prudence, from a language of climate justice (which motivates only those who are already religiously, ethically, or publicly inclined to change their behavior for the betterment of the least well-off) to a language of personal and local well-being, and from a language of the other to a language of the self. Climate justice is a language of disproportionate impact and blame, and it is not the best vehicle for instigating broad social change. Those who are already immune to the urgencies of human suffering, and to the appeals of social justice *sui generis*, will not be moved by the addition of a threat multiplier like climate change. In fact, as noted at the beginning of this chapter, the language of climate justice can trigger defensive retrenchment in some audiences and ethical back peddling into the mindset of "misfortune."

The early modern idea of "enlightened self-interest" can be instructive here. In *The Passions and the Interests*, Albert Hirschman explored the way early modern thinkers used the idea of "enlightened self-love" as a motivational proxy for altruism (Hirschman, 1977). This helps explain why individuals sometimes sacrifice immediate benefits for social or collective ends though they are not motivated by social or collective reasons to do so. Starting with a realist's concession that only a small subset of us will be reliably motivated by altruism, selfish individuals nevertheless can produce pro-social ends by exercising prudence, or foresight, to pursue interests that coincide with the interests of others; in a selfish world, private and public can still converge. Prudence can restrain present gratification for greater gratification in the future.

The idea of enlightening selfishness is obviously not new to the climate change discussion. The use of carrots rather than sticks to motivate climate-friendly behavior is one of California's great, though sometimes controversial, success stories ("cap-and-trade") for the ethical compromises incentives always entail.

But along these lines of thinking about carrots, why not also think creatively about another powerful motivation: the personal health risks of climate change. Maria Neira suggests (see Chap. 8 in this book) that we frame climate change not as an environmental crisis but as a public health crisis. This section concludes with five thoughts about this.

First, we need to convince people and politicians where their true interests lie. This requires a pervasive shift in our norms (Griskevicius et al., 2008). "Mitigating climate change" needs to become synonymous in people's minds with "preserving health and well-being"—for me, and for all. According to a 2017 Gallup poll, only 42% of Americans presently believe that climate change will pose a serious threat in their lifetime (Saad, 2017). We need to increase public knowledge about very concrete and broad public health threats in order to enlighten self-interest.

Second, relatedly, we need to "equalize impacts" as a matter of practical ethics, to stimulate a more generalized sense of public urgency. Focusing on health disparities for the poor does not get us there. We need to talk about the relation of air quality, for example, to stroke, cardiovascular disease, respiratory disease, and so forth. This is a powerful equalizing tool. Advancing the idea that we are *all* in peril, regardless of our wealth and social position, expansion of the circle of potential victimhood is a powerful equalizing tool. Again, those of us committed to climate justice need to make compromises like this. We are distraught over the disparities in our world. But the language of disparity and ethical duty to others does not move people who are already numb to global justice. We need to be strategic.

Third, we need to think about who is best situated to communicate public health risks. The issue of public trust here is essential. In our brave new fact-free world, politicians, journalists, scientists, and intellectuals have all been castigated as biased, corrupt, detached from reality, and untrustworthy. This may be a very American (and indeed increasingly European) phenomenon, but these are the populations we are struggling to reach. Faith leaders have been effective messengers in some places, framing the perils of climate change through scriptural inspiration, respect for Mother Earth, and divine compassion for our fellow creatures. Religious communities can also help to promote new social norms among their ranks—an important dimension of sustainable behavioral shift, since norms always have a social regulatory component. The argument here is that public health might help us intervene at even broader scales, *motivating through different channels,* by tapping into very conventional concerns about personal health and well-being, and moving people from what they love—not Mother Earth, not common humanity, but their well-being and the well-being of those they love.

To accomplish this, fourth, health practitioners will need to get on board. For the most part they have been less publicly tarnished than the other relevant professions. And health practitioners are also capable of engaging people where they are: in the intimacy of an examination room. It may well be that the general practitioner (GP) the pediatrician, the obstetrician/gynecologist, the midwife, and the nurse practitioner are our most important soldiers on the ground right now and potentially our most effective climate educators.

This entails, fifth, an important shift in the way we train our health professionals today—practitioners as well as researchers, policy makers, and community-based activists—in the public health arena. Medical schools and public health institutes will need to become increasingly more collaborative, interdisciplinary, community engaged, and better insulated from corporate agendas. They will need to commit themselves to integrating environmental health education, and policy agendas more explicitly into their mission.

Take, for example, the Community Stations at the University of California San Diego, a network of research and teaching stations (located in four disadvantaged neighborhoods on both sides of the US–Mexico border) that focus on community-engaged climate action and environmental health education.[2] At these sites, teaching and research are conducted collaboratively with local environmental nonprofit organizations. The Community Stations are committed to three main public health agendas: (1) providing community-based environmental health education for university students, teaching them to become climate communicators and educators; (2) increasing environmental health literacy among adults and especially children in partner communities (with the idea that educating children is our most essential task); and (3) stewarding participatory climate action at the neighborhood scale— from small-scale activities (like raising awareness of consumption habits and stimulating a walking and cycling culture) to designing and financing zero–net energy solutions and retrofits of homes, schools, and businesses.

One of these stations, the EarthLab Community Station, is located in Encanto, San Diego's most challenged inner-city neighborhood, and situated near Chollas Creek, the city's most polluted waterway.[3] EarthLab is a 4-acre outdoor environmental classroom with community gardens, solar houses, water-harvesting facili-

---

[2] Led by the author in collaboration with Teddy Cruz, Professor of Public Culture and Urbanism at the University of California, San Diego.

[3] EarthLab is a partnership between the University of California, San Diego, the Encanto-based environmental nonprofit organization Groundwork San Diego, and the San Diego Unified School District. For further discussion, see Forman et al. (2016).

ties, an energy "nanogrid," and other green infrastructure designed by university researchers as learning tools for the six public schools within walking distance of the site. Thousands of low-income youth and their families circulate through EarthLab each year, learning about the public health effects of climate change and environmental degradation in their own neighborhoods and participating in climate action in concrete ways.

This is highly replicable and scalable activity. This project began as an experimental coalition between a university, a school district, and a community-based environmental nonprofit organization—precisely the kind of integral action Pope Francis advocated in *Laudato Si'*. Universities everywhere, large and small, are positioned to facilitate this kind of local collaborative work for the benefit of students, local communities, and the planet.

Conclusion

This chapter acknowledges the urgent and disproportionate impact of climate change on the bottom three billion among us. But it argues that the language of climate justice is not the most effective framework for communicating about climate change to the larger public. Those who are not receptive to social justice *sui generis* will not be motivated to change their attitudes and behaviors about climate change on the grounds that it disproportionately harms the vulnerable. The chapter argues that most people are motivated most often by what most impacts themselves, their families, and their neighborhoods. This suggests that a more generalizable language about the health impacts of climate change might be a more effective way of communicating with those who are not motivated by the narrative of social justice to change their attitudes and behaviors. The chapter closes with some reflections about who is best positioned to communicate these impacts. The takeaway from these reflections is that those of us who are committed to mitigating climate change to lessen the impacts on the most vulnerable ought to consider that narratives about generalizable health impacts might just be a more expeditious path toward the ends we are seeking.

References

Barber, B. (2013). *If mayors ruled the world: Dysfunctional nations, rising cities*. New Haven: Yale University Press.

Barber, B. (2017). *Cool cities: Urban sovereignty and the fix for global warming*. New Haven: Yale University Press.

Barnett, J., & Adger, W. N. (2007). Climate change, human security, and violent conflict. *Climate Change and Conflict, 26*, 639–655. https://doi.org/10.1016/j.polgeo.2007.03.003

Biermann, F., & Boas, I. (2008). Protecting climate refugees: The case for a global proto-col. *Environment: Science and Policy for Sustainable Development, 50*, 8–17. https://doi.org/10.3200/ENVT.50.6.8-17

Cosgrove, T. (2012). Memo on "Sea Level Rise Focus Groups," Union of Concerned Scientists.

Dasgupta, P., & Ramanathan, V. (2014). Pursuit of the common good: Religious institutions may mobilize public opinion and action. *Science, 345*. https://doi.org/10.1126/science.1259406

Davis, M. (2007). *Planet of slums*. London: Verso.

Forman, F. (2018). Social norms and the cross-border citizen: From Adam Smith to Antanas Mockus. In C. Tognato (Ed.), *Cultural agents reloaded: The legacy of Antanas Mockus* (pp. 333–356). Cambridge, MA: Harvard University Press.

Forman, F., & Cruz, T. (2017). Latin America and a new political leadership: Experimental acts of co-existence. In J. Burton, S. Jackson, & D. Wilsdon (Eds.), *Public servants: Art and the crisis of the common good* (pp. 71–90). Boston, MA: MIT Press.

Forman, F., & Cruz, T. (2018). Global justice at the municipal scale: The case of Medellín, Colombia. In L. Cabrera (Ed.), *Institutional cosmopolitanism* (pp. 189–215). New York: Oxford University Press.

Forman, F., & Ramanathan, V. (2018). Climate change, mass migration and sustainability: A prob-abilistic case for urgent action. In M. Suarez-Orozco (Ed.), *Humanitarianism and mass migra-tion* (pp. 42–59). Oakland, CA: University of California Press.

Forman, F., Solomon, G., Morello-Forsch, R., & Pezzoli, K. (2016). Bending the curve and closing the gap: Climate justice and public health. *Collabra, 2*, 22. https://doi.org/10.1525/collabra.67

Furth, I., & Gantwerk, H. (2013). *Citizen dialogues on sea level rise: Start with impacts/end with action*. San Diego, CA: Viewpoint Learning. Retrieved February 9, 2020 from https://climate-access.org/system/files/Viewpoint%20UCS_sea%20level%20dialogues.pdf

Gadanho, P. (2014). *Uneven growth: Tactical urbanisms for expanding megacities*. New York, NY: Museum of Modern Art.

Griskevicius, V., Cialdini, R. B., & Goldstein, N. J. (2008). Social norms: An underestimated and underemployed lever for managing climate change. *International Journal for Sustainability Communication, 3*, 5–13.

Hirschman, A. O. (1977). *The passions and the interests: Arguments for capitalism before its tri-umph*. Princeton: Princeton University Press.

IOM (International Organization for Migration) (2008). *Migration and climate change* (IOM migration research series) (31st ed.). Geneva: IOM.

IPCC (Intergovernmental Panel on Climate Change) (1990). *Climate change: The IPCC impacts assessment* (pp. 5–11). Canberra: IPCC.

Kelley, C. P., Mohtadib, S., Canec, M. A., Seagerc, R., & Kush-Nirc, Y. (2015). Climate change in the Fertile Crescent and implications of the recent Syrian drought. *Proceedings of the National Academy of Sciences of the United States of America, 112*, 3241–3246.

Lancet Commissions (2015). Health and climate change: policy responses to protect public health. *Lancet, 386*, 1861–1914.

Mary Robinson Foundation-Climate Justice (MRFCJ) (n.d). Retrieved February 9, 2020 from https://www.mrfcj.org/principles-of-climate-justice/share-benefits-and-burdens-equitably/

McGuirk, J. (2014). *Radical cities: Across Latin America in search of a new architecture*. London: Verso.

McLeman, R. A. (2014). *Climate and human migration: Past experiences, future challenges*. Cambridge, UK: Cambridge University Press. Retrieved February 9, 2020 from www.cam-bridge.org/core/books/climate-and-humanmigration/0D0D207E8BA3D169452A84FEB7D779CB

Myers, N. (2005). Environmental refugees: An emergent security issue, 13th Economic Forum, May 2005, Prague.

NRC/IDMC (Norwegian Refugee Council and Internal Displacement Monitoring Centre) (2015). *Global estimates 2015: People displaced by disasters*. Geneva: IDMC.

Peñalosa, E. (2014). Why buses represent democracy in action (TED talk). Retrieved February 9, 2020 from https://www.ted.com/talks/enrique_penalosa_why_buses_represent_democracy_in_action

Pianigiani, G. (2015). At Vatican, mayors pledge climate change fight. *New York Times*, July 21, 2015.

Pincetl, S. (2010). Implementing municipal tree planting: Los Angeles million-tree initiative. *Environmental Management, 45*, 227–238.

Pope, F. (2015) *Laudato Si (Encyclical letter on care of our common home)*. Retrieved February 9, 2020 from http://www.vatican.va/content/francesco/en/encyclicals/documents/papa-francesco_20150524_enciclica-laudato-si.html

Ramanathan, V., Allison, J., Auffhammer, M., Auston, D., Barnosky, A. D., Chiang, L., et al. (2015). Chapter 1. Bending the curve: Ten scalable solutions for carbon neutrality and climate stability. *Collabra, 2*, 15. Oakland: University of California Press. Retrieved February 9, 2020 from https://doi.org/10.1525/collabra.55

Saad, L. (2017). Global warming concern at three-decade high in US. *Gallup*. Retrieved February 9, 2020 from https://news.gallup.com/poll/206030/global-warming-concern-three-decade-high.aspx

Sen, A. (2009). *The idea of justice*. Belknap: Harvard.

Shklar, J. (1990). *Faces of injustice*. New Haven: Yale.

UNHCR (2009). *Climate change could become the biggest driver of displacement: UNHCR Chief*. Retrieved February 9, 2020 from http://www.unhcr.org/en-us/news/latest/2009/12/4b2910239/climate-change-become-biggest-driver-displacement-unhcr-chief.html

UNICEF (United Nations Children's Fund) (2015). *Children will bear the brunt of climate change*. Retrieved February 9, 2020 from www.unicef.org/media/media_86347.html

CHAPTER 19

Health of People, Health of the Planet, Health of Migrants

Marcelo Suárez-Orozco

You shall neither vex a stranger, nor oppress him: for you were strangers in the land of Egypt.

—*Exodus 22:21*

Summary Migration is as old as mankind. Modern humans are the children of immigration (Antón, Potts, & Aiello, 2014). Today's migrations unfold in complex ecologies involving demographic factors, economic variables, social practices, political processes, historical relationships, the environment itself, and various combinations thereof (McLeman, 2014). In the twenty-first century, mass migration is the human face of globalization—the sounds, colors, and aromas of a miniaturized, interconnected, and ever-fragile world. Above all, migration is a condition of all humanity.

Catastrophic Migrations of the Twenty-First Century: A New Cartography

Mass migrations are increasingly defined by the slow-motion disintegration of failing states with feeble institutions, war and terror, unchecked climate change, environmental degradation, and demographic imbalances (IDMC, 2016; McLeman, 2014). Symbiotically, these forces are the drivers of the catastrophic migrations of the twenty-first century (Betts, 2010; Chapter 2 in Suárez-Orozco, 2019) (Fig. 19.1).

In the first quarter of the twenty-first century, the world witnessed forcible displacement of the largest number of human beings in history: while precise numbers

M. Suárez-Orozco (✉)
UCLA Graduate School of Education and Information Studies, Los Angeles, CA, USA
e-mail: mms-o@gseis.ucla.edu

© The Author(s) 2020   251
W. K. Al-Delaimy, V. Ramanathan, M. Sánchez Sorondo (eds.), *Health of People, Health of Planet and Our Responsibility*, https://doi.org/10.1007/978-3-030-31125-4_19

	New displacements Jan – Dec 2016	Total number of IDPs as of the end of 2016
CONFLICT	6.9 million	40.3 million
DISASTERS	24.2 million	?

are both elusive and changing (see "Data on Movements of Refugees and Migrants are Flawed," 2017), United Nations (UN) data show that more than 68 million people—the equivalent of every man, woman, and child in Lagos, Sao Paulo, Seoul, London, Lima, New York, and Guadalajara combined—have escaped home into the unknown (UNHCR, 2017). The majority of those seeking shelter are internally displaced persons (IDPs), not formal refugees displaced across international borders (Suárez-Orozco, 2019). In addition, approximately nine in ten international asylum seekers remain in a neighboring country; Asians stay in Asia, Africans in Africa, Americans in the Americas (Fig. 19.2).

While migration is normative, it is increasingly catastrophic: "The majority of new displacements in 2016 took place in environments characterized by a high exposure to natural and human-made hazards, high levels of socioeconomic vulnerability, and low coping capacity of both institutions and infrastructure" (IDMC, 2017). By the end of 2016, there were 31.1 million *new* internal displacements due to conflict and violence (6.9 million) and disasters (24.2 million), the equivalent of "20 people [being] forcibly displaced every minute" (IDMC, 2017), and it was noted that "an unknown number remain displaced as a result of disasters that occurred in and prior to 2016" (IDMC, 2017). Internal displacement associated with war and terror has been growing since the beginning of the millennium, and by 2017, most "displacement [had] occurred in sub-Saharan Africa, with the Democratic Republic of the Congo (DRC) overtaking Syria in the top ranking." Today the number of internally displaced persons is significantly larger than the number of refugees; in 2017 it was estimated that worldwide, there were 22.5 million refugees under the United Nations High Commissioner for Refugees (UNHCR) mandate (UNHCR, 2017c).

In sum, catastrophic migrations unfold at the interstices of war and terror, "fossil fuel use, the pollution of the atmosphere and the oceans, climate change, [and crises in] public health, the health of ecosystems and sustainability" (Pontifical Academy of Sciences, 2017).

Nine out of 10 refugees across the world find asylum within their region of origin
Number of refugees by region of origin and destination, 2015 (in millions)

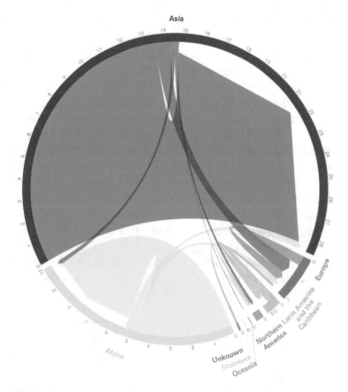

Fig. 19.2 Refugees by region (Source: United Nations High Commissioner for Refugees, Global Trends: Forced Displacement in 2015, UNHCR, Geneva, 2016 and United Nations Relief and Works Agency for Palestine Refugees in the Near East, Annual Operational Report 2015 for the Reporting Period, 1 January–31 December 2015, UNRWA, 2016)

Environmental Dystopia

Forman and Ramanathan (Chapter 1 in Suárez-Orozco, 2019) argue that unchecked climate change and geophysical hazards increase morbidity and mortality, disrupt production, decrease agricultural yields, decimate livestock, and forcefully displace millions the world over (see also McLeman, 2014). According to the *Global Report on Internal Displacement*, South and East Asia were the regions most affected in 2016, as in previous years (IDMC, 2017). The 2016 data continued an alarming trend in extreme weather patterns and weather-related hazards: in 2015, rising sea levels and floods, droughts, high-intensity cyclones, monsoons, hurricanes, heat waves, and forest fires triggered 14.7 million displacements, and "4.5 million displacements were brought on by large-scale geophysical hazards." The 2017 *Global Report on Internal Displacement* stated that "over the past eight years, 203.4 million displacements have been recorded, an average of 25.4 million each year"

(IDMC, 2016). By 2017, the majority of new displacements had occurred in "low- and lower-middle-income countries and as a result of large-scale weather events, and [had] predominantly [occurred] in South and East Asia. While China, the Philippines and India [had] the highest absolute numbers, small island states [suf- fered] disproportionally once population size [was] taken into account. Slow-onset disasters, existing vulnerabilities[,] and conflict also [continued] to converge into explosive tipping points for displacement" (IDMC, 2017).

Documented displacements due to environmental factors took place in 113 coun- tries across all regions of the world. Floods, storms, cyclones, monsoons, hurricanes, earthquakes, volcanic eruptions, wildfires, landslides, and extreme temperatures dis- placed millions of people in 2015 (Chapter 1 in Suárez-Orozco, 2019). India, China, and Nepal accounted for the largest numbers of people displaced, "with totals of 3.7 million, 3.6 million, and 2.6 million[,] respectively." The Philippines experienced three massive storms, which together displaced two million people (IDMC, 2016). In Myanmar, Cyclone Komen "displaced more than 1.6 million people[,] . . . the fifth highest figure worldwide in absolute terms. . . . Twelve of the country's fourteen states and regions suffered widespread destruction". In Central America, millions have been affected by environmental factors (Durham, 1979; Chapter 2 in Suárez- Orozco, 2019). By 2017, the world witnessed ferocious hurricanes in the Atlantic that devastated entire regions of the Caribbean, including Antigua and Barbuda. According to Prime Minister Gaston Alphonso Browne, after the largest storm ever in the Atlantic Ocean in September 2017, "the island of Barbuda [was] decimated[,] its entire population left homeless[,] and its buildings reduced to empty shells" (UN News Centre, 2017a, 2017b). The entire island of Puerto Rico was left without power. A month earlier (in August 2017), devastating monsoons in South Asia killed more than 1200 people; forced millions from their homes in India, Nepal, and Bangladesh; and shut 1.8 million children out of school.

The UNHCR predicts that climate change will perhaps become the biggest driver of population displacements, both inside and across national borders. Though there is a general consensus that quantitative estimates are presently unreliable, Forman and Ramanathan (Chapter 1 in Suárez-Orozco, 2019) make a plea for an ethical global policy response to the emerging climate migration crisis. They argue that we simply cannot await reliable metrics. International cooperation on climate mitiga- tion is more urgent than ever as the USA, under President Trump's leadership, is moving toward an ever more retrograde agenda on climate issues. Establishing international protocols that outline the rights of climate refugees and the responsi- bilities of industrialized nations toward them cannot wait.

Jeffrey Sachs (2017) has argued that in addition to the physical environment, demography itself is a main driver of mass migrations. Africa and the Middle East are a case in point. In the 1950s, Europe had twice the combined popula- tions of the Middle East and all of Africa, so migration to Europe was not a *problématique* of significance; with labor shortages and the need to rebuild after World War II, immigration was a solution, not a problem. In an epic reversal, the Middle East and Africa now have twice the population of Europe. Europe now has about 740 million people. The Middle East and Africa combined have about

1.4 billion people. Furthermore, according to UN forecasts, Europe's population will remain level because of aging and low fertility rates, whereas the population of the Middle East and Africa combined is on its way to four billion people by 2100 (Sachs, 2017).

War and Terror

War and terror are pushing millions of human beings from home. Millions of people linger in camps far away from the wealthy cities of Asia, Europe, North America, and Australia. Indeed, the world is witnessing what Sánchez Terán (2017) calls the great "forced confinement crisis" of the twenty-first century. Millions have been internally displaced, millions are awaiting asylum, and millions more are living in the shadow of the law as irregular or unauthorized immigrants.

In the aftermath of antigovernment uprisings beginning in 2010, the Middle East and North Africa had the largest number of human beings displaced by war and terror. But by the end of 2016, sub-Saharan Africa led the way, with the Democratic Republic of the Congo (DRC) overtaking Syria in the top ranking with "the most new displacements by conflict and violence." Ongoing conflict "in North and South Kivu and an increase in intercommunal clashes in southern and central regions such as Tanganyika, Kasai, Kasai-Oriental, Ituri[,] and Uele caused more than 922,000 new displacements in total during the year. Some people were forced to flee more than once." War and terror in Syria resulted in more than 800,000 new displacements recorded during 2016. In Iraq, almost 680,000 new displacements occurred as a result of nine military campaigns. In Yemen, at least 478,000 new displacements took place against the backdrop of a persistently dynamic and volatile security situation (IDMC, 2017).

In Syria, an estimated 12 million people have fled their homes since 2011. By 2016, more than half of the Syrian population lived in displacement either across borders or within their own country. "In the sixth year of war, 13.5 million [were] in need of humanitarian assistance within the country. Among those escaping the conflict, the majority [had] sought refuge in neighboring countries or within Syria itself. According to the United Nations High Commissioner for Refugees, 4.8 million [had] fled to Turkey, Lebanon, Jordan, Egypt, and Iraq, and 6.6 million [were] internally displaced within Syria. Meanwhile, about one million [had] requested asylum [in] Europe. Germany, with more than 300,000 accumulated applications, and Sweden, with 100,000, [were] the EU's top receiving countries" (UNHCR, 2017c).

In 2017, just three countries—Syria, Iraq, and Yemen—accounted for more than half of all internally displaced persons (IDMC, 2016). Likewise, in 2017, more than half of all international refugees under the UNHCR mandate originated in four states: Syria (approximately 5.5 million), Afghanistan (2.5 million), South Sudan (1.6 million), and Somalia (900,000). The conflicts in these countries [have been] disparate and incommensurable in nature but have shared a chronic, protracted quality. By 2017, Syria's descent into a Dantesque inferno had been six years in the making; the Afghanistan conflict had gone on for twice as long. In Somalia, "more than two million Somalis [were] displaced by a conflict that [had] lasted over two decades. An estimated 1.5 million people [were] internally displaced in Somalia[,] and nearly 900,000 [were] refugees in the near region, including some 308,700 in Kenya, 255,600 in Yemen[,] and 246,700 in Ethiopia" (UNHCR, 2017b). In Sudan, war and terror displaced almost a million people in 2016 alone. These conflicts [had] endured longer than World War I and World War II. In each case, environmental dystopia and extreme weather patterns anteceded and accentuated the catastrophic movement of people.

Syria has continued to represent "the world's largest refugee crisis" (Suárez-Orozco, 2019) and, in its collapse, has embodied the noxious synergies among the environment, war and terror, and mass human displacement. A report published in 2016 cited NASA data showing that Syria was experiencing the driest drought on record. The NASA scientists found that "estimating uncertainties using a resampling approach, [they could] conclude that there [was] an 89 percent likelihood that this drought [was] drier than any comparable period [in] the last 900 years and a 98 percent likelihood that it [was] drier than the last 500 years" (Cook, Anchukaitis, Touchan, Meko, & Cook, 2016). According to UN data, the drought caused "75 percent of Syria's farms to fail and 85 percent of livestock to die between 2006 and 2011. The collapse in crop yields forced as many as 1.5 million Syrians to migrate to urban centers like Homs and Damascus" (Stokes, 2016).

Long-term conflicts, unchecked climate change, extreme weather patterns, and environmental degradation in Africa are generating massive forced migrations. "Four countries in Africa—Nigeria, the Democratic Republic of the Congo, the Central African Republic, and South Sudan—were among the top ten globally for new violence-induced internal displacements in 2015. . . . In total, more than 12 million people [had] been internally displaced by conflict and violence within Africa—more than twice the number of African refugees" (UNICEF, 2016).

In 2017, the UNHCR observed that in South Sudan, "some 1.9 million people [had been] displaced internally, while outside the country there [were] 1.6 million South Sudanese refugees [who had been] uprooted, mainly in Ethiopia, Sudan, and Uganda" (UNHCR, 2017a). Again , environmental concerns loomed large: "Drought and environmental degradation, and a food crisis that became a famine because of government neglect and changing regional demographics" were behind the collapse in Sudan (IDMC, 2016). The UN noted that "a famine produced by the vicious combination of fighting and drought [was] now driving the world's fastest growing refugee crisis. . . . The rate of new displacement [was] alarming, representing an impossible burden on a region that [was] significantly poorer [than other African regions] and [was] fast running short of resources to cope. Refugees from South Sudan [were] crossing the borders to the neighboring countries. The majority of them [were going] to Uganda[,] where new arrivals spiked from 2,000 per day to 6,000 per day in February [2017] and averaged more than 2,800 people per day" (UNHCR News Centre, 2017a). The UN World Food Program estimated that by 2017, 4.9 million people (40% of South Sudan's population) were facing famine (UNHCR News Centre, 2017b).

Indeed, famine lurked as a macabre specter. In 2017 the UNHCR stated that:

In all, more than 20 million people in Nigeria, South Sudan, Somalia[,] and Yemen are experiencing famine or are at risk. The regions in which these countries sit, including the Lake Chad basin, Great Lakes, East, Horn of Africa[,] and Yemen[,] together host well over 4 million refugees and asylum seekers. Consecutive harvests have failed, conflict in South Sudan coupled with drought is leading to famine and outflows of refugees, insecurity in Somalia is leading to rising internal displacement, and rates of malnutrition are high, especially among children and lactating mothers. In the Dollo Ado area of southeast Ethiopia[,] for example, acute malnutrition rates among newly arriving Somali refugee children aged between six months and five years are now running at between 50 [and] 79 percent. (UNHCR, 2017b)

By large margins, African refugees [have stayed] on the continent: "Some 86 percent . . . [have found] asylum in other African countries. Five of the largest refugee populations in the world are hosted in Africa, led by Ethiopia, Kenya, and Uganda. The protracted nature of crises in sending countries means that some of these host countries have shouldered responsibilities for more than two decades. Generations of displaced children have been born in some of the longest standing camps" (Dryden-Peterson, chapter 10 in Suárez-Orozco, 2019).

Only small numbers of refugees have made it to the high- and middle-income countries. Europe is a case in point. By the end of 2015, Europe had approximately one in nine of all refugees under the UNHCR's mandate, a total of 1.8 million people (UNICEF, 2016). Of these, most were "divided in nearly equal measure among Germany, the Russian Federation[,] and France (17, 17[,] and 15 percent of refugees in Europe, respectively). In 2015, more than one-third of the refugees living in Germany were from the Syrian Arab Republic, with smaller proportions from Iraq and Afghanistan (38, 17[,] and 10 percent, respectively). Nearly all of the 315,000 refugees hosted in the Russian Federation by the end of 2015 were from Ukraine. By the end of 2015, Germany had become the world's largest recipient of new individual applications for asylum—receiving more than twice as many as the next closest country" (UNICEF, 2016).

In the Americas, a new migration map has also taken shape. First, by 2015, Mexican migration to the USA, the largest flow of international migration in US history, was at its lowest in over a quarter of a century. Second, for the first time in recent history, more Mexicans were returning (voluntarily and involuntarily) to their country than were migrating to the USA. According to data analyzed by the Pew Hispanic Center,

more Mexican immigrants [had] returned to Mexico from the [United States] than [had] migrated here since the end of the Great Recession . . . . The same data sources also show the overall flow of Mexican immigrants between the two countries [was] at its smallest since the 1990s, mostly due to a drop in the number of Mexican immigrants coming to the [United States].
    From 2009 to 2014, one million Mexicans and their families (including US-born children) left the [United States] for Mexico, according to data from the 2014 Mexican National Survey of Demographic Dynamics. (ENADID, 2014)

Third, as Mexican migration decreased, uncontrolled criminality (Chapter 2 in Suárez-Orozco, 2019), terror, and environmental dystopia put Central Americans at the center of the new map. Indeed, the Americas gave the new immigration map a new nomenclature: *mass unauthorized immigration* (Pew Research Center, 2016) and *unaccompanied minors*.

The sources of the forced movements of people in Central America have disparate and complex histories, finding their distal origins in the Cold War, inequality, uncontrolled criminality, and environmental malfeasance. In the case of Central America, 1998 began a new cycle of catastrophic migrations. That is the year Hurricane Mitch hit Honduras and the rest of the region. Hurricane Mitch was the second-deadliest Atlantic hurricane on record, causing over 11,000 fatalities in Central America, with over 7,000 occurring in Honduras alone because of the catastrophic flooding due to the slow motion of the storm. The hurricane left severe environmental and psychosocial scars. Data from the School of Medicine at Brown University estimated that of the total of 3.3 million adult (15 years of age or older) inhabitants of Honduras, more than 49,000 suffered posttraumatic stress disorder (PTSD) due to Hurricane Mitch. The near-savage deforestation of Honduras left a country with weak institutional capacity extremely vulnerable to devastation in the wake of the hurricanes. Hondurans then began an ecological exodus north.

A generation before, *La guerra del fútbol*—the so-called Soccer War of 1969 between El Salvador and Honduras—had more to do with environmental factors flowing from extraordinary inequality in land holdings than with the region's beloved game. Running out of cultivable land, some 300,000 Salvadoreans packed up and migrated over the border to Honduras. The ensuing war lasted 100 h and forecasted the noxious synergies between environmental malfeasance, war and terror, and mass migrations (Durham, 1979).

The New Face of Catastrophic Migrations

Crying children are the face of the catastrophic migrations of the twenty-first century. Worldwide, one in every 200 children is a refugee, almost twice the number of a decade ago (UNICEF, 2017). According to UN data, in 2016 there were 28 million forcefully displaced children. Another 20 million children were international migrants (UNICEF, 2017). Their total number is now larger than the populations of Canada and Sweden combined. Millions of children are internal migrants. In China alone there were an estimated 35 million migrant children in 2010 and a staggering 61 million children who were left behind in the countryside as their parents migrated to the coastal cities.

> [Children] are a sign of hope, a sign of life, but also a 'diagnostic' sign, a marker indicating the health of families, society[,] and the entire world. Wherever children are accepted, loved, cared for[,] and protected, the family is healthy, society is more healthy[,] and the world is more human. (Pope Francis, 2014)

Yet, few forcefully displaced children ever make it to safety in the high- or middle-income countries. In 2016, UNICEF (2016) reported that 900,000 children had been forcefully displaced within South Sudan and more than 13,000 had been reported as "missing" or separated from their families. The vast majority of children seeking refuge remain internally displaced or settle in a neighboring country. Jacqueline Bhabha (chapter 3 in Suárez-Orozco, 2019) notes that of the more than 600,000 South Sudanese refugees currently sheltered in Uganda, some 300,000 are under age 18, and the majority are girls and women.

By 2015, the world had witnessed a record number of unaccompanied or separated children, with 98,400 formal asylum applicants—mainly Afghans, Eritreans, Syrians, and Somalis—being lodged in 78 countries. "This was the highest number on record since UNHCR started collecting such data in 2006" (UNHCR, 2016). By the end of 2016, a new record had been set, with at least "300,000 unaccompanied and separated children moving across borders[. They] were registered in 80 countries in 2015–16—a near fivefold increase from 66,000 in 2010–11. The total number of unaccompanied and separated children on the move worldwide [was] likely much higher" (UNICEF, 2017).

Conclusions

Mass migration and demographic change are, under the best of circumstances, destabilizing and generate disequilibrium. Catastrophic migrations produce multiple additional layers of distress. The forcefully displaced undergo violent separations and carry the wounds of trauma. Millions of human beings are caught in permanent limbo, living in zones of confinement. In these zones "humiliation is re-created in the camp environment when individuals are not allowed to work, grow food, or make money" (Mollica, Chapter 5 in Suárez-Orozco, 2019). Catastrophic migrations assault the structure and coherence of families in terms of their legislative, social, and symbolic functions.

The outright rejection of unwanted refugees, asylum seekers, and unauthorized immigrants compounds trauma. In many countries of immigration, too, we have identified zones of confinement where de facto and de jure policies are forcing millions of immigrant and refugee families to live in the shadow of the law. In the USA, the country with the largest number of immigrants, millions are separated, millions are deported, millions are incarcerated, and millions more inhabit a subterranean world of illegality (Chapter 4 in Suárez-Orozco, 2019).

Catastrophic migrations and violent family separations disrupt the essential developmental functions necessary for children to establish basic trust, feel secure, and have a healthy orientation toward the world and the future. Catastrophic migrations tear children from their families and communities. Furthermore, physical, sexual, and psychological abuse are normative features of forced migrations, especially when they involve human trafficking and subhuman conditions prevailing in many migrant camps. The health of children in such contexts is the existential crisis of our times (Chapter 7 and Chapter 5 in Suárez-Orozco, 2019).

Catastrophic migrations remove children and youth from the prescribed pathways that enable them to reach and master culturally determined developmental milestones in the biological, socioemotional, cognitive, and moral realms required to successfully make the transition to adulthood. Catastrophic migrations are life thwarting, harming children's physical, psychological, moral, and social well-being by placing them in contexts that are inherently dangerous.

When immigrants and refugees manage to settle in new societies, they bring new kinship systems, cultural sensibilities (including racial, linguistic, and religious sensibilities), and identities to the forefront. These may misalign with (and even contravene) taken-for-granted cultural schemas and social practices in the receiving societies. The world over, immigrants and refugees are arousing suspicion, fear, and xenophobia. Immigration is the frontier pushing against the limits of cosmopolitan tolerance. Immigration intensifies the general crisis of connection and flight from the pursuit of our inherent humanitarian obligations concerning the welfare of others (Chapter 14 in Suárez-Orozco, 2019).

Reimagining the narrative of belonging, reclaiming the humanitarian call, and recalibrating the institutions of the nation-state are a *sine qua non* to move beyond the current immigration malaise the world over. In the long term, we must retrain hearts and minds, especially younger ones, for democracy in the context of demographic change and superdiversity. We need to convert a dread of the unfamiliar "Other" into empathy, solidarity, and a democratizing desire for cultural difference. We must endeavor to cultivate the humanistic ideal to find oneself "in Another" (Ricoeur, 1992/1995) in the refugee, in the asylum seeker, and in the forcefully displaced.

References

Anton, S. C., Potts, R., & Aiello, L. C. (2014). Evolution of early homo: An integrated biological perspective. *Science, 345*, 1236828. https://doi.org/10.1126/science.1236828

Betts, A. (2010). Survival migration: A new protection framework. *Global Governance, 16*, 361–382.

Cook, B. I., Anchukaitis, K. J., Touchan, R., Meko, D. M., & Cook, E. R. (2016). Spatiotemporal drought variability in the Mediterranean over the last 900 years. *Journal of Geophysical Research, 121*, 2060–2074. https://doi.org/10.1002/2015JD023929

Data on movements of refugees and migrants are flawed. (2017). *Nature, 543*, 5–6.

Durham, W. H. (1979). *Scarcity and survival in Central America: Ecological origins of the soccer war*. Stanford, CA: Stanford University Press.

ENADID (Instituto Nacional de Estadística y Geografía) (2014). *National survey of demographic dynamics 2014*. Retrieved February 8, 2020 from http://en.www.inegi.org.mx/proyectos/enchogares/especiales/enadid/2014/

Internal Displacement Monitoring Centre (IDMC) (2016). *Global report on internal displacement*. Norway: IDMC. Retrieved February 8, 2020 from http://internal-displacement.org/globalreport2016/#home2016

Internal Displacement Monitoring Centre (IDMC) (2017). *Global report on internal displacement*. Norway: IDMC. Retrieved February 8, 2020 from https://reliefweb.int/sites/reliefweb.int/files/resources/2017-GRID.pdf. IOM (International Organization for Migration). United Nations.

McLeman, R. A. (2014). *Climate and human migration: Past experiences, future challenges*. Cambridge: Cambridge University Press.

Pew Research Center Hispanic Trends (2016). *Unauthorized immigrant population trends for states, birth countries, and regions.* Retrieved February 9, 2020 from http://www.pewhispanic.org/interactives/unauthorized-trends/

Pontifical Academy of Sciences (2017). *Health of people and planet: Our responsibility.* Retrieved February 9, 2020 from http://www.casinapioiv.va/content/accademia/en/events/2017/health.html

Pope Francis (2014). *Homily of Pope Francis: "Pilgrimage to the Holy Land on the Occasion of the Fiftieth Anniversary of the Meeting between Pope Paul VI and Patriarch Athenagoras in Jerusalem".* Retrieved February 9, 2020 from https://w2.vatican.va/content/francesco/en/homilies/2014/documents/papa-francesco_20140525_terra-santa-omelia-bethlehem.html

Ricoeur, P. (1992). *See oneself as another* (Reprint 1995). Chicago: University of Chicago Press.

Sachs, J. (2017). *Humanism and the forced confinement crisis.* Paper presented at the Workshop on Humanitarianism and Mass Migration, UCLA, Los Angeles, January 18–19, 2017.

Sánchez Terán, G. (2017). *Humanism and the forced confinement crisis.* Paper presented at the Workshop on Humanitarianism and Mass Migration, UCLA, Los Angeles, January 18–19, 2017.

Stokes, E. (2016). The drought that preceded Syria's civil war was likely the worst in 900 years. *Vice News.* Retrieved February 9, 2020 from https://news.vice.com/article/the-drought-that-preceded-syrias-civil-war-was-likely-the-worst-in-900-years

Suárez-Orozco, M. (Ed.). (2019). *Humanitarianism and mass migration: Confronting the world crisis.* Oakland, CA: University of California Press.

UN News Centre (2017a). *Hurricane Irma erased 'footprints of an entire civilization' on Barbuda, prime minister tells UN.* Retrieved February 9, 2020 from https://news.un.org/en/story/2017/09/566372-hurricane-irma-erased-footprints-entire-civilization-barbuda-prime-minister.

UN News Centre (2017b). *South Sudan now world's fastest growing refugee crisis.* Retrieved February 9, 2020 from http://www.un.org/apps/news/story.asp?NewsID=56367#.WN6wTBiZOqB

UNHCR (2016). *Global trends forced displacement in 2015.* Retrieved February 9, 2020 from http://www.unhcr.org/en-us/statistics/unhcrstats/576408cd7/unhcr-global-trends-2015.html

UNHCR (2017a). *South Sudan situation.* Retrieved February 9, 2020 from http://data.unhcr.org/SouthSudan/country.php?id=251

UNHCR (2017b). *Somalia situation.* Retrieved February 9, 2020 from http://www.unhcr.org/591ae0e17.pdf

UNHCR (2017c). *Syrian refugees: A snapshot of the crisis in the Middle East and Europe.* Florence: Migration Policy Centre.

UNHCR News Centre (2017a). *South Sudan now world's fastest growing refugee crisis.* Retrieved February 9, 2020 from http://www.unhcr.org/en-us/news/latest/2017/3/58cbfa304/refugee-crisis-south-sudan-worlds-fastest-growing.html

UNHCR News Centre (2017b). UNHCR says death risk from starvation in horn of Africa, Yemen, Nigeria growing, displacement already rising. Retrieved February 9, 2020 from http://www.unhcr.org/en-us/news/briefing/2017/4/58ec9d464/unhcr-says-death-risk-starvation-horn-africa-yemen-nigeria-growing-displacement.html

UNICEF (2017). *A child is a child: Protecting children on the move from violence, abuse, and exploitation.* Retrieved February 9, 2020 from https://www.unicef.org/publications/index_95956.html

UNICEF (United Nations Children's Fund) (2016). *Uprooted: The growing crisis for refugee and migrant children.* Retrieved February 9, 2020 from https://www.unicef.org/publications/files/Uprooted_growing_crisis_for_refugee_and_migrant_children.pdf

Part VI
Overarching Solutions: The Role of Religion

CHAPTER 20
Faith in God and the Health of People

Leith Anderson

Summary Engaging religious groups in environmental stewardship for human health begins with their theological presuppositions and is most persuasive when influenced by those leaders who share their faith.

Faith in God and the health of people are universal. Eighty-four percent of the world's 7.3 billion people are affiliated with some religion (Christians 31.2%; Muslims 24.1%; unaffiliated 16%; Hindus 15.1%; Buddhists 6.9%; folk religions 5.7%; other religions 0.8%; Jews 0.2%) (Hackett & McClendon, 2017). While not everyone in the world is religious, about eight in ten are religious and the rest are affected by the religious beliefs of the many around them. However, the health of humans involves everyone. One hundred percent of the world's population are somewhere on the continuum of good health to poor health.

While religion and health usually are studied under different academic disciplines, practical experience combines them in most lives. As Pope Francis asserted in *Laudato Si'*, "everything is interrelated" (Catholic Ecology, 2015). We cannot conflate religion and human health, but we can seek to connect them. It is a challenging task because there is generally greater consensus on health than on religion. Moreover, there are different starting points. In broad terms, religion tends to start with God and faith, while science starts with research and predictability. That said, Christians are more likely to begin with God and move on to health and the environment, while "some non-Western religious traditions, like Hinduism, make a tighter link between representations of the divine and nature" (Ecklund & Scheitle, 2018).

L. Anderson (✉)
National Association of Evangelicals, Washington, DC, USA
e-mail: leith.anderson@wooddale.org

W. K. Al-Delaimy, V. Ramanathan, M. Sánchez Sorondo (eds.), *Health of People, Health of Planet and Our Responsibility*, https://doi.org/10.1007/978-3-030-31125-4_20

The Protestant Evangelical Perspective

As an American Protestant Evangelical Christian, my perspective is rooted in one segment of Christian faith and one nation's current engagement in the "Health of People, Health of Planet and Our Responsibility" (Pontifical Academy of Sciences, 2017). Evangelicals worldwide total around 600 million, with nearly 100 million in the USA (about 30% of the population of the USA) (Wikipedia, n.d.). It is more of a movement than an institution; there are thousands of independent subsets with diverse sub-beliefs, organizations, and ethnicities without direct knowledge of or connection to one another. They all share four Christian beliefs, which mark their identity: the Bible, Jesus Christ, salvation, and activism/evangelism (Anderson & Stetzer, 2016).

Beginning with God and the Bible, there are foundational beliefs that God is eternal and sovereign, and that God created the Earth and all on the Earth. The pinnacle of God's creation is humans, who were created in God's image of rational intellect, emotion, and volition (Biblica Inc., 1973a). Since God created the Earth, he is the owner, who has delegated stewardship/management responsibility to his human creatures. We are the beneficiaries of God's creation but also accountable to God for the care of his creation (Biblica Inc., 1973b). This becomes the theological basis for Christian environmentalism—humans are responsible to God for the preservation, protection, and thriving of all God has created, including air, land, water, animals, plants, and especially our fellow humans. There are great challenges in fulfilling this calling, but the calling is to be considered an honor and a privilege to serve as agents of God.

The Bible does not start with environmental stewardship. It starts with God and is therefore theocentric. In other words, environmental stewardship is not an end in itself but grows out of who God is, what God has created, and all God has assigned for us to do in the care of his creation. Loving and serving God is always first. When Jesus was asked what our highest human responsibility is, he said, "Love the Lord your God with all your heart and with all your soul and with all your mind. This is the first and greatest commandment" (Biblica Inc., 1973c). On the basis of the primacy of God and the authority of the Bible, Christians are also called to love our fellow humans. After the first commandment to love God, Jesus added, "And the second is like it: 'Love your neighbor as yourself'" (Biblica Inc., 1973d). These core expectations of Christians did not originate with Jesus; they also are written in the Hebrew scriptures of the Old Testament (Biblica Inc., 1973e). On the basis of these principles, the moral code of Jesus in the New Testament includes care for the poor and vulnerable (Biblica Inc., 1973f).

Although not all Christians have adhered to the teachings of the Bible on the care of God's creation or the moral imperatives of Jesus on the care of the poor and vulnerable, these are foundational doctrines and expectations. As millions of Christians take these callings seriously, they are immediately challenged with prioritization. What if environmental stewardship hurts the poor? What if helping the poor hurts the environment?

When Hurricane Maria devastated the island Commonwealth of Puerto Rico, 97% of the electrical power grid was knocked out. Transportation was curtailed; hospitals ran on diesel generators or closed; patients on ventilators or dialysis died. Puerto Rico is primarily powered by fossil fuels (U.S. Energy Information Administration, 2017). Generators and fossil fuels were shipped, airlifted, and carried to sustain human life and to improve living conditions. Of course, cataclysmic crises should not be the standard for deciding between protecting the environment from pollution and protecting human life, but, in everyday choices, decisions are often made between good competing values. Sadly, too many decide their priorities first and then unnecessarily choose to harm one and help another when options are available that will harm neither.

In 2010, Evangelical Christian leaders from 198 countries met in Cape Town, South Africa, under the auspices of the Lausanne Movement and issued the Cape Town Commitment (Lausanne Movement, 2010). Included in Part II of the lengthy Commitment is a statement on caring for God's creation and caring for the poor and suffering:

> All human beings are to be stewards of the rich abundance of God's good creation. We are authorized to exercise godly dominion in using it for the sake of human welfare and needs, for example in farming, fishing, mining, energy generation, engineering, construction, trade, or medicine. As we do so, we are also commanded to care for the earth and all its creatures, because the earth belongs to God, not to us. We do this for the sake of the Lord Jesus Christ who is the creator, owner, sustainer, redeemer and heir of all creation.
>
> We lament over the widespread abuse and destruction of the earth's resources, including its bio-diversity. Probably the most serious and urgent challenge faced by the physical world now is the threat of climate change. This will disproportionately affect those in poorer countries, for it is there that climate extremes will be most severe and where there is little capability to adapt to them. World poverty and climate change need to be addressed together and with equal urgency.
>
> We encourage Christians worldwide to:
> A) Adopt lifestyles that renounce habits of consumption that are destructive or polluting;
> B) Exert legitimate means to persuade governments to put moral imperatives above political expediency on issues of environmental destruction and potential climate change;
> C) Recognize and encourage the missional calling both of (i) Christians who engage in the proper use of the earth's resources for human need and welfare through agriculture, industry and medicine, and (ii) Christians who engage in the protection and restoration of the earth's habitats and species through conservation and advocacy. Both share the same goal for both serve the same Creator, Provider and Redeemer. (The Lausanne Movement, 2011)

This statement was subsequently adopted by the National Association of Evangelicals (USA) in 2015 (National Association of Evangelicals, 2015).

Engaging in the Issue

In the USA, issues related to climate change have become politically polarized, with politics often eclipsing scientific, religious, and moral teaching. Many explanations are offered to interpret this polarization, including the divide between those who trust government and institutions and those who do not share that trust. There are so many complex presuppositions and group identities involved that simple answers are hard to find and are often rejected, especially during a time of political transition and disruption. It is not that sources of political polarization are unimportant, but our discussion here focuses on description and possible paths to solutions.

Religion Versus Science

What Religious People Really Think, written by Elaine Howard Ecklund and Christopher Scheitle (New York: Oxford University Press), reports surveys of 10,000 Americans and interviews with nonreligious persons, religious persons, and science professionals. Funded by the John Templeton Foundation, in collaboration with the American Association for the Advancement of Science, and based at Rice University, this extensive research included questions on climate change.

Consider the breakdown of responses to just two of the many questions asked:

"Please tell me how interested you are in the following things: the environment ('Very interested')." All respondents 32.0%; Evangelical Protestants 27.8%; Mainline Protestants 31.2%; Catholics 32.0%; Jews 34.3% (Ecklund & Scheitle, 2018).

"The climate is changing and human actions are a significant cause of the change." All respondents 41.8%; Evangelical Protestants 34.7%; Mainline Protestants 41.6%; Catholics 42.4%; Jews 43.8%; non-Western 55.4%; unaffiliated, atheists, agnostics 49.9% (Ecklund & Scheitle, 2018).

The differences are statistically significant but fall short of a majority on both questions in all groups except for adherents to non-Western religions, with 55.4% agreeing that the climate is changing with human action as a significant cause. Overall, most Americans of all religious affiliations say they are not "very interested" in the environment and they do not affirm that the climate is changing and human actions are a significant cause of the change. While it is worth noting that Evangelical Protestants are on the lower end of "very interested" in the environment and on the lower end of affirming climate change caused by human action, they are among the majority of all respondents in both categories. This indicates that raising concern and commitment on these issues relates to the majority of Americans although mostly so among Evangelical Protestants.

Ecklund and Scheitle give suggestions on how to engage different religious communities in the issues of environment and climate change. Here are a few that specifically relate to American Evangelical Protestants (because they are 30% of the US population and less likely to engage in these issues):

- Evangelicals are also more likely to say either that the climate is not changing at all or that the climate is changing but not because of human actions. So, while politics does contribute to lower levels of acceptance of the scientific consensus in climate change among evangelicals, politics does not appear to be the sole factor. So, for those conducting outreach to evangelical communities on this issue, the message needs to reach beyond the political issues to address the distinct faith concerns of evangelicals.

- Attitudes on climate change are also strongly influenced by factors like political ideology. Over and over, we found, particularly among evangelical Christians, that environmentalism was tied in their minds to political ideologies they want to distinguish themselves from. Evangelicals are the religious groups most likely to be suspicious of scientists, so they need to hear about climate change from voices they trust. Religious leaders can help here, making climate change action about theology not politics.
- Congregations not only can draw connections between theology and environmental care, but they also can provide opportunities for members to practice and reinforce environmentally conscious behaviors.
- Scientists can help make environmental issues more salient for religious believers by focusing on the human impacts of climate change, and religious leaders can use those impacts to provide people of faith with a religious rationale for climate change action. (Ecklund & Scheitle, 2018)

Practical steps to address faith, poverty, suffering, human health, and the care of the Earth's environment begin with the presuppositions of faith and lead to actions based on faith.

- Faith communities may organize study groups, meeting over several weeks. Provide study guides that begin with Bible teaching on God's creation and God's call to care for his creation. Next, study Bible teachings on God's call to protect and provide for the poor, suffering, and vulnerable. Introduce information on the impacts of climate change that degrade God's creation and contribute to suffering and poverty. Provide stories of positive outcomes where the care of God's creation and mitigation of climate change contribute to human health and flourishing. Discuss individual and collective actions that fulfill the mandates of faith in the context of a changing climate.
- Provide persons of faith and communities of faith with theological and scientific information that comes from respected leaders who share their faith. Prioritize scientists who have strong faith commitments and are active in local faith communities. Utilize theologians, pastors, authors, and other known faith leaders who write and speak about the care of creation and climate change. Connecting with these trusted persons of influence from both the theological and scientific perspectives may be multisourced by engaging them as guest speakers in churches, authors of books and brochures, bloggers, and social media communicators, and also via other modes of communication. This approach builds on already established trust, which is leveraged to communicate information and specific practical calls to individual and collective action.
- Encourage grassroots decisions rather than prescribing personal or political action lists. This is not to preclude suggestions and recommendations. Stories, examples, and action ideas are powerful tools of influence. However, groups brainstorming how to act out their faith can lead to bottom-up movements that are often not possible when there is a concern over top-down external prescriptions and pressure.

Conclusions

Interfaith dialogue on the health of people, the health of the planet, and our responsibility is most valuable and important when focused on humanitarian needs. Collaboration and cooperation are best suited to helping those suffering from floods, droughts, crop failures, famines, migration, and other cataclysmic events, regardless of the causes. When crises come to countries and communities of varied religions, all people of faith should step forward to rescue and restore together.

Interfaith dialogue and cooperation may be least valuable when one seeks to inform and persuade each religion's constituents. The lesser value is rooted in different theologies, presuppositions and historical ignorance, and lack of trust of each other's religious faith. While dialogue among leaders is frequently valued and practiced, it has minimal predictability of influence among the masses in foreseeable generations. In other words, cooperation in practical humanitarian actions is good in that it helps those in need, but the trajectory from practical meeting of needs to persuasion of differing religious groups is best accomplished unilaterally rather than multilaterally. It comes down to matters of prioritization and deployment of resources. The priority of each religious community on issues of the environment and humanitarianism and the use of resources will produce greater effects when largely focused on separate religious traditions and communities.

This most applies to the continuum from individuals and communities of faith to individuals and communities of agnostics, atheists, and the unaffiliated. The belief systems and worldviews are at such a distance that they are a substantial barrier and distraction to dialogue and collaboration on environmental concerns. The same may be said of the distance between Western and non-Western religions. However, this conclusion is less applicable to groups within Christianity. For example, many of the doctrinal presuppositions and social teachings of the Roman Catholic Church and American Evangelicalism align (e.g., cardinal Christian doctrines and prolife social teaching).

The strategies for increasing engagement and action within each religious tradition are most likely to succeed when (1) leaders give voice to those who trust them as leaders; (2) persuasion begins with the tenets of the faith and not just current events or scientific consensus; (3) environmental catastrophes dominate the news and create opportunities for engagement and persuasion (preparing plans for communication before such catastrophes occur and quickly seizing the news cycle); and (4) scientists within the faith have disproportionate influence because of their understanding and terminology of both their faith and their science. And, finally, all of this takes a long time.

References

Anderson, L., & Stetzer, Ed. (2016, March 2). Defining evangelicals in an election year. *Christianity Today Magazine.*

Biblica, Inc. The Holy Bible, New International Version®, Niv® Copyright © 1973a, 1978, 1984, 2011 Biblica, Inc., Genesis 1–3.

Biblica, Inc. The Holy Bible, New International Version®, Niv® Copyright © 1973b, 1978, 1984, 2011 Biblica, Inc., Genesis 1:27–31.

Biblica, Inc. The Holy Bible, New International Version®, Niv® Copyright © 1973c, 1978, 1984, 2011 Biblica, Inc., Matthew 22:37–38.

Biblica, Inc. The Holy Bible, New International Version®, Niv® Copyright © 1973d, 1978, 1984, 2011 Biblica, Inc., Matthew 22:39.

Biblica, Inc. The Holy Bible, New International Version®, Niv® Copyright © 1973e, 1978, 1984, 2011 Biblica, Inc., Deuteronomy 6:4–5 and Leviticus 19:17–18.

Biblica, Inc. The Holy Bible, New International Version®, Niv® Copyright © 1973f, 1978, 1984, 2011 Biblica, Inc., Matthew 25:36–44 (God's judgment), Luke 10:25–37 (Parable of the Good Samaritan) and approximately 2000 references in the Bible to advocacy and care of the poor.

Catholic Ecology. (2015, June 18). *With Laudato Si, "Everything is connected".* Catholic Ecology. Retrieved September 30, 2017, from https://catholicecology.net/blog/laudato-i-everything-connected

Ecklund, E. H., & Scheitle, C. P. (2018). *Religion vs. science: What religious people really think.* Oxford, UK: Oxford University Press. Chapter 6.

Hackett, C., & McClendon, D. (2017, April 5). *Christians remain world's largest religious group, but they are declining in Europe.* Pew Research Center. Retrieved September 30, 2017, from http://www.pewresearch.org/fact-tank/2017/04/05/christians-remain-worlds-largest-religious-group-but-they-are-declining-in-europe/

National Association of Evangelicals (2015). *Caring for God's creation: A call to action.* National Association of Evangelicals. Retrieved October 1, 2017, from https://www.nae.net/caring-for-gods-creation/

Pontifical Academy of Sciences (2017, November 2–4). *Health of people, health of planet and our responsibility.* The workshop of the Pontifical Academy of Sciences, Vatican.

The Lausanne Movement (2010, October 17–24). *Cape Town 2010: The third Lausanne Congress on World evangelization.* Lausanne Movement. Retrieved October 1, 2017, https://www.lausanne.org/gatherings/congress/cape-town-2010-3

The Lausanne Movement (2011). *The Cape Town Commitment, Section IIB, 6.* The Lausanne Movement. Retrieved October 1, 2017, from https://www.lausanne.org/content/ctc/ctcommitment

U.S. Energy Information Administration (2017, September 21). *Puerto Rico.* U.S. Energy Information Administration. Retrieved September 30, 2017, from https://www.eia.gov/state/?sid=RQ

Wikipedia. (n.d.). *Evangelicalism.* Wikipedia. Retrieved September 30, 2017, from https://en.wikipedia.org/wiki/Evangelicalism

CHAPTER 21
Caring for Creation: The Evangelical's Guide

Mitchell C. Hescox

Summary Understanding creation care as prolife helps the Evangelical and Catholic communities relate to environmental concerns as more than interest in fauna and flora, and as a primary matter of life for our children, the Majority World's poor, and even many of the economically disadvantaged in the United States, whose homes border some of our country's most toxic air, foulest water, and most polluted soil.

Introduction

The religious landscape is changing in the United States. Today, only 43% of Americans identify as white and Christian, in comparison with 81% in 1976. Yet even amid these shifting demographics, white Evangelicals still comprise 35% of the Republican Party (Cox & Jones, 2017). Together, Evangelicals and Catholics represent approximately 28% of the American populace, yet they consistently vote in disproportionately high numbers. Over 80% of white Evangelicals voted for Mr. Trump in 2016, representing 28% of actual voters. More broadly, the last three US presidential elections saw prolife voters average 45% of the electorate (Evangelicals 26%, Catholics 19%), and for many of these voters, their antiabortion (prolife) stance overrode all other electoral considerations.

With the American populace more conservative than progressive, it is not surprising that climate change continues to be a polarizing issue. Exacerbating the issue is the failure to recognize that values are the driving force behind these differing ideologies. Most attempts to mobilize Americans for a bright future utilizing clean energy and addressing climate change utilize language and value appeals that are inconsistent with the core values of conservatives. *The Righteous Mind: Why*

M. C. Hescox (✉)
Evangelical Environmental Network, New Freedom, PA, USA
e-mail: mitch@creationcare.org

Good People Are Divided by Politics and Religion by Jonathan Haidt provides insight into how people formulate their moral foundations (Haidt, 2012).

Haidt's research on moral frames compares favorably with our experience in using messages geared toward the Evangelical community. These frames—tested in over 1000 presentations in churches, town halls, and Christian colleges around the nation—have been used to drive our education, which has resulted in over 3,500,000 prolife Evangelical Christians acting on climate, clean energy, and other related initiatives within the past 3 years.

Values/Morality-Based Messaging

Attempting to prove the science and pushing for government-based solutions as the major goal remain problematic for our community. While anecdotal, the following encounter described by one of our Creation Care Champions typifies our hypothesis.

> *Yesterday, I had an experience that was simultaneously so strange and yet so emblematic when it comes to encountering fellow Evangelical Christians on the issue of climate change.*
>
> *The context of the situation was our town's monthly "First Friday" event, which is staged April through October, in Warrenton, VA, to stage a kind of block party between 5:00 and 9:00 p.m. Local craft and food vendors and all sorts of civic organizations set up booths along the street to advertise and sell their products or to publicize their organizations and causes. For the past several years we have participated in First Friday.*
>
> *While manning the booth, I engaged a local pastor in conversation over climate change. The man leads a well-established local congregation and as we entered a conversation he resorted to many of the highly predictable tropes of climate change denial, e.g., "the science isn't settled," "there's no consensus," "it's a liberal political plot," "God granted us dominion over creation," and "the oceans aren't warming." On each point I summoned substantial and substantive reference to either publicly verifiable empirical data or actual reference to Scripture to counter his propositions, at which point he would immediately move on to the next denial trope without any rebuttal to the response I'd given him.*
>
> *But, when we got to the matter of the warming oceans, why they are warming, the role that warming plays in modifying global climate patterns, and the mountain of empirical data that directly and unequivocally links this warming to the rise in atmospheric greenhouse gases resulting from increasing consumption of hydrocarbon fuels, things started down a path into certifiable weirdness.*
>
> *At first, he flatly denied the cause of the atypical rise in ocean temperature as a thoroughly documented unnatural phenomenon with more than 70 years of direct observations correlating it with a commensurate rise in atmospheric CO_2, which is further correlated with increased hydrocarbon fuel consumption—i.e., a verifiable "given 'A,' then 'B,' then 'C'" chain of observation leading to causation. He then demonstrated no understanding of the relationship between climate and weather, citing that it was "hot" when he visited Africa and "it's always been hot in Africa" so there was no truth to what I just told him regarding what we're seeing in the oceans and the linkage to what we're seeing play out in climate. Then, as I had to busy myself with other tasks at the booth, he delivered his final assessment that, contrary to what he just asserted, the oceans are in fact warming after all, but the cause is that Hell is expanding and that is what is heating up the oceans, and that he could factually substantiate that from Scripture.*

Some have asserted that simply being more winsome and compelling in communicating the science is the key to overcoming the climate conundrum in the United States. However, as exemplified in the story from our grassroots champion, this assumes both that science is well understood by the general population and that scientific evidence has an equally weighted value in our decision-making processes. However, that supposition is not correct for many in the Evangelical community.

Instead, reframing climate change and clean energy within an existing moral framework and adding individual action help to engage a much wider audience (Wolsko, Ariceaga, & Seiden, 2016). When solutions are presented as involving multiple sectors and actions, and are framed according to existing values and priorities, many previously disengaged individuals willingly engage and act on climate change. Change depends on engaging existing faith and moral frameworks. These frames must include children's health (unborn and born), the potential harm to future generations, business opportunities, efficient and limited government action, and hope. Solutions must also include practical and meaningful individual engagement.

Sanctity (the sacredness of life) and purity (being morally untainted) remain the top moral values for the conservative community. Therefore, any meaningful communication and education in our community must focus on these two primary concerns. And the best and most common messages for our community, as well as in the Catholic Communion, center on life (Feinberg & Willer, 2012).

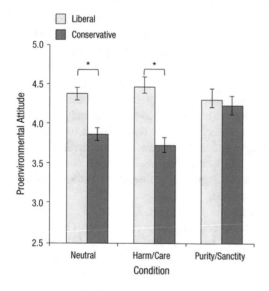

Source: Feinberg and Willer (2012)

Creation Care: It's a Matter of Life

Modern man [sic] has been upsetting the balance of nature and the problem is drastic and urgent. It is not just a matter of aesthetics, nor is the problem only future; the quality of life has already diminished for many modern men. (Schaeffer, 1970)

Most people live lives as a seesaw, trying to balance work, family, values, and faith. Individuals change only when everyday life balance and values are impacted in such a way as to upset that tipping point; only as issues touch the center of our being and who we care most about are changes initiated and actions taken. When issues like climate change arise, it's very easy to dismiss them or deny them because it feels like one more additional stress on our already hectic everyday lives.

Most Americans understand that climate change is real and are concerned about it. But most still see climate change as a faraway threat, in both time and place, and as something that threatens the future of polar bears but not necessarily people. (The Medical Society Consortium on Climate and Health, 2017)

In introducing our theme, "Creation Care: It's A Matter of Life," one must be cognisant that for many US Evangelicals, there is a psychological distance between the impacts of climate change and their everyday lives. And recognizing this reality helps to explain why previous education attempts have failed and have allowed denial to exacerbate climate change inaction, as described by Naomi Oreskes and Erik M. Convay's book, *Merchants of Doubt*.

The Evangelical Climate Initiative (2006), The Cape Town Commitment (2010) by the Lausanne Movement (founded by Billy Graham & John Stott), and the *Encyclical Letter, Laudato Si' of the Holy Father Francis on Care for Our Common Home* (2015) all make persuasive biblical and moral arguments for addressing climate change. These documents remain central to a biblical understanding of creation care. However, all three documents most strongly highlight climate impacts in the Majority World. None would disagree that most impacts are already occurring in the developing world, and these statements address what should be a biblical priority; however, for conservative Evangelical Christians, it has provided an easy way to disassociate themselves from the threat. The perception of distance with respect to climate impacts allows climate change to be disregarded as irrelevant to one's family or to one's immediate concerns.

Polling yields conflicting results on the impacts of *Laudato Si'*. The higher impact end suggests increases in climate concern among American Catholics of 15% and perhaps as much as 13% among American Evangelicals (Mills, Rabe, & Borick, 2015). However, other polling indicates little or no impact of the encyclical on US Catholic and Evangelical thought. Perhaps the most telling survey came from the Center for Applied Research in the Apostolate at Georgetown University, Washington, DC, which discovered that 11 months after *Laudato Si'* was published, only 32% of Catholics and 28% of Evangelicals had ever heard of or read the encyclical (CARA, 2016).

What matters is not the movement or the polling but how climate change threats and opportunities are made real to Evangelicals and others in the conservative faith community. And for that we must return to the message of life.

The Evangelical Environmental Network understands that a consistent, holistic prolife position requires care and concern for all life, and this perspective reflects others in our community as well. As *Focus on The Family* put it recently, "pro-life is not a political statement, it's a way of life" (The Dignity of Life, n.d.). In a National Association of Evangelicals (NAE) statement on end-of-life concerns, the NAE states, "we are pro-life from womb to the tomb" (National Association of Evangelicals, 2014). This whole-life view expresses the best of our theological tradition, and it certainly reflects Catholic social teaching as well. The unborn child is very important to us, but so is each child of God at every stage of life.

If we believe Jesus's words in John 10:10—"I have come that they may have life, and have it to the full"—or, as other translations phrase it, "abundant life" or "life abundantly"—how can Christians not be concerned about quality of life? Who would not pray for our children or grandchildren to have a high quality of life? An abundant life includes health, faith, family, and a fulfilling career in serving God through loving our neighbors. Without doubt, prolife should be about not ending pregnancy, but it must also be about assuring the right to have the opportunity for abundant life. We must have a whole-life theology that cares for the unborn and born alike.

As prolife Evangelicals, we want children to be born healthy, unhindered by the ravages of pollution even before they take their first breath. The medical community has long known the environmental impacts on our unborn children. The once-held belief that a pregnant mother gives her developing child complete chemical protection is untrue. One of the body's protective shields against brain damage, called "the blood–brain barrier," is not fully developed until after the first 3 years of life. Thus, in the unborn child, toxins can cross this incomplete barrier and accumulate in the brain, causing developmental disabilities and brain damage, resulting in lowered intelligence and learning problems. One study found that "the resulting loss of intelligence causes diminished economic productivity that persists over the entire lifetime of these children" (Trasande, Landrigan, & Schechter, 2005). In economic terms, the poisoning of our unborn children's brains costs between US$60 billion and US$106 billion in the United States every year (Trasande & Liu, 2011).

Besides the critical neurodevelopment that occurs before birth, the other major reason babies in the womb are so vulnerable is bioaccumulation. Chemicals readily pass from the mother through the placenta. Unfortunately, a developing fetus, unlike the mother, cannot eliminate toxins through normal biological processes. In fact, one method by which a pregnant woman's body removes chemicals is by passing them into her uterus.

Recent studies have shown that smog, air toxins, and volatile organic compounds (VOCs) like hydrocarbons, benzene, and formaldehyde have a disproportionate impact upon life in the womb. Stacy et al. at the University Of Pittsburgh found evidence that low birth weight in babies was associated with proximity to unconventional natural gas wells in Butler County, PA (Stacy et al., 2015), and McKenzie et al. at the Colorado School of Public Health published peer-reviewed medical findings linking birth defects to methane production (McKenzie et al., 2014).

The American Lung Association's *State of the Air 2019* report stated that more than four in 10 Americans, approximately 43.3% of the population, live in counties that have monitored unhealthy ozone and/or particle pollution, and 141 million people live in counties with unhealthy levels of pollution. The latest childhood epidemics with strong links to petrochemicals and fossil fuel energy—that is, asthma, autism, attention deficit–hyperactivity disorder (ADHD), and allergies—impact as many as one in three children in the United States (Bock & Stauth, 2007). Dr. Philippe Grandjean states, "We are facing massive prevalence of brain dysfunction, autism, and many other signs of ill health due to development insults. Because the exposures to toxic chemicals happen worldwide, the adverse effects are appearing now as a silent pandemic" (Grandjean, 2013).

In the 1960s, a woman living in the United States had a 1:20 chance of developing breast cancer in her lifetime; now the frequency is 1:8. Unfortunately, the breast cancer news is not good. According to Dr. Philip Rosenberg at the National Cancer Institute, breast cancer rates are expected to increase by 30% by 2030 (Rosenberg, Barker, & Anderson, 2015). The number of women diagnosed with breast cancer will soar (in the United States) from 283,000 in 2011 to 441,000 in 2030. The one glimmer of light in the research shows a slightly lower mortality rate due to more effective treatment.

While the modern medical field understands a great deal about breast cancer, much is still unknown. Doctors know that breast cancer rates are higher in the developed world than in the Majority World. They also know that only 30% of women with breast cancer have known risk factors such as genetics, late menopause, or having children later in life. The causes of 70% of breast cancer diagnoses are, as yet, unclear. Nevertheless, a growing body of research points to the environment, and especially suspect are chemicals and plastics that act like hormones in the human body. A large body of plastics such as bisphenol-A (BPA), high-density polyethylene (HDPE) and a host of other resins used in plastic bags and packaging (including bottles labeled BPA free), other plastics, and common fertilizers and pesticides all mimic estrogen (Yang, Yaniger, Jordan, Klein, & Bittner, 2011). These same chemicals have also been linked to potential male reproductive issues, including low sperm counts, malformed genitalia, and an increased frequency of undescended testicles (Jeng, 2014).

We are poisoning ourselves and our children, but, unfortunately, many in our pews have not yet made the connection. Until our communities understand and identify the problems using the language of their existing values, our communities have no way to internalize, accept, and act on the threats facing our children.

In a world where our children are poisoned by pollution, climate change is adding insult to injury. With rising temperatures, smog will get worse—impacting those with asthma and making life more difficult. Dengue fever (also known as bonebreak fever because of the pain it causes)—a mosquito-borne disease never native to the United States—is now present in Florida, Texas, and Hawaii, at least in part because of our changing climate and warming temperatures. And in Pennsylvania, my home state, Lyme disease has become almost epidemic as a result of earlier springs and later autumns. Just recently, the Centers for Disease Control stated that the number of counties with high rates of Lyme disease—including counties in Pennsylvania, New York, Maryland, Connecticut, and Massachusetts—increased by more than 320%.

Lyme disease and other health outcomes offer the best opportunity for awareness and engagement in our Evangelical community. One could add related outcomes from extreme weather or any number of health catalysts. The critical takeaway is this: for the threat of climate change to be accepted, it must become real and personal. As numerous studies and simple human nature testify, until people recognize local impacts, the reality of crises in other regions or countries is easily discounted.

The Bible as the Source of Moral Value

If prolife values and children's health are the emotional key to opening the door for addressing creation care and climate, the Bible remains the door itself. Throughout the history of Protestantism, and especially in its Evangelical wing, the Bible has always been our primary rule of faith. Sola Scriptura remains our central theological statement. All Christians have open access to the Bible. While doctrine, tradition, experience, and scholarship play important roles depending on individual denominations or congregations, sharing Scripture's concern for creation care is paramount. Although it must be admitted that biblical interpretations do cause a fair amount of chaos, the Bible remains the basis for life in Christ. And, as such, scriptural biblical knowledge among US Evangelicals remains woefully lacking.

Some time ago, I led a community men's morning bible study in a conservative area of the United States. The teaching was based on the following:

> For in him all things were created: things in heaven and on earth, visible and invisible, whether thrones or powers or rulers or authorities; all things have been created through him and for him. He is before all things, and in him all things hold together. And he is the head of the body, the church; he is the beginning and the firstborn from among the dead, so that in everything he might have the supremacy. For God was pleased to have all his fullness dwell in him, and through him to reconcile to himself all things, whether things on earth or things in heaven, by making peace through his blood, shed on the cross. (Colossians 1:16–20 (New International Version))

The key verse is "For in him all things were created: things in heaven and on earth, visible and invisible, whether thrones or powers or rulers or authorities; all things have been created through him and for him." This Scripture tells Christians that we are not the owners of the Earth. The Earth, God's creation, was formed by and for God—by and for Jesus. Unfortunately, too many Christians, especially Evangelicals, do not understand the imperative to "tend the garden."

As a case in point, immediately after sharing my thoughts and during a time set aside for reflection and discussion, one gentleman said, "I've read the Bible all my life, and I never saw this Scripture in the light of caring for the Earth."

Humanity has been given a precious gift, a planet that can provide for all our needs if we only follow God and use it wisely. Just as we are called to love our neighbor, not subjugate him or her, the same applies to creation. Never does the Bible support the Earth being trashed or misused. Genesis states just the opposite. The Earth supplies the necessities for biological life; God designed creation for exactly this purpose. God created the garden and was the first gardener. For life to prosper, humans are to empower the garden to flourish. We have been clearly given the responsibility, being created in God's image, to reflect his image—God's presence—by caring for creation.

The sad reality is that our stewardship reflects our relationship with God. Upon a close reading of Genesis 3, we understand that original sin was the temptation to be god-like, to be in control. When we look back at human history, our principal failing always seems to be the desire to be in charge, combined with the inability to live within God-given limits. The Genesis account describes a universal order with God as the loving and very good creator, humans cast in his image as partners in maintaining creation, and all creation living in a sustainable relationship.

Our desire to be in control, however, breaks the order, attempts to bypass the limits, and injures our relationship with God, leading to a broken and unsustainable world. Each time we use more than we need or consume greater than our share, we perpetuate our brokenness, support our vanity, and continue disregarding God's limits. This distorts the creation and impacts all.

Throughout the Old Testament, God defines and provides deliberate instructions for tending the Earth. Although most Christians have not made the connection, the Bible provides definitive mandates to live in a reciprocal relationship with the nonhuman creation. In Deuteronomy, Numbers, and Leviticus, God gives clear instruction for Sabbath rest for the land, indications of crop rotation, and animal husbandry. There are strict ordinances regarding farming, livestock management, and land use in general. These conditions define the parameters for living in relationship with God, people, and the Earth in an integrated approach to abun

> If you follow my decrees and are careful to obey my commands, I will send you rain in its season, and the ground will yield its crops and the trees their fruit. (Leviticus 26:3–4 (New International Version))

Overcoming our sin, our failure, requires a renewed study of Scripture and the resources to help both clergy and laity grasp the critical importance of creation care as an act of discipleship in following our Risen Lord. A renewed understanding of the Biblical mandate for creation care is central to our relationship with God, each other, and all creation. As John Stott—one of the great Evangelical leaders of the twentieth and early twenty-first centuries—wrote in his last book, *The Radical Disciple*, creation care is one of the "neglected aspects of our calling" (Stott, 2010).

Hope

May the God of hope fill you with all joy and peace as you trust in him, so that you may overflow with hope by the power of the Holy Spirit. (Romans 15:13 (New International Version))

Hope should be the paramount aspiration for any Christian. We believe in a God who healed the sick, who proclaimed that the greatest commandment was to love, and who overcame death to restore humanity and all creation. However, hope—while solidly based in faith—must have tangible and physical realities. One need only consider the well-known Biblical passage of "Doubting Thomas" as an example.

Compounding our inaction is the doom so readily served up by many in the climate space. This is not to minimize climate threats or impacts, but gloom instills doom, and combining the two in the human heart often leads to paralyzing fear. Climate impacts are already bad, but we must not make them worse by spreading apathy and cynicism. Strides have already been made in addressing climate, but to mobilize the world, a real prescription of hope is required (Ojala, 2012).

Climate change is the greatest moral challenge of our generation, as each of God's children worldwide is impacted. However, properly addressing climate solutions provides the greatest opportunity for hope. As we address the threats posed by a changing climate, the potential exists to turn energy poverty into energy prosperity and to replace resource scarcity with sustainable economies.

Clean energy is the foundation for a sustainable world, and that energy transition is well underway. Remarkably, the pace and scale are greater than experts have forecasted. According to Michael Liebreich at Bloomberg New Energy Finance (BNEF), from 2016 to 2017 the United States saw wind installations increase by 262%; solar installations by utilities increase by 4645%; solar installations by homeowners and businesses increase by 143%; utility power purchase agreements for solar and wind increase by 83% and 71%, respectively; and investment in energy efficiency grow by 100%. Moreover, households paid 20% less for electricity and natural gas. Climate pollution in the United States is down 23%, and our 2025 Paris Climate Agreement Goals have been halfway reached (https://about.bnef.com/summit/event/new-york/new-york-highlights/).

While not wishing to oversell technological advances, the decline of fossil fuels is happening quicker than many anticipated. In a recent meeting with a family-owned fossil fuel company, the chairperson stated, "Our family has benefited economically for three generations in the petroleum business. However, our business will not survive into a fourth generation, and we are doing everything possible to divest and reorganize our family's holdings."

A close second to renewable energy in the work toward a sustainable, abundant future is sustainable and nutritious food. As *National Geographic* reported in 2017, the Netherlands is already meeting the sustainable food challenge (http://www.nationalgeographic.com/magazine/2017/09/holland-agriculture-sustainable-farming/). It is the world's second largest "exporter of food as measured by value, second only to the United States, which has 270 times its landmass," and has reduced its water needs for growth of key crops by up to 90%, eliminated pesticide use almost entirely, and reduced antibiotics in poultry and livestock by up to 60%.

The above are just a few examples of the hope that is possible as we continue to decouple fossil fuels from our economy and build a sustainable world that offers abundant life for all creation. Hope alone does not change our future, but without hope we will never rise to overcome the challenges. In our book, *Caring For Creation: The Evangelical's Guide to Climate Change*, my coauthor, Paul Douglas, writes, "We are not hopeless and we are not helpless."

Conclusion

Securing a total commitment to solving the global climate crisis requires the engagement of the United States. A primary reason for the lack of leadership to date has been the inability to motivate Evangelicals. As Evangelicals remain the largest single political force in conservative American political life, it is paramount that a substantial portion of our community become climate champions.

While recent polling suggests that over 60% of Evangelicals understand climate science, the failure to properly communicate climate change within the community's existing value sets keeps climate action from becoming a priority. This is not a surprise. Very few systemic changes have ever occurred without a personally perceived threat to one's values or way of life. US engagement in World War II would likely have never occurred had the naval base at Pearl Harbor never been attacked.

Facing the fear of losing one's way of life (i.e., a fossil fuel–based economy), coupled with conservative philosophical concern regarding big government overreach, makes it easy to understand how denial and confusion campaigns have worked so extremely well in the Evangelical community.

However, the Evangelical *Titanic* is turning. By using values consistent with our community and by helping Evangelicals understand climate impacts more personally, we are opening hearts to realize the danger, to understand the clear biblical message, and to hope for a better future for our children.

References

Bock, K., & Stauth, C. (2007). *Healing the new childhood epidemics: Autism, ADHD, asthma, and allergies: The groundbreaking program for the 4-A disorders*. New York, NY: Random House.
CARA Catholic Poll (CCP) 2016: Attitudes about climate change. (2016, May). Washington, DC: Center for Applied Research in the Apostolate. Retrieved February 13, 2020 from http://cara.georgetown.edu/climate%20summary.pdf
Cox, D., & Jones, R. P. (2017). *America's changing religious identity*. Washington, DC: Public Religion Research Institute.

Feinberg, M., & Willer, R. (2012). The moral roots of environmental attitudes. *Psychological Science, 24*, 56–62.

Grandjean, P. (2013). *Only once chance, how environmental pollution impairs brain development & how to protect the brains of the next generation* (p. xiii). New York, NY: Oxford University Press.

Haidt, J. (2012). *The righteous mind: Why good people are divided by politics and religion.* New York: Pantheon Books.

Jeng, H. A. (2014). Exposure to endocrine disrupting chemicals and male reproductive health. *Frontiers in Public Health, 2*, 55. https://doi.org/10.3389/fpubh.2014.00055

McKenzie, L. M., Guo, R., Witter, R. Z., Savitz, D. A., Newman, L. S., & John Adgate, L. (2014). Birth outcomes and maternal residential proximity to natural gas development in rural Colorado. *Environmental Health Perspectives, 122*, 412–417. https://doi.org/10.1289/ehp.1306722

Mills, S. B., Rabe, B. G., & Borick, C. (2015). *Acceptance of global warming rising for Americans of all religious beliefs.* Ann Arbor, MI: The Center for Local, State, and Urban Policy at the Gerald R. Ford School of Public Policy, University of Michigan. Retrieved April 18, 2020 from http://closup.umich.edu/files/ieep-nsee-2015-fall-religion.pdf

National Association of Evangelicals (2014). Allowing natural death. *Resolution.* Retrieved February 13, 2020 from http://nae.net/611/

Ojala, M. (2012). Hope and climate change: The importance of hope for environmental engagement among young people. *Environmental Education Research, 18*, 625–642.

Rosenberg, P. S., Barker, K. A., & Anderson, W. F. (2015). *Estrogen receptor status and the future burden of invasive and in-situ breast cancers in the United States, presentation.* Philadelphia, PA: American Association for Cancer Research. Retrieved February 13, 2020 from http://mb.cision.com/Public/3069/9755232/81b414b4ec298479.pdf

Schaeffer, F. (1970). *Pollution and the death of man.* Wheaton, IL: Tyndale House. (Reprinted by Crossway, Wheaton, IL, 2011, p. 21).

Stacy, S. L., Brink, L. L., Larkin, J. C., Sadovsky, Y., Goldstein, B. D., Pitt, B. R., et al. (2015). Perinatal outcomes and unconventional natural gas operations in Southwest Pennsylvania. *PLoS One, 10*(6), e0126425. https://doi.org/10.1371/journal.pone.0126425

Stott, J. (2010). *The radical disciple.* Downers Grove, IL: InterVarsity Press.

The Dignity of Life (n.d.). Focus of Family Video. Retrieved February 13, 2020 from https://www.youtube.com/watch?v=Y63Ksd8yHa4

The Medical Society Consortium on Climate and Health (2017). *MEDICAL ALERT! Climate change is harming our health.* The Medical Society Consortium on Climate and Health, (American Academy of Asthma, Allergy, Immunology, American Academy of Family Physicians, American Academy of Pediatrics, American Congress of Obstetricians and Gynecologists, American College of Physicians, American College of Preventive Medicine, American Podiatric Medical Association, National Medical Association Society of General Internal Medicine).

Trasande, L., Landrigan, P. J., & Schechter, C. (2005). Public health and economic consequences of methyl mercury toxicity to the developing brain. *Environmental Health Perspectives, 113*, 590–596.

Trasande, L., & Liu, Y. (2011). Reducing the staggering costs of environmental disease in children, estimated at $76.6 billion in 2008. *Health Affairs, 30*, 863–870. https://doi.org/10.1377/hlthaff.2010.1239

Wolsko, C., Ariceaga, H., & Seiden, J. (2016). Red, white, and blue enough to be green: Effects of moral framing on climate change attitudes and conservation behaviors. *Journal of Experimental Social Psychology, 65*, 7–19. https://doi.org/10.1016/j.jesp.2016.02.005

Yang, C. Z., Yaniger, S. I., Jordan, V. C., Klein, D. J., & Bittner, G. D. (2011). Most plastic products release estrogenic chemicals: A potential health problem that can be solved. *Environmental Health Perspectives, 119*, 989–996. Retrieved February 13, 2020 from http://www.ncbi.nlm.nih.gov/pmc/articles/PMC3222987/pdf/ehp.1003220.pdf

CHAPTER 22
Call to Action from Faith Leaders

Alastair Redfern

Summary This chapter considers the responsibilities of faith leaders in pursuing the agenda of the health of people and the health of the planet—not least because of the particular opportunities and contributions available to faith communities at this significant moment in our global history. Consideration is given to two key areas: first, the context in which faith leaders are challenged to contribute to this crucial agenda, and second, some possible responses, challenges and perspectives that faith leaders might offer.

The Context of Faith

Other chapters in this book highlight the amazing variety of expert insights into climate change, health, and their interrelationship. Many contributions in this area of concern display the marks of what might be called missionary fervour. It could be that climate change and health professionals are being called to be the evangelists of our times. Further, we can discern two contrasting models of such a mission. One calls for small-scale, local, measurable engagement to offer demonstrable signs to encourage and influence others: a micro-plus approach. In my terms, this is a model often associated with a Protestant understanding of the church, small local groups being the key areas for action.

By contrast, other contributors have emphasized "structural sin" and call for systemic universal action—a more "Catholic" approach to the making of the church, seeking complete fulfilment while acknowledging the reality and inevitability of some interpretations and differences of pace and enthusiasm: a macro-minus approach.

A. Redfern (✉)
Bishop Alastair Redfern, Church of England, London, UK

© The Author(s) 2020
W. K. Al-Delaimy, V. Ramanathan, M. Sánchez Sorondo (eds.), *Health of People, Health of Planet and Our Responsibility*, https://doi.org/10.1007/978-3-030-31125-4_22

Of course, as with the mission of Christian faith more generally, both approaches are valuable, and an important endeavour is to discern and develop the right balance between local, personal engagement and the broadest vision and ideals. Citizens need a kingdom, and a kingdom needs citizens in their local contextuality.

A second feature of the context that should be carefully considered revolves around indications that the Enlightenment liberal project could be unravelling. Freedom for and within society has become translated into freedom for the individual, thus creating huge issues about the possibility of cohesion and connectivity. We retreat into self-reinforcing "echo chambers" of our own choosing. Pope Francis describes this phenomenon as "the globalization of indifference", a "liberal" toleration of others subtly moving into a way of living that does not even recognize the "other", let alone feel any responsibility towards them.

"Freedom" now seems to deliver an expectation of so many different, and often apparently contradictory, possibilities and therefore has tended to become focused upon power being manifested through education and progressiveness. Power becomes knowledge and resides with liberal elites, whose rhetoric about freedom and rights masks increasing poverty and exclusion. This is a challenging political context—an ever-sharpening reality within which responses to health and climate change need to be crafted.

There is a contemporary stream of critique that accuses politicians of not knowing enough, of being too easily swayed by wealthy vested interests. Of course, there is some truth in these accusations, but from my own experience in the English Parliament, I want to be clear that politics is a vocation for some, and there are thus important allies within the structures of policy making who need to be sought out and encouraged.

Another element of the context into which we seek to bring change is an increasing concern for what is sometimes called the power of populism. Nietzsche was clear that popular agitations, beginning in his time in the nineteenth century with the development of mass media, created "moments" of public concern and connectivity, which were in fact "headless"—that is, with no tradition or detailed vision for the future (Hansard, 2017). Such "moments" or emotional expressions were dangerous because of the way they fuelled incoherence. The task in such times is to turn "moments" into movements, with a sense of tradition/continuity and a coherent vision for future well-being. Such wisdom will be very important for the future of our own aspirations. Faith aims to convert moments into movements, which is an urgent and vital contribution needed to counter increasing confusion and division.

Two other factors are important in helping us understand our context in a way that will enable a more effective contribution. In my work on modern slavery it is clear that the ever-expanding trade in children for sexual exploitation is a sign of what one American law enforcement officer described as a world that now has no "moral stops". Until recently, the liberal agenda of free choice was checked by an instinct to preserve and protect children. There is much evidence that even this "moral stop" is disappearing fast—which creates a challenging environment for the common discourse in this area of global concern for a "moral case" to be made for radical change in relation to climate change and health care.

Finally, in terms of the context into which we are aiming to reach, there are strong signs of what Hannah Arendt termed "the banality of evil". She used this phrase in relation to Eichman and the Holocaust—pointing to the sense in which, despite the enormously evil outcomes, he and his colleagues could feel that they were simply doing their jobs, cogs in a system for which others were ultimately responsible (Arendt, 1963). This mentality could be part of what the Holy Father has called the globalization of indifference.

The Contribution of Faith

Given such a complex and challenging context, what might be the contributions of faith and of faith leaders? There are a number of key contributions.

At the most fundamental level, faith calls all people to recognize the common ground upon which we all exist. A strong theme in our concern for the health of people and the health of the planet is the issue of air pollution. In the Bible, life begins by God breathing into a human form. The "breath" of God is the mystery that gives life to every creature. It is universal and inclusive—in faith terms, "Catholic". And such a reality reminds every human creature that to live is to be dependent—on God, on creation, on others. This is an important reality for measuring our own efforts and confidence.

Further, faith leaders have a great ability to convene people to work together around significant issues. In the work I do with Bishop Marcelo Sánchez Sorondo in fighting the evil crime of modern slavery, through the Global Sustainability Network, we can gather people from different faiths or none, and from significant sectors such as government, business, media, academia and nongovernmental organizations (NGOs). At these gatherings, Bishop Sánchez Sorondo sometimes reminds participants that there is something in every human heart that can recognize and respond to the values of the Beatitudes—the importance of being pure, meek, humble and thirsty to offer justice, and the need to reach out to any who are poor, suffering or in mourning. These values are the basis of the golden rule to do to others as you would have them do to you.

Faith leaders can convene people around such values because we are generally credited with being on the side of goodness but lacking a particular expertise or agenda in terms of detailed responses to the challenges facing society. Thus, faith leaders can invite others to make their various and necessary contributions in a space of mutual commitment to goodness and mutual flourishing. This is a unique kind of convening power.

This potential is recognized in the chapter by Professor Dasgupta, calling our attention to recognize the importance of what is "within us". This is the deep desire for goodness that embraces others as well as ourselves and yet needs the wisdom and expertise of numerous disciplines in order to proceed effectively.

Another important contribution of faith is the primary focus on "salvation". This word comes from *salvus*, which means health. Salvation and health are the same thing—for creation, for creatures, for our endeavours in this crucial and urgent project of preserving the planet. There is an interplay between health being properly understood and pursued and our care for the planet and its inhabitants, which are two sides of the same coin.

Sin, which is both structural and personal, is recognition that human efforts often miss the mark and fall short because of our selfishness and short-sightedness. Sin requires judgment and a robust response, not least in terms of tackling corporate corruption and individual greed. But the Christian understanding of sin also involves an invitation to recognize the miracle of forgiveness and the gift of new life. This mystery needs to inform our relationship with others, especially those who might seem to be perpetrators of the trends we so urgently need to challenge regarding global warming and approaches to health care. In a sinful world, all parties need to know our need for God's grace through forgiveness and new life. Profound humility and self-criticalness must inform a faith approach. In his chapter, Governor Brown eloquently captures this deep truth in his powerful call for "transformation".

The Journey of Faith

A key New Testament word used to describe the ministry and mission of Jesus is *hodos* (the way). Faith calls us in a direction—a journey during which we will continue to learn new things, and within which none of us is immune from making mistakes. This should provide the comfort that no one set of tools or answers will suffice. Often, new insight or energy can come from unexpected sources. Thus, faith can provide a context that is both creative and critical, a common ground inhabited with humility, openness to others and graciousness towards fellow sinners who may hold different visions and values. In this way, faith groups are often more open to diversity and debate than "purist" organizations offering "perfect" solutions. The latter approach is important in offering a prophetic element but always needs embedding in the complex and sometimes seemingly contradictory agendas so prevalent amongst God's diverse children.

Moreover, the Gospels make clear that the prism through which faith pursues its calling is not one of perfection (as can be the dangers of an Enlightenment project that has become reduced to the power of knowledge). Rather, Christian faith is a call to see and act through and with the poorest: those suffering, threatened and excluded. Faith is a power that unfolds through the unpowerful. Worship is an enactment of this reality about human dependency and imperfection called into greater wholeness or health. Jesus is clear that wisdom resides with the unlearned, not the worldly wise.

In this way, very much reinforcing the emphasis of climate change concern upon the link between air pollution and health, faith invites us into the narrative of the breath. Life begins with breathing: the life of God breathed into the creature to give life within creation. Breath gives life and needs to be pure and holy/healthy. But, for Christians, when on the cross, Jesus, who represents every creature, gives up His breath—a process that will happen to each of us and to every human being. Life, inspired or breathed into by God, will end in this mortal setting. The miracle and mystery is that God breathes new life into the mortality of human life: resurrection. And on the day of Pentecost the breath of the Holy (healthy or whole) Spirit is breathed into the disciples to enliven the seeds of eternity in souls touched by this faith in the resurrection of the body.

This miracle is not an excuse to give up on our challenges and leave everything to God; rather, faith recognizes this mystery of the breath of resurrection as a power to sustain and nourish human life through adversity, failure and challenge. Thus, faith invites engagement with all the complexities of the penultimacy of our endeavours—an arena needing hope and love to sustain and encourage, even when the direction seems less than perfectly clear and plans struggle to be adequately comprehensive.

In this way, faith can help to create an atmosphere that enables ideals and the frustrating realities of limitation to exist creatively together in hope as much as in demonstrable "answers", bound together in the greater purposes of a love that can flow through us and between us, connecting more closely with the fuller purposes of creation in its proper healthiness (or holiness/wholeness).

Of most obvious significance regarding the role of faith is the fact that all believers are citizens and consumers. The potential "power" of mobilizing believers and faith communities to practise and support appropriate action for climate change and proper approaches to health is enormous. The congruence between the teaching of faith about salvation/health in this world as stewardship and preparation for a greater wholeness (health), with the values and visions so powerfully expressed in this book, should provide a proper challenge to faith leaders to pursue these possibilities. The fact of a common ground (breath) for all creatures and the convening power of faith to call different people together in the service of goodness, reliant on the expertise of others, need to be replicated in local contexts too. Faith leaders can initiate raising awareness of key issues related to climate change and health, and encourage appropriate action as witnesses to the Gospel of salvation.

This strategy does not need to wait until appropriate "answers" have been fully formed and agreed on; it can proceed while the "movement" we are trying to form continues "on the way".

The Paris Agreement and the Sustainable Development Goals (SDGs) of the United Nations provide important and internationally agreed frameworks. Faith can give energy, perspective, commitment and essential humility for the journey—a hope that will set out in the right direction ahead of "conclusive" evidence, modelling appropriate lifestyles and embracing the wisdom or perspective of the "unlearned" by human standards. Perhaps most significantly, faith empowers its adherents to give themselves for the good of others, sacrificially, at cost to the self—for the health (salvation) of all.

This is the Catholic doctrine for a deeply connected creation, inviting recognition of and participation in the purposes of the Creator and the fulfilment of creation. Moreover, to act in faith is to trust that just a small contribution can play a key part; the fraction at the Eucharist reminds us that broken bits can be joined to give health to creation.

Given the urgency of the issues highlighted in this book and a deep recognition of the interconnectivity of so many factors that together can make or break our health on this planet, faith leaders have a huge responsibility to practise what has been preached for 2000 years by Christian churches.

The Role of Faith Leaders

In response to this context, and the contribution and journey that faith might offer, there are a number of key roles that faith leaders need to consider in terms of our contribution to the health of people and the health of the planet.

First is the importance of authoritative guidance. Many people do not believe that climate change is an urgent matter, and evidence shows that it has a very low priority for voters. Those in positions of leadership need to provide appropriate guidance and challenge—always substantiated with credible evidence and clearly connected to the teaching and values of the faith.

Second, places of worship and congregations can play a key role in the formation of community life, providing sophisticated networks for mobilizing views and values, and possessing the ability to organize potential action at both the policy and grass-roots levels. Careful consideration needs to be given to how this resource can be utilized in a way that does not seem merely idealistic, while inviting people to make appropriate and manageable responses. Important areas include the transition to a low-carbon economy and recognition of the inalienable link between climate change and human health.

Current research shows that there are key groups who are relatively less engaged with the matter of climate change. These include older citizens, parents, right-wing voters and the less scientifically educated. Any response of faith leaders and groups needs to be particularly targeted in terms of work with these sectors.

Finally, in terms of targets, any strategy must highlight the reality of perceived social norms, the need for creating higher political priorities in this area and the skill to present opportunities rather than burdens to those who might respond. Faith leaders and communities can also have considerable influence in the world of industry/business, particularly in convening spaces for conversation and a commitment to progress that is realistic about pace and possible outcomes. There is also huge potential for faith values to be translated into citizen action through voting and participating in political debate. Faith leaders could be more proactive in convening spaces for discussion and debate.

Faith creates "households of commitment", domestically and institutionally. The ultimate human household is Planet Earth. Our values, commitments and care need to be fully expressed for the best ordering and stewardship of this place of residence. It is a particularly human responsibility and one that faith leaders need to help present as a divine mandate too. We are stewards of the environment into which life is born, not owners or proprietors. Humility, wonder and thankfulness should energize our efforts, not simply calculation and self-interest.

The health of all God's people is directly and deeply related to the health of the planet we are privileged to inhabit. Faith in the future owns this privilege and pledges a responsible stewardship in return.

This challenge and call need leadership to best identify and articulate a response appropriate for each "faith" so that cooperation and companionship can be developed and human flourishing can be joined to a common concern for the well-being of our shared household. Such witness will offer encouragement and an invitation to all people of goodwill and good sense to play their part too—a common human cause for a common human flourishing in a common human household.

This is not an option but a divinely given duty. We each need to take a deep breath.

We have to recognize the responsibility of stewardship of the household we inhabit, join our teaching and our actions in a common work of praising our Creator and cherishing creation, and invite others to discover this call and to join in our endeavour.

References

Hansard Column 385 (2017, January 19). Vol 778. Retrieved January, 16, 2019, from https://hansard.parliament.uk/lords/2017-01-19/debates/D7DB9D5E-0078-4ED2-81D2-106C40E59DA4/PopulismAndNationalism

Arendt, H. (1963). *Eichmann in Jerusalem: A report on the banality of evil* (Volume 165 of Compass books). New York: Viking Press.

Part VII
Overarching Solutions: The Role of Science and Technology

CHAPTER 23
Public Health Co-benefits of Reducing Greenhouse Gas Emissions

Qiyong Liu and Jinghong Gao

Summary The public health co-benefits that curbing climate change would have may make greenhouse gas (GHG) mitigation strategies more attractive and increase their implementation. The primary purpose of this chapter is to review the evidence on GHG mitigation measures and the related health co-benefits; identify potential mechanisms, uncertainties, and knowledge gaps; and provide recommendations to promote further development and implementation of climate change response policies at both national and global levels. Evidence of the effects of GHG abatement measures and related health co-benefits has been observed at regional, national, and global levels, involving both low- and high-income societies. GHG mitigation actions have mainly been taken in five sectors—energy generation, transport, food and agriculture, households, and industry—consistent with the main sources of GHG emissions. GHGs and air pollutants to a large extent stem from the same sources and are inseparable in terms of their atmospheric evolution and effects on ecosystems; thus, reductions in GHG emissions are usually, although not always, estimated to have cost-effective co-benefits for public health. Some integrated mitigation strategies involving multiple sectors, which tend to create greater health benefits, have also been investigated, and this chapter discusses the pros and cons of different mitigation measures, issues with existing knowledge, priorities for

Q. Liu (✉)
State Key Laboratory of Infectious Disease Prevention and Control, Collaborative Innovation Center for Diagnosis and Treatment of Infectious Diseases, National Institute for Communicable Disease Control and Prevention, Chinese Center for Disease Control and Prevention, Beijing, China

Shandong University Climate Change and Health Center, School of Public Health, Shandong University, Jinan, Shandong, China
e-mail: liuqiyong@icdc.cn

J. Gao
The First Affiliated Hospital of Zhengzhou University, Zhengzhou, Henan, China

National Engineering Laboratory for Internet Medical Systems and Applications, Zhengzhou, Henan, China
e-mail: fccgaojh@zzu.edu.cn

© The Author(s) 2020
W. K. Al-Delaimy, V. Ramanathan, M. Sánchez Sorondo (eds.), *Health of People, Health of Planet and Our Responsibility*, https://doi.org/10.1007/978-3-030-31125-4_23

research, and policy implications. Findings from this study can play a role not only in motivating large GHG emitters to make decisive changes in GHG emissions, but also in facilitating cooperation at international, national, and regional levels to promote GHG mitigation policies that protect public health from climate change and air pollution simultaneously.

Climate Change

There is robust evidence that climate change is occurring and that anthropogenic greenhouse gas (GHG) emissions, primarily from human activity–related burning of fossil fuels, are the main drivers (Stocker et al., 2013). According to the Fifth Assessment Report (AR5) from the Intergovernmental Panel on Climate Change (IPCC), the evidence of climate change is unequivocal. During the period 1880–2012, there was a warming of 0.85 °C in the global average surface temperature (Pachauri et al., 2014a; Stocker et al., 2013). Without further mitigation actions, the average temperature may rise by 2.6–4.8 °C by the end of this century (Watts et al., 2015). In addition, it has been suggested that even if CO_2 emissions abruptly ceased, climate change would continue for hundreds of years because of the inertia in the global climate system (Solomon et al., 2009).

Anthropogenic GHG emissions—primarily from human activity–related energy generation, transport, food and agriculture, household, and industrial processes—are considered the main driver of climate change (Pachauri et al., 2014a; Stocker et al., 2013). In order to hold the increase in the global average temperature to less than 2 °C relative to preindustrial levels to avoid the risk of potentially catastrophic climate change impacts, it was reported that total anthropogenic CO_2 emissions needed to be kept below 2900 billion tonnes (Gt) by the end of this century (Watts et al., 2015; Whitmee et al., 2015). However, in the years 2003–2011, an average global annual emissions growth rate of 3% per year was observed, whereas the growth figure over the 1980–2002 period was 1.2% annually (Netherlands Environmental Assessment Agency, 2007). In 2014, emissions from the combustion of fossil fuels and industrial processes totaled 35.7 Gt of CO_2, with current trends expected to exceed the required emissions target over the next 15–30 years (Watts et al., 2015).

Health Effects of Climate Change

The health of human beings is sensitive to shifts in weather patterns and other changes in climate systems (e.g., temperature, precipitation, and occurrence of extreme weather events) (Smith et al., 2014). To date, converging evidence from different lines of research generally suggests that climate change, both directly and indirectly, has already started to damage human health and is expected to cause increasingly adverse impacts in the future (Field et al., 2014; Pachauri et al., 2014b;

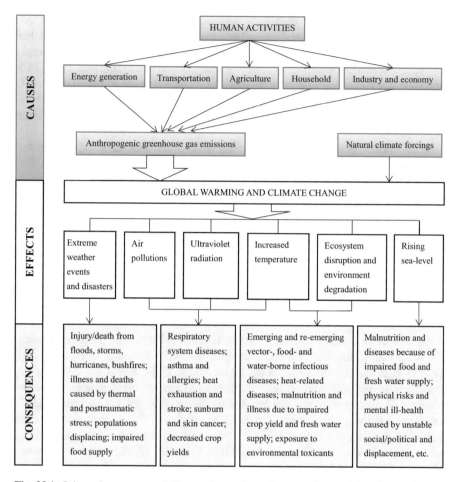

Fig. 23.1 Schematic summary of climate change determinants and potential pathways through which climate change affects human health

Patz et al., 2008; Smith et al., 2014). Climate change can affect public health via various pathways (Fig. 23.1) (Field et al., 2014). Directly, climate change can increase the frequency and intensity of extreme weather events such as heat waves, floods, droughts, and other weather-related natural disasters, which will lead to increased injury, morbidity, and mortality, especially for vulnerable populations (Field et al., 2014; IPCC, 2007; Smith et al., 2014). Indirectly, health impacts may result from climate change–related ecosystem alterations and environmental degradation, as well as the corresponding decline in air quality and impairment of fresh water and food supplies, which, in turn, influence the distribution and incidence of water-, food-, and vector-borne infectious diseases and respiratory system diseases, and can degrade nutritional status (Barros et al., 2014; Field et al., 2014; Patz et al., 2008; Smith et al., 2014). In addition, climate change may play a role in other less direct health outcomes by mediating societal systems. These outcomes include physical risks and mental illness caused by unstable social or political status, violent conflict, and population displacement associated with rising sea levels (Barros et al., 2014; IPCC, 2007; Pachauri et al., 2014b); malnutrition due to impaired crop yields and food insecurity due to ecosystem disruption and rising sea levels (Barros et al., 2014; Pachauri et al., 2014b); and loss of workforce, economy, and health care systems due to extreme weather events (Field et al., 2014; Pachauri et al., 2014a; Smith et al., 2014).

If no further climate change mitigation actions are undertaken, the combined effects of the selected impacts on the global annual gross domestic product (GDP) are expected to rise over time to likely levels of 1.0–3.3% by 2060, with the largest negative economic consequences being suffered by regions in Africa and Asia (Organization for Economic Co-operation and Development (OECD), 2015). Climate change has been described as the biggest global threat confronting public health in the twenty-first century (Costello et al., 2009).

Existing Response and Challenges

Although there are now several policy initiatives, lifestyle and other recommendations, and technology instruments that can help mitigate climate change (Pachauri et al., 2014b; Smith et al., 2014; Watts, 2009; Xia et al., 2015), many countries (especially developing countries) remain reluctant to make decisive changes (Edenhofer et al., 2014a). Of the worldwide efforts to reduce GHG emissions, active participation and decisive actions of developing countries are essential to limit the increase in GHG concentrations and prevent dangerous anthropogenic interference with the climate system (Pachauri et al., 2014a). However, developing nations are concentrating so much on economic development, air quality improvement, and poverty reduction that they have limited economic resources and allocate a low priority on their political agenda to tackling the challenges posed by GHG emissions and climate change (Barros et al., 2014; Field et al., 2014). Additionally, developing countries often insist that the "common but shared responsibility" principle of the 1992 United Nations Framework Convention on Climate Change (UNFCCC) should be applied, meaning they should not have the same obligations to reduce GHG emissions as developed countries until they have achieved certain level of human development (Costa, Rybski, & Kropp, 2011). Low- and middle-income countries point out that they are only retracing the same development path taken in the past by present-day high-income countries. This highlights the worldwide challenge of balancing environmental and public health concerns against economic growth. Thus, the issue of how to balance countries' rights to sustainable development and economic growth—especially the rights of developing nations—with enforceable reductions in GHG emissions may be the key challenge of ambitious global mitigation action (Watts et al., 2015).

The Potential Opportunity: Public Health Co-benefits of Reducing GHG Emissions

According to the IPCC, mitigation strategies not only act to curb the emissions of climate-warming pollutants (mainly GHGs) but also, if well chosen and implemented, deliver substantial simultaneous improvements in public health, independent of the effects on climate change, with most of these impacts being beneficial (Field et al., 2014; Gao et al., 2018a; Pachauri et al., 2014a; Smith et al., 2014). A series of studies published in *The Lancet* has also shown that appropriate climate change mitigation strategies (mainly targeting reductions in GHG emissions) can have additional, independent, and largely beneficial effects on public health (Watts, 2009). For example, actions like reducing fossil fuel combustion and improving energy efficiency, aimed primarily at cutting GHG emissions, can also produce ancillary health benefits from decreased air pollution (West et al., 2013). One of the mechanisms of these so-called health co-benefits of mitigation measures is that GHGs and air pollutants are, to a large extent, emitted from the same sources and are interlinked in terms of their atmospheric behavior and effects on the ecosystem and human beings (Fig. 23.2) (Haines et al., 2009; Lelieveld et al., 2015; Pachauri et al., 2014a; Stocker et al., 2013). Moreover, some air pollutants such as black carbon (BC) and ozone (O_3) are also greenhouse gas pollutants (climate-warming agents) with even higher radiative forcing (RF) per unit than CO_2 (Stocker et al., 2013).

These so-called co-benefits from simultaneously curbing climate change and improving ancillary public health may make GHG mitigation strategies more attractive to developed and developing countries, and may encourage their implementation (Edenhofer et al., 2014a; Field et al., 2014; Shindell et al., 2012). They also bridge the development gap between high- and low-income countries and thus can play an important role in future international negotiations on the climate convention (e.g., the Conference of the Parties) (Haines et al., 2009; Watts, 2009). At the very least, co-benefits can reduce the costs of taking actions against climate change (under certain conditions, the value of health gains may be comparable to or exceed

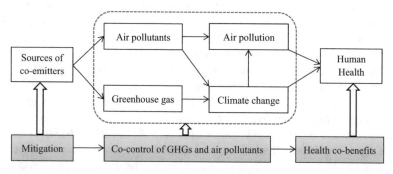

Fig. 23.2 Potential mechanisms and pathways through which reductions in greenhouse gas (GHG) emissions result in public health co-benefits

abatement costs) and therefore can strengthen the case for climate change mitigation policies in the face of scientific uncertainty (Nemet, Holloway, & Meier, 2010).

Thus, with a view to filling some of the knowledge gaps on the topic of GHG emissions and related mitigation measures, the main purposes of this study are to (1) synthesize the current evidence of the public health co-benefits of reducing GHG emissions to improve our understanding of the economic sectors involved in GHG mitigation measures, how and through which pathways reductions in GHG emissions can bring ancillary health benefits (mechanisms), and the relevant uncertainties and knowledge gaps associated with the process of assessing health co-benefits; and (2) discuss the potential policy implications.

Public Health Co-benefits of Measures to Mitigate GHG Emissions

Here, we define public health co-benefits as measures to reduce the emissions of climate-warming pollutants (mainly GHGs), which also hold the potential to simultaneously deliver significant improvements in human health, independent of the effects on climate change. Although there are intersections and overlaps, the associations between reductions in GHG emissions and health co-benefits are based almost entirely on modeling studies and are drawn primarily from five sectors: energy generation and provision, transportation, agriculture and food, households, and industrial and economic processes. According to studies conducted in both high- and low-income countries, GHG mitigation policies in the five domains are often (but not always) estimated to have net co-benefits in terms of public health, and comprehensive measures across various sectors tend to provide greater health benefits. In some cases, the positive health consequences seem to be substantial, cost effective, and attractive to multiple parties (Haines et al., 2009).

The health co-benefits of mitigation measures appear in a number of forms, depending on the sources of the GHG emissions being reduced. The main health co-benefits of reducing GHG emissions in the energy generation sector are linked to the corresponding reductions in common air pollutants (Fig. 23.2), and the most significant pollutants impacted are particulate matter (PM), black carbon (BC), SO_2, and nitrogen oxides (NOx) (Crawford-Brown, Barker, Anger, & Dessens, 2012). In the transportation sector, in addition to the co-control of GHGs and air pollutant emissions, and the health gains from improved air quality (Fig. 23.3), GHG mitigation activities such as active travel can also increase physical activity, social contact, and the opportunity to interact with the natural environment, which may reduce the risks of a range of diseases (e.g., cardiovascular disease, type 2 diabetes, colon and breast cancer, and depression) (Woodcock et al., 2009; Xia et al., 2015). A combination of agricultural technological improvements and reductions in consumption of foods from animal sources in high-consumption populations could provide an effective contribution to meeting targets to reduce GHG emissions and substantially benefit public health—for example, via reductions in type 2 diabetes, ischemic heart

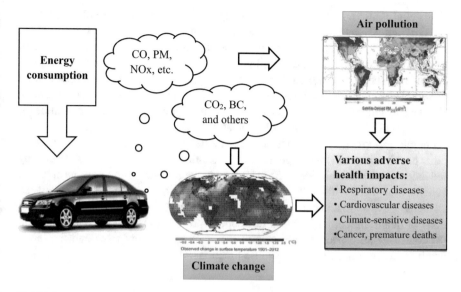

Fig. 23.3 Co-control of greenhouse gas (GHG) and air pollutant emissions and the corresponding health gains due to GHG mitigation measures in the transportation sector (adapted from the work of Chong, Yim, Barrett, & Boies, 2014). *BC* black carbon, *NOx* nitrogen oxides, *PM* particulate matter

disease, and the prevalence of obesity (Friel et al., 2009; Macdiarmid, 2013). Mitigation actions in the residential and household sector—such as improvements in combustion energy efficiency (Dora, Röbbel, & Fletcher, 2011), substitution of traditional cooking and space-heating practices with clean fuel technology and lower-emission household appliances (Venkataraman, Sagar, Habib, Lam, & Smith, 2010; Wilkinson et al., 2009), and energy saving through improvements in fabrics, fuel switching, behavioral changes, etc. (Wilkinson et al., 2009)—could bring about cost-effective health co-benefits (especially for women and children) in addition to reductions in GHG emissions (Anenberg et al., 2013). With regard to industrial and economic processes, co-benefits of GHG abatement, air quality improvements, and health gains could be expected through a series of measures such as improving energy efficiency, promoting the use of clean and renewable energy, and adjusting the industrial energy structure (Crawford-Brown et al., 2012; Gao et al., 2018b).

Challenges and Uncertainties in Estimating Health Co-benefits

Estimating the full range of the health co-benefits of reducing GHG emissions presents several common challenges to conventional epidemiological approaches and assessment studies (Haines et al., 2009; Pachauri et al., 2014a; Patz et al., 2008;

Watts et al., 2015), including the following: (1) development of credible scenarios for GHG emissions under "business-as-usual" and mitigation projections over the relevant time course; (2) the fact that rapid development of the energy structure, transportation patterns, land use, building construction, technology innovation, levels of exposure to health drivers, and demographic characteristics (population growth, baseline mortality rates, and value of statistical life) in some societies can change substantially in a short time with major implications for public health; (3) the fact that different subgroups of populations (e.g., age, gender, racial, or socioeconomic groups) may face disproportionate health impacts from air pollution or other health drivers; (4) the large number of health outcomes potentially affected by reductions in GHG emissions; (5) the short- to medium-term and long-term health benefits associated with GHG mitigation actions; (6) the varying lag times between changes in exposure and changes in health outcomes; (7) different economic valuations of health outcomes between developed and nonindustrialized countries; and (8) controversial aspects of key parameters such as discount rates and the terms involved in the concentration–response functions.

Knowledge Gaps

On the basis of a review of the current literature involving GHG emissions, mitigation strategies, and the related health co-benefits, several knowledge gaps have been identified. First, although several studies have attempted to model or quantify the associations between reductions in GHG emissions and health co-benefits, few studies have tried to establish a thorough performance appraisal system for the evaluation of environmental, socioeconomic, and public health co-benefits in relation to GHG mitigation measures and low-carbon policies (Gao et al., 2018a; Haines et al., 2009; Pachauri et al., 2014b; Smith et al., 2014). It has been indicated that an integrated performance appraisal system, including preimplementation analyses of interventions and follow-up cost–benefit appraisals of program implementation and related results, is crucial to comprehensively assess the performance of specific GHG abatement strategies and could help enhance the efficiency of decision-making processes and help programs to compare and prioritize potential options (McMichael, Barnett, & McMichael, 2012). Second, to date, most of the health benefit assessments have been performed in developed societies, with insufficient research having been carried out in developing regions, especially in areas like Africa and Asia, where the least GHG emissions are generated while the most severe climate change consequences are suffered (Haines et al., 2009; OECD, 2015; Pachauri et al., 2014b; Smith et al., 2014). These already susceptible areas—affected by conflicts, unstable politics, and impaired water supplies, as well as poor health infrastructure and limited economic resources—may become more vulnerable because of further climate change and the projected increasing frequency and intensity of extreme weather events (Barros et al., 2014; IPCC, 2007; OECD, 2015; Pachauri et al., 2014b; Shindell et al., 2012; West, Fiore, & Horowitz, 2012; Whitmee et al., 2015).

In addition, according to the present review, to date there is little evidence in the scientific literature of cost-effectiveness analysis, in practice, of health-promoting interventions to reduce GHG emissions (Haines et al., 2009). Most of the studies are descriptive or modeling investigations, and a conspicuous gap in the scientific research on the health co-benefits of GHG mitigation is the lack of intervention studies and assessments based on actual surveillance data (Pachauri et al., 2014a). Finally, despite potential health benefits of urban green space having been suggested by some studies (Salmond et al., 2016), the quantitative relationships between reductions in GHG emissions associated with green space and human health gains have not been fully evaluated. In order to take advantage of the public health opportunities offered by climate change mitigation measures (Wang & Horton, 2015), all of these areas for future research need to be explored.

Policy Implications

To date, although health co-benefits of GHG mitigation strategies in different sectors and countries have been modeled or quantified to a certain extent, their potential to provide cost-beneficial solutions has not been widely recognized, especially in developing nations, and this may hamper international cooperation for global reductions in GHG emissions (Haines et al., 2009; Pachauri et al., 2014a). Therefore, it is crucial to promote education and raise awareness on this topic among policy makers, health professionals, and other stakeholders, particularly in countries and regions confronted with multiple challenges from development, urbanization, population growth, air pollution, and climate change (Haines et al., 2009; Jiang et al., 2013).

Given the complexities of the challenges posed by climate change, major medical associations and nongovernmental health organizations have been calling for integrated multisectoral and multidisciplinary policies and actions to protect human health from dangerous climate change (Edenhofer et al., 2014b; Gao et al., 2018b; Haines et al., 2009). The existing evidence has suggested that comprehensive mitigation strategies generally provide greater health gains and larger reductions in GHG emissions (Friel et al., 2009; Shindell et al., 2012; West et al., 2012; Woodcock et al., 2009). Thus, interdisciplinary mitigation frameworks linking energy generation, transport, agriculture, household, and industrial processes are essential for curbing GHG emissions and improving public health simultaneously. However, we argue that in order to work toward the target of GHG mitigation, joint actions and collaborations—not only by different governmental departments at the national level but also between high- and low-income countries at the international level— are crucial (Edenhofer et al., 2014b; IPCC, 2007; Pachauri et al., 2014b; Whitmee et al., 2015).

In addition, studies have suggested that health professionals could play an important role in combating climate change and GHG emissions. For instance, they could advocate for comprehensive local, national, and international policies to reduce

GHG emissions and achieve the corresponding health co-benefits (Haines et al., 2009). Professionals can also serve as role models for practices in their own workplaces, communities, and even regions, to help inform and educate the local and national public and policy makers about the health risks posed by climate change and the health co-benefits of GHG mitigation (Watts, 2009).

Although numerous studies have focused on the roles of governments and of various economic sectors in mitigating climate change, relatively limited attention has been paid to the effects of individual behavioral change on reductions in GHG emissions (Gao et al., 2018a; Macdiarmid, 2013). In terms of the potential health co-benefits resulting from behavioral changes—such as limiting car trips in favor of active travel, limiting consumption of foods from animal sources, using lower-emission stoves, and reducing energy use—the collective impact of small behavioral changes may result in a considerable reduction in global GHG emissions (Friel et al., 2009; Haines et al., 2009; Venkataraman et al., 2010; Xia et al., 2015).

Conclusion

This chapter has summarized the available evidence on quantitative associations between reductions in GHG emissions and health benefits. The results generally suggest that GHG mitigation strategies in the energy generation, transport, agriculture and food, household, and industrial sectors could bring ancillary health benefits at the same time, while comprehensive measures across various sectors would tend to provide greater health gains. In addition to raising awareness, the findings of this review can provide valuable information for central and local governments, nongovernmental organizations, and policy makers, as well as for other relevant stakeholders concerned with the development and implementation of low-carbon technologies and policies. Besides, the anticipated cost-effective health co-benefits produced by GHG mitigation actions could make climate change mitigation policies more appealing and true "no-regrets" options for policy makers and GHG emitters, especially in low-income countries with finite economic resources. This could play a key role in motivating large GHG emitters to make voluntary and decisive changes that would help reduce emissions while prioritizing mitigation measures with optimal health gains. It would help in facilitating cooperation and co-action at the international, national, and regional levels on the basis of traditional situations, social expectations, and resource availability to protect human health from dangerous climate change and air pollution simultaneously.

Acknowledgments External funding for this study was obtained from the China Prosperity Strategic Programme Fund (SPF) 2015-16 (Project Code: 15LCI1) and the National Basic Research Program of China (973 Program) (Grant No. 2012CB955504). The funders played no role in the design, development, or interpretation of the present work. The views expressed in this chapter are those of the authors and do not necessarily reflect the position of the funding bodies. The authors also want to thank Alistair Woodward, Sotiris Vardoulakis, Sari Kovats, Paul Wilkinson, Jing Li, Shaohua Gu, and Xiaobo Liu for their kind and constructive suggestions and comments on this chapter.

References

Anenberg, S. C., Balakrishnan, K., Jetter, J., Masera, O., Mehta, S., Moss, J., et al. (2013). Cleaner cooking solutions to achieve health, climate, and economic cobenefits. *Environmental Science & Technology, 47*, 3944–3952.

Barros, V. R., Field, C. B., Dokken, D. J., Mastrandrea, M. D., Mach, K. J., Bilir, T. E., et al. (2014). *IPCC 2014: Climate change 2014: Impacts, adaptation, and vulnerability. Part B: Regional aspects. Contribution of Working Group II to the Fifth Assessment Report of the Intergovernmental Panel on climate change* (p. 688). Cambridge, UK and New York, NY: Cambridge University Press.

Chong, U., Yim, S. H., Barrett, S. R., & Boies, A. M. (2014). Air quality and climate impacts of alternative bus technologies in greater London. *Environmental Science & Technology, 48*, 4613–4622.

Costa, L., Rybski, D., & Kropp, J. P. (2011). A human development framework for CO2 reductions. *PLoS One, 6*, e29262.

Costello, A., Abbas, M., Allen, A., Ball, S., Bell, S., Bellamy, R., et al. (2009). Managing the health effects of climate change: Lancet and University College London Institute for Global Health Commission. *The Lancet, 373*, 1693–1733.

Crawford-Brown, D., Barker, T., Anger, A., & Dessens, O. (2012). Ozone and PM related health co-benefits of climate change policies in Mexico. *Environmental Science & Policy, 17*, 33–40.

Dora, C., Röbbel, N., & Fletcher, E. (2011). *Health co-benefits of climate change mitigation: Housing sector.* Geneva, Switzerland: Public Health & Environment Department, Health Security & Environment Cluster, World Health Organization.

Edenhofer, O., Pichs-Madruga, R., Sokona, Y., Farahani, E., Kadner, S., Seyboth, K., et al. (2014a). *Climate change 2014: Mitigation of climate change. Contribution of Working Group III to the Fifth Assessment Report of the Intergovernmental Panel on Climate Change* (pp. 511–597). Geneva, Switzerland: IPCC.

Edenhofer, O., Pichs-Madruga, R., Sokona, Y., Farahani, E., Kadner, S., Seyboth, K., et al. (2014b). *IPCC 2014: Climate change 2014: Mitigation of climate change. Contribution of working group III to the fifth assessment report of the Intergovernmental Panel on Climate Change.* Cambridge, UK and New York, NY: Cambridge University Press.

Field, C. B., Barros, V. R., Dokken, D. J., Mach, K. J., Mastrandrea, M. D., Bilir, T. E., et al. (2014). *IPCC 2014: Climate change 2014: Impacts, adaptation, and vulnerability. Part A: Global and sectoral aspects. Contribution of Working Group II to the Fifth Assessment Report of the Intergovernmental Panel on Climate Change* (p. 1132). Cambridge, UK and New York, NY: Cambridge University Press.

Friel, S., Dangour, A. D., Garnett, T., Lock, K., Chalabi, Z., Roberts, I., et al. (2009). Public health benefits of strategies to reduce greenhouse-gas emissions: Food and agriculture. *Lancet, 374*, 2016–2025.

Gao, J., Hou, H., Zhai, Y., Woodward, A., Vardoulakis, S., Kovats, S., et al. (2018a). Greenhouse gas emissions reduction in different economic sectors: Mitigation measures, health co-benefits, knowledge gaps, and policy implications. *Environmental Pollution, 240*, 683–698.

Gao, J., Kovats, S., Vardoulakis, S., Wilkinson, P., Woodward, A., Li, J., et al. (2018b). Public health co-benefits of greenhouse gas emissions reduction: A systematic review. *The Science of the Total Environment, 627*, 388–402.

Haines, A., McMichael, A. J., Smith, K. R., Roberts, I., Woodcock, J., Markandya, A., et al. (2009). Health and climate change 6 public health benefits of strategies to reduce greenhouse-gas emissions: Overview and implications for policy makers. *Lancet, 374*, 2104–2114.

IPCC (2007). *Intergovernmental Panel on Climate Change: Fourth assessment report: Climate change 2007: Synthesis report.* Geneva, Switzerland: IPCC.

Jiang, P., Chen, Y., Geng, Y., Dong, W., Xue, B., Xu, B., et al. (2013). Analysis of the co-benefits of climate change mitigation and air pollution reduction in China. *Journal of Cleaner Production, 58*, 130–137.

Lelieveld, J., Evans, J. S., Fnais, M., Giannadaki, D., & Pozzer, A. (2015). The contribution of outdoor air pollution sources to premature mortality on a global scale. *Nature, 525*, 367–371.

Macdiarmid, J. I. (2013). Is a healthy diet an environmentally sustainable diet? *Proceedings of the Nutrition Society, 72*, 13–20.

McMichael, C., Barnett, J., & McMichael, A. J. (2012). An ill wind? Climate change, migration, and health. *Environmental Health Perspectives, 120*, 646–654.

Netherlands Environmental Assessment Agency. (2007). *China now no. 1 in CO2 emissions; USA in second position.* Bilthoven, The Netherlands: PBL Netherlands Environmental Assessment Agency. Retrieved February 23, 2017, from http://www.pbl.nl/en/dossiers/Climatechange/Chinanowno1inCO2emissionsUSAinsecondposition

Nemet, G., Holloway, T., & Meier, P. (2010). Implications of incorporating air-quality co-benefits into climate change policymaking. *Environmental Research Letters, 5*, 014007.

Organization for Economic Co-operation and Development (OECD). (2015). *The economic consequences of climate change.* OECD iLibrary: OECD Publishing.

Pachauri, R. K., Allen, M., Barros, V., Broome, J., Cramer, W., Christ, R., et al. (2014a). *IPCC 2014: Climate change 2014: Synthesis report. Contribution of Working Groups I, II and III to the Fifth Assessment Report of the Intergovernmental Panel on Climate Change* (p. 151). Geneva, Switzerland: IPCC.

Pachauri, R. K., Allen, M., Barros, V., Broome, J., Cramer, W., Christ, R., et al. (2014b). *Climate Change 2014: Synthesis report. Contribution of Working Groups I, II and III to the Fifth Assessment Report of the Intergovernmental Panel on Climate Change.* Geneva, Switzerland: IPCC.

Patz, J., Campbell-Lendrum, D., Gibbs, H., & Woodruff, R. (2008). Health impact assessment of global climate change: Expanding on comparative risk assessment approaches for policy making. *Annual Review of Public Health, 29*, 27–39.

Salmond, J. A., Tadaki, M., Vardoulakis, S., Arbuthnott, K., Coutts, A., Demuzere, M., et al. (2016). Health and climate related ecosystem services provided by street trees in the urban environment. *Environmental Health, 15*, 36.

Shindell, D., Kuylenstierna, J. C., Vignati, E., van Dingenen, R., Amann, M., Klimont, Z., et al. (2012). Simultaneously mitigating near-term climate change and improving human health and food security. *Science, 335*, 183–189.

Smith, K. R., Woodward, A., Campbell-Lendrum, D., Chadee, D., Honda, Y., Liu, Q., et al. (2014). Human health: Impacts, adaptation and co-benefits. *Climate Change*, 709–754.

Solomon, S., Plattner, G.-K., Knutti, R., & Friedlingstein, P. (2009). Irreversible climate change due to carbon dioxide emissions. *Proceedings of the National Academy of Sciences of the United States of America, 106*, 1704–1709.

Stocker, T., Qin, D., Plattner, G., Tignor, M., Allen, S., Boschung, J., et al. (2013). *IPCC 2013: Climate change 2013: The physical science basis. Contribution of Working Group I to the fifth assessment report of the Intergovernmental Panel on Climate Change* (p. 1535). Cambridge, UK and New York, NY: Cambridge University Press.

Venkataraman, C., Sagar, A., Habib, G., Lam, N., & Smith, K. (2010). The Indian national initiative for advanced biomass cookstoves: The benefits of clean combustion. *Energy for Sustainable Development, 14*, 63–72.

Wang, H., & Horton, R. (2015). Tackling climate change: The greatest opportunity for global health. *The Lancet, 386*, 1798–1799.

Watts, G. (2009). The health benefits of tackling climate change: An executive summary. *The Lancet Series 2009*, 1–8. https://doi.org/10.1016/S0140-6736(09)61759-1.2

Watts, N., Adger, W. N., Agnolucci, P., Blackstock, J., Byass, P., Cai, W., et al. (2015). Health and climate change: Policy responses to protect public health. *Lancet, 386*, 1861–1914.

West, J. J., Fiore, A. M., & Horowitz, L. W. (2012). Scenarios of methane emission reductions to 2030: Abatement costs and co-benefits to ozone air quality and human mortality. *Climatic Change, 114*, 441–461.

West, J. J., Smith, S. J., Silva, R. A., Naik, V., Zhang, Y., Adelman, Z., et al. (2013). Co-benefits of mitigating global greenhouse gas emissions for future air quality and human health. *Nature Climate Change, 3*, 885–889.

Whitmee, S., Haines, A., Beyrer, C., Boltz, F., Capon, A. G., de Souza Dias, B. F., et al. (2015). Safeguarding human health in the Anthropocene epoch: Report of the Rockefeller Foundation-lancet commission on planetary health. *Lancet, 386*, 1973–2028.

Wilkinson, P., Smith, K. R., Davies, M., Adair, H., Armstrong, B. G., Barrett, M., et al. (2009). Public health benefits of strategies to reduce greenhouse-gas emissions: Household energy. *Lancet, 374*, 1917–1929.

Woodcock, J., Edwards, P., Tonne, C., Armstrong, B. G., Ashiru, O., Banister, D., et al. (2009). Public health benefits of strategies to reduce greenhouse-gas emissions: Urban land transport. *Lancet, 374*, 1930–1943.

Xia, T., Nitschke, M., Zhang, Y., Shah, P., Crabb, S., & Hansen, A. (2015). Traffic-related air pollution and health co-benefits of alternative transport in Adelaide, South Australia. *Environment International, 74*, 281–290.

CHAPTER 24
Good Health in the Anthropocene Epoch: Potential for Transformative Solutions

Andy Haines

Summary Unprecedented environmental changes threaten to reverse progress on health and development unless they are addressed by decisive actions. The implementation of policies to reduce the environmental impact of human societies can reduce risks and also bring a range of near-term health benefits—for example, from lowered emissions of pollutants co-emitted with greenhouse gases. This chapter outlines potential transformative policies in sectors such as energy, transport, urban development and food systems, which can reduce their environmental footprint and improve health.

Introduction

Humanity faces multiple environmental challenges dominated by climate change but also encompassing ocean acidification, air pollution, biodiversity loss, declining fisheries, deforestation and depletion of freshwater supplies, which together characterise the Anthropocene epoch. The Anthropocene reflects the dominance of one species, our own, over the Earth's natural systems and is leading to rapid changes in these systems with potentially far-reaching implications for human health and development. Humanity flourished in the Holocene epoch, a relatively stable climatic period of about 11,500 years, and there is concern that the dramatic gains in life expectancy over recent decades could be undermined by these multiple environmental stressors acting in concert. Although this chapter largely addresses the challenge of climate change, it does so within the broader context of "planetary health", which acknowledges that the health of human civilisation depends on the state of underpinning natural systems (Whitmee et al., 2015).

Transformative approaches imply that health and development must increasingly be achieved at much lower environmental impacts than the current fossil fuel–intensive and inequitable world economy.

A. Haines (✉)
London School of Hygiene and Tropical Medicine, London, UK
e-mail: andy.haines@lshtm.ac.uk

The Imperative of Transformative Change

In order to attain the aims of the Paris Agreement, it will be necessary to decarbonise the world economy and implement deep cuts in short-lived climate pollutants (SLCPs) as a matter of urgency. In the absence of deep cuts in both CO_2 and SLCP emissions, temperature increases are likely to exceed 1.5 °C during the 2030s and exceed 2 °C by mid-century (Ramanathan & Xu, 2010; Shindell et al., 2012). After 3 years of stable greenhouse gas (GHG) emissions, in 2017 global GHG emissions from energy and industry increased (UNEP Emissions Gap Report, 2018).

The perception that transformative change towards a clean economy based on non-polluting renewable sources of energy is prohibitively expensive and politically unachievable in the near term is a major barrier to progress. An alternative view is that it is both inevitable and welcome because it brings many benefits in the near term as well as creating the opportunity to promote a sustainable and productive economy for the future. Health co-benefits of the "zero-carbon" economy can arise from a range of policies in sectors such as transport, energy, housing, and food and agriculture, and through several pathways, including reduced air pollution exposure, healthy low-environmental-impact diets and increased physical activity (Haines et al., 2009). These policies must be complemented by others that halt and reverse the profound damage to a range of natural systems whilst safeguarding health and development (Whitmee et al., 2015).

Health Co-benefits of Reduced Air Pollution from Clean Renewable Energy Sources

A growing body of evidence attests to the major health co-benefits that accompany policies to reduce GHG/SLCP emissions. A major pathway is through the reduction in exposure to fine particulate air pollution from a combination of ambient and household air pollution, and from tropospheric ozone. The sources of ambient air pollution vary by country, as shown in Fig. 24.1, which compares the USA and India. Ambient air pollution has been estimated to cause over 4.5 million premature deaths annually worldwide (Lelieveld, Haines, & Pozzer, 2018).

The Global Burden of Disease (GBD) study estimated that 2.9 million deaths in 2015 were associated with household air pollution (Forouzanfar, Alexander, Anderson, et al., 2015), whereas the World Health Organization (WHO) estimated that there were 4.3 million related deaths in 2012 (WHO, 2014). This difference is partly due to the approaches used to quantify exposure–outcome associations (see Landrigan et al., 2018, for further discussion of differences between the GBD and WHO approaches). The magnitude of these health effects shows the potential improvements from universal access to clean renewable energy sources, which can be further augmented by indirect health benefits—for example, from reduced time collecting firewood for domestic use (a task that falls disproportionately on women and girls, exposing them to risks of sexual violence).

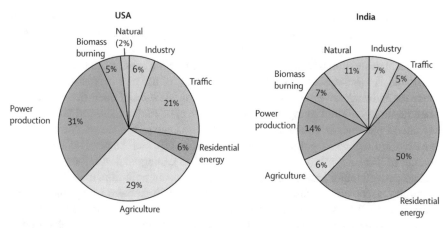

Fig. 24.1 Sources of mortality from air pollution. The percentages are the proportions of deaths attributable to ambient air pollution. Excess deaths attributable to air pollution in 2015 were estimated at 120,000 (95% confidence interval 81,000–156,000) in the USA and 967,000 (753,000–1,150,000) in India. "Natural" refers to natural sources of air pollution, predominantly aeolian dust. (Source: Lelieveld et al., 2018, with permission from *The Lancet Planetary Health*)

A growing number of studies have estimated the health co-benefits of policies to scale up access to clean energy from low-carbon and renewable sources. The International Energy Agency estimated that a 7% increase in total energy investment over the period to 2040 in a clean air scenario would lead to a dramatic improvement in health in comparison with a main reference scenario. Premature deaths from outdoor air pollution would be 1.7 million lower in 2040, and those from household pollution would be 1.6 million lower. Furthermore, there would be additional benefits in terms of energy security, CO_2 emissions and import bills (IEA, 2016).

Co-benefits may be particularly large in countries with a large burden of disease from air pollution; for example, the implementation of climate mitigation strategies in China could result in about a 50% reduction in months of life lost as a result of air pollution exposure by 2050 (Markandya et al., 2009).

Implementing a range of SLCP mitigation actions could result in between 700,000 and 4.7 million annually avoided premature deaths, mainly from reductions in black carbon, together with increases in annual crop yields of 30–135 million metric tons due to ozone reductions in 2030 and beyond (Shindell et al., 2012). The measures to reduce black carbon emissions include reduced agricultural burning, replacement of domestic solid fuels with clean sources of energy and use of improved-efficiency brick kilns. Methane reductions could be achieved by reducing leakage of natural gas, improved rice paddy management and so on. A rapid phase-out of fossil fuels to reduce CO emissions, together with policies to cut SLCPs, offer the best prospect of achieving the targets in the Paris Agreement and will provide both near- and long-term benefits to health (Shindell et al., 2017).

There are a number of potential approaches to reducing exposure to household air pollution, including improved-efficiency cookstoves, electrification (preferably from low-carbon sources) and use of liquefied petroleum gas (LPG). LPG is a fossil fuel and therefore would result in a small increase in CO_2 emissions but much lower particulate air pollution exposure (Pachauri, 2014).

Careful evaluation of effects is necessary because the anticipated benefits are not inevitable. A large-scale trial evaluating the effects of use of a cleaner-burning bio-mass-fuelled cookstove and solar panel versus continuation of open-fire cooking on pneumonia in children living in two rural districts of Malawi showed no effect on childhood pneumonia, probably because of low usage of the stoves, the failure of the stoves to reduce air pollution emissions sufficiently and the persistence of other sources of local air pollution (Mortimer et al., 2017). Integrated solutions are needed that address all major local sources of fine particulate air pollution, using affordable, culturally acceptable technologies that can be easily repaired and maintained. Access to electricity generated from clean renewable sources or stoves that use clean renewable fuels such as ethanol and biogas can bring large climate change and SLCP benefits (WHO/CCAC, 2016). The replacement of kerosene lamps with solar lamps reduces emissions of black carbon, resulting in benefits to the climate and to health, as well as reducing poor households' expenditure on kerosene (Climate and Clean Air Coalition, 2014).

Valuation of the Air Pollution Co-benefits of Mitigation Strategies

A study comparing a reference scenario with a global GHG mitigation scenario projected that the latter would avoid 0.5 ± 0.2 million and 1.3 ± 0.5 million premature deaths in 2030 and 2050, respectively. The global average marginal co-benefits of avoided mortality were estimated to be valued at US$50–380 per metric ton of CO_2 (using the value of a statistical life), which would exceed marginal abatement costs in 2030 and 2050 (West et al., 2013) and would tend to be largest in locations where the current air pollution health burdens are greatest. A comparison of a climate change mitigation policy scenario that is consistent with the European Commission's 2° target and a baseline medium-high emissions scenario showed major health and economic co-benefits (Holland et al., 2011).

The monetised human health benefits associated with air quality improvements from reduced ozone and fine particulate pollution resulting from GHG mitigation policies could offset 26–1050% of the cost of US carbon policies. Larger net co-benefits could result from policies that minimised costs, such as cap-and-trade standards, than from specific sectoral policies (electricity and transportation) (Thompson, Rausch, Saari, & Selin, 2014). An assessment of fine particulate and ozone reductions from a US emissions trajectory consistent with 2 °C warming from energy and vehicles showed that by 2030, clean energy policies could prevent about 175,000 premature deaths, with 22,000 (95% confidence interval 11,000–96,000) fewer annually thereafter (Shindell, Lee, & Faluvegi, 2016). Furthermore, clean transportation could prevent ~120,000 premature air pollution–related deaths by 2030 and subsequently 14,000 (9000–52,000) annually.

A retrospective analysis of the effects of implementation of US state–level renewable energy (RE) portfolio standards in 2013 showed that in addition to reductions in life cycle GHG emissions estimated at 59 million metric tons of carbon dioxide–equivalent (CO_2e), there were substantial air pollution, health and environmental benefits (Wiser et al., 2016).

Cities

Most of the world's population now lives in cities, and an additional 2.5 billion people are projected to live in cities by 2050, largely in sub-Saharan Africa and in Asia (United Nations Environment Programme, 2013). Cities provide multiple and interconnected routes for realising environmental and health benefits. These include lower GHG emissions and air pollution, and non-communicable disease (NCD) reduction through improvement of the urban and built environment, including modifications to housing, transport and energy systems (Milner, Davies, & Wilkinson, 2012; Wilkinson & Haines, 2015).

A large prospective study in Ontario, Canada (Creatore et al., 2016), showed that greater neighbourhood walkability was associated with a decreased prevalence of overweight and obesity and a decreased incidence of diabetes between 2001 and 2012. A large-scale cross-sectional study (Flint & Cummins, 2016) and a nationally representative longitudinal study in the UK (Martin, Panter, Suhrcke, & Ogilvie, 2015) both demonstrated health benefits of active travel, including a reduced risk of overweight and obesity.

The modelled effects of low-emission vehicles and increased active travel in London and Delhi showed major benefits to health, which outweighed the increased risk of exposure to road injuries (Woodcock et al., 2009). In the case of London, the main health benefits were projected to result from increased active travel, although the GHG benefits resulted largely from the introduction of low-emission vehicles.

In cities with high air pollution levels, such as Delhi, a larger proportion of the health benefits is likely to be due to air pollution reduction. However, even where concentrations of particulates with a diameter of 2.5 μm ($PM_{2.5}$ concentrations) reach 100 μg/m³, harms from increased inhalation of air pollution would exceed the benefits only if individuals exceeded 1 h 30 min of cycling per day or more than 10 h of walking per day in comparison with staying at home. In comparison with driving, the benefits of physical activity would exceed harms from air pollution at up to 3 h 30 min of cycling per day (Tainio et al., 2016). There could be substantial benefits to national health services; for example, if urban populations in England and Wales had patterns of walking and cycling similar to those achieved in Copenhagen, about £17 billion of costs to the National Health Service (NHS) could be averted over 20 years, increasing beyond that period because of the lag between increased physical activity and disease prevention (Jarrett et al., 2012).

The introduction of very-low-emission vehicles (e.g. electric or electric hybrid vehicles) would result in increased health benefits from reductions in air pollution. Autonomous vehicles would bring the added benefit of reduced accidents. Electric bicycles may increase the range of travel and encourage older, unfit or physically disabled people to take up cycling (Johnson & Rose, 2015).

Provision of accessible and secure green space may provide a number of benefits to health and the environment, including improved mental health (Bowler, Buyung-Ali, Knight, & Pullin, 2010); reduced risks of fatal and non-fatal cardiovascular disease, obesity and diabetes; and improved pregnancy outcomes (WHO, 2016). The mechanisms may include reduced psychological stress; increased physical activity; and reduced air pollution, noise and heat exposure. Urban sprawl and less mixed land use (which poses a higher risk) are consistently associated with obesity in North American studies, but elsewhere the associations are more heterogeneous (Mackenbach et al., 2014). In the UK, a study of over 400,000 people showed that below 1800 housing units per square kilometre higher residential density was associated with higher adiposity, whereas above 1800 units per square kilometre, residential density was associated with significantly lower body mass index and waist circumference values and with decreased odds of obesity (Sarkar, Webster, & Gallacher, 2017). Areas of high residential density may promote greater physical activity because of the proximity of shops and other facilities, and perhaps because they provide greater social support and community engagement. Urban sprawl may also be associated with increased risks of urban heat exposure (Stone, Hess, & Frumkin, 2010) and with higher transport GHG emissions. A study of US cities found that doubling population-weighted density was associated with a 48% reduction in CO_2 emissions from household travel and a 35% reduction in CO_2 emissions from residential energy use (Lee & Lee, 2014).

Health in the Circular Economy

There is growing interest in transforming the economy from the current, inefficient, resource-intensive, polluting approach to one based on recycling, reuse and remanufacturing, encouraging the use of renewable and biodegradable resources. Such approaches can greatly reduce exposure to hazardous waste and inefficient use of resources, ensuring that human progress is achieved at much lower levels of environmental damage than today's wasteful and inequitable economy. One example would be the introduction of policies to reduce food waste; currently about 30% of agricultural land is used to grow food that is never eaten but contributes to climate and other environmental changes. Another priority is to reduce exposure of mainly poor populations to hazardous waste dumps (Landrigan et al., 2018). The circular economy—if designed to address health, development and environmental sustainability in an integrated fashion—could both address pollution and help mitigate climate change whilst promoting decent work and livelihoods.

A recent paper, using data from 637 Chinese cities, demonstrated the potential to combine the circular economy and climate change mitigation by capitalising on urban–industrial symbiosis, including the use of waste heat from industry in the commercial and residential sectors (Ramaswami et al., 2017). Industrial cities have high direct CO_2 emissions intensity and are hot spots for interventions to reduce emissions and capitalise on the potential symbiosis.

Food and Agriculture

There are a range of potential solutions to the threats posed by environmental change to crop yields, food security and nutrition, including reduction of food waste, biofortification of crops, sustainable intensification of agriculture, sustainable aquaculture, more efficient use of water and fertiliser, and promotion of low-environmental-impact, healthy diets (Whitmee et al., 2015). A systematic review has shown the potential to reduce GHG emissions, together with land and water use, from shifting current intakes to environmentally more sustainable diets, suggesting that median reductions of 20–30% across these indicators are possible in high-income settings (Aleksandrowicz, Green, Joy, Smith, & Haines, 2016). The per-calorie environmental impacts were largest with consumption of ruminant meat, followed by other animal products, and smallest with consumption of plant-based foods (Green et al., 2015). Dietary shifts also resulted in modest estimated reductions in the all-cause mortality risk (Milner et al., 2017). More evidence from low-income settings is needed.

A study using data from 555 life cycle analyses on 82 types of crops and animals showed that in comparison with the global-average income-dependent diet projected for 2050, three comparator diets—Mediterranean, pescatarian and vegetarian—resulted in 30%, 45% and 55% reductions in per capita GHG emissions, respectively (Tilman & Clark, 2014), and required much less cropland than the income-dependent diet. Compared with conventional omnivorous diets, the three alternative diets were estimated to reduce incidence rates of type 2 diabetes by 16–41% and ischaemic heart disease by 20–26%. Sustainable aquaculture needs scaling up to meet increased fish requirements, as much current aquaculture depends on inputs of antibiotics, pesticides and food from unsustainable sources such as krill.

Pricing Greenhouse Gas Emissions to Reduce Climate Change and Improve Health

Removing fossil fuel subsidies and implementing carbon taxes could, if properly designed, improve health, reduce GHG emissions, redistribute wealth and stimulate employment (Cuevas & Haines, 2015), but recent disturbances in France show the adverse effects of policies that increase inequalities. Estimated nationally efficient carbon prices based on self-interest (i.e. including consideration of the resulting domestic co-benefits) in 2010 prices averaged around US\$57 per ton for the top 20 emitting nations (Parry, Heine, Lis, & Li, 2014), but there are wide differences between countries, depending on their mix of energy sources. Implementing taxation that reflects the costs of carbon dioxide emissions, local air pollution and additional transport-related externalities (such as congestion and accidental injury) from burning fossil fuels could yield additional revenues, equivalent to about 2.6% of global gross domestic product (GDP), while simultaneously reducing carbon dioxide emissions by 23% and pollution-related mortality by 63% (Parry et al., 2014).

By 2050, food-related GHG emissions could take up half the remaining GHG emissions to keep the world on track for 2 °C warming in the absence of decisive actions to address the food system. An analysis of the role of GHG-related food taxes showed that food-related GHG emissions would be reduced by 1.0 $GtCO_2e$ (9.3%) under full tax coverage and by 919 $MtCO_2e$ (8.6%) in a regionally optimised tax scenario. About two thirds of the emissions reductions were due to reduced consumption of beef and one quarter were due to reduced milk consumption (Springmann, Godfray, Rayner, & Scarborough, 2016). However, care would be needed to avoid potential negative health effects, particularly in low-income settings, by omitting food groups known to have health benefits from the tax regime or subsidising healthy low-emission food groups. The health benefits identified in this study (100,000–500,000 avoided deaths globally) were broadly similar to the potential health benefits of 500,000 ± 200,000 avoided premature deaths per year by 2030 that global climate policies could have on air pollution associated with coal-fired power generation (West et al., 2013).

Potential Adverse Effects of Policies to Address Environmental Change

Assessments of health co-benefits of policies to mitigate environmental change should specifically consider the potential for unintended adverse consequences (Haines et al., 2009). Some types of biofuels, such as corn alcohol, compete directly with important food crops and may contribute to food price increases, but others, such as sugar cane alcohol, do not compete directly. Increasing the energy efficiency of houses by better insulation and draught proofing may increase exposure to household air pollution unless it is accompanied by improved ventilation. Diesel engines have been promoted in some countries because of their reduced GHG emissions but have higher emissions of fine particulates and nitrogen oxides (NOx) (Wilkinson & Haines, 2015).

The Post-2015 Development Agenda

The post-2015 development agenda represents a major opportunity to improve health through the lens of sustainable development (Dora et al., 2015; Haines, Alleyne, Kickbusch, & Dora, 2012). The Sustainable Development Goals (SDGs) of the United Nations will shape the development agenda until 2030, and a recent paper has outlined the links between interventions and policies to reduce the emissions of short-lived climate pollutants and the SDGs (Haines et al., 2017).

There are opportunities to harness some of the climate funding of US$100 billion annually that was previously pledged to be provided by 2020 for mitigation and adaptation (http://www.greenclimate.fund/). This funding could help countries capitalise on the health co-benefits of climate change mitigation policies and reduce climate change impacts on health by preventing dangerous climate change.

SDG target 3.7 aims to "ensure universal access to sexual and reproductive health-care services, including for family planning, information and education, and the integration of reproductive health into national strategies and programmes." This, together with improved education of women (SDG goal 4), can improve pregnancy outcomes and the prospects for child survival through birth spacing. Although per capita GHG emissions in regions such as sub-Saharan Africa are currently low, these will grow if development is powered by fossil fuels, and population growth in vulnerable regions makes adaptation to climate change more difficult.

Conclusion

Whilst global environmental change poses major challenges to human health, there are opportunities to address disease prevention and environmental sustainability through integrated transformative policies. However, in order to realise this potential, unprecedented collaboration across sectors will be required in the coming decades. The economic and social drivers responsible for the challenges faced by humanity in the Anthropocene epoch must be addressed, including by removal of harmful subsidies, realignment of taxes to focus on health-damaging externalities, and reduction of inequities in health and wealth. Better health cannot be achieved and sustained without the recognition that human health ultimately depends on the integrity of natural systems.

References

Aleksandrowicz, L., Green, R., Joy, E. J. M., Smith, P., & Haines, A. (2016). The impacts of dietary change on greenhouse gas emissions, land use, water use, and health: A systematic review. *PLoS One, 11*, e0165797. https://doi.org/10.1371/journal.pone.0165797

Bowler, D. E., Buyung-Ali, L. M., Knight, T. M., & Pullin, A. S. (2010). A systematic review of evidence for the added benefits to health of exposure to natural environments. *BMC Public Health, 10*, 456.

Climate and Clean Air Coalition (2014). *Scientific advisory panel briefing: Kerosene lamps and SLCPs*. https://ccacoalition.org/en/news/kerosene-lamps-are-important-target-reducing-indoor-air-pollution-and-climate-emissions (retrieved October 1st 2019)

Creatore, M. I., Glazier, R. H., Moineddin, R., Fazli, G. S., Johns, A., et al. (2016). Association of neighborhood walkability with change in overweight, obesity, and diabetes. *JAMA, 315*, 2211. https://doi.org/10.1001/jama.2016.5898

Cuevas, S., & Haines, A. (2015). The health benefits of a carbon tax. *Lancet, 386*, 7–9. Retrieved on February 16, 2020 from https://www.thelancet.com/journals/lancet/article/PIIS0140-6736(15)00994-0/fulltext

Dora, C., Haines, A., Balbus, J., Fletcher, E., Adair-Rohani, H., Alabaster, G., et al. (2015). Indicators linking health and sustainability in the post-2015 development agenda. *The Lancet, 385*, 380–391.

Flint, E., & Cummins, S. (2016). Active commuting and obesity in mid-life: Cross-sectional, observational evidence from UK Biobank. *The Lancet Diabetes and Endocrinology, 4*, 420–435.

Forouzanfar, M. H., Alexander, L., Anderson, H. R., et al. (2015). Global, regional, and national comparative risk assessment of 79 behavioural, environmental and occupational, and metabolic risks or clusters of risks in 188 countries, 1990–2013: A systematic analysis for the Global Burden of Disease Study 2013. *Lancet, 386*, 2287–2323.

Green, R., Milner, J., Dangour, A. D., Haines, A., Chalabi, Z., Markandya, A., et al. (2015). The potential to reduce greenhouse gas emissions in the UK through healthy and realistic dietary change. *Climatic Change, 129*, 253–265. https://link.springer.com/article/10.1007%2Fs10584-015-1329-y

Haines, A., Alleyne, G., Kickbusch, I., & Dora, C. (2012). From the Earth Summit to Rio+20: Integration of health and sustainable development. *Lancet, 379*, 2189–2197.

Haines, A., Amann, M., Borgford-Parnell, N., Leonard, S., Kuylenstierna, J., & Shindell, D. (2017). Short-lived climate pollutant mitigation and the sustainable development goals. *Nature Climate Change, 7*, 863–869.

Haines, A., McMichael, A. J., Smith, K. R., Roberts, I., Woodcock, J., Markandya, A., et al. (2009). Public health benefits of strategies to reduce greenhouse-gas emissions: Overview and implications for policy makers. *Lancet, 374*, 2104–2114.

Holland, M., Amann, M., Heyes, C., Rafaj, P., Schöpp, W., Hunt, A., et al. (2011). The reduction in air quality impacts and associated economic benefits of mitigation policy. Summary of results from the EC RTD ClimateCost Project. In P. Watkiss (Ed.), *The ClimateCost Project. Final report. Volume 1: Europe*. Sweden: Stockholm Environment Institute. ISBN: 978-91-86125-35-6.

International Energy Agency (2016). *World Energy Outlook (WEO) special report 2016 energy and air pollution*. Paris, France: International Energy Agency.

Jarrett, J., Woodcock, J., Griffiths, U. K., Chalabi, Z., Edwards, P., Roberts, I., et al. (2012). Effect of increasing active travel in urban England and Wales on National Health Service costs. *Lancet, 379*, 2198–2205.

Johnson, M., & Rose, G. (2015). Extending life on the bike: Electric bike use by older Australians. *Journal of Transport & Health, 2*, 276–283.

Landrigan, P. J., Fuller, R., Acosta, N. J. R., Adeyi, O., Arnold, R., et al. (2018). Lancet commission on pollution and health. *Lancet, 391*, 462–512. https://doi.org/10.1016/S0140-6736(17)32345-0

Lee, S., & Lee, B. (2014). The influence of urban form on GHG emissions in the U.S. household sector. *Energy Policy, 68*, 534–549.

Lelieveld, J., Haines, A., & Pozzer, A. (2018). Age-dependent health risk from ambient air pollution: A modelling and data analysis of childhood mortality in middle and low-income countries. *Lancet Planetary Health, 2*, e292–e300.

Mackenbach, J., Rutter, H., Compernolle, S., Glonti, K., Oppert, J. M., et al. (2014). Obesogenic environments: A systematic review of the association between the physical environment and adult weight status, the SPOTLIGHT project. *BMC Public Health, 14*, 233. http://www.biomedcentral.com/1471-2458/14/233

Markandya, A., Armstrong, B. G., Hales, S., Chiabai, A., Criqui, P., Mima, S., et al. (2009). Public health benefits of strategies to reduce greenhouse-gas emissions: Low-carbon electricity generation. *Lancet, 374*, 2006–2015.

Martin, A., Panter, J., Suhrcke, M., & Ogilvie, D. (2015). Impact of changes in mode of travel to work on changes in body mass index: Evidence from the British Household Panel Survey. *Journal of Epidemiology and Community Health, 69*, 753–761.

Milner, J., Davies, M., & Wilkinson, P. (2012). Urban energy, carbon management (low carbon cities) and co-benefits for human health. *Current Opinion in Environmental Sustainability, 4*, 398–404.

Milner, J., Joy, E. J. M., Green, R., Harris, F., Aleksandrowicz, L., Agrawal, S., et al. (2017). Dangour A Projected health effects of realistic dietary changes to address freshwater constraints in India: A modelling study. *Lancet Planetary Health, 1*, e 26–e 32.

Mortimer, K., Ndamala, C. B., Naunje, A. W., Malava, J., Katundu, C., et al. (2017). A cleaner burning biomass-fuelled cookstove intervention to prevent pneumonia in children under 5 years old in rural Malawi (the Cooking and Pneumonia Study): A cluster randomised controlled trial. *The Lancet, 389*, 167–175.

Pachauri, S. (2014). Household electricity access a trivial contributor to CO2 emissions growth in India. *Nature Climate Change, 4*, 1073–1076.

Parry, I. W., Heine, M. D., Lis, E., & Li, S. (2014). *Getting energy prices right: From principle to practice*. Washington, DC: International Monetary Fund.

Ramanathan, V., & Xu, Y. (2010). The Copenhagen Accord for limiting global warming: Criteria, constraints, and available avenues. *Proceedings of the National Academy of Sciences of the United States of America, 107*, 8055–8062. https://doi.org/10.1073/pnas.1002293107

Ramaswami, A., Tong, K., Fang, A., Lal, R. M., Nagpure, A. S., et al. (2017). Change urban cross-sector actions for carbon mitigation with local health co-benefits in China. *Nature Climate Change, 7*, 736–742. https://doi.org/10.1038/nclimate3373

Sarkar, C., Webster, C., & Gallacher, J. (2017). Association between adiposity outcomes and residential density: A full-data, cross-sectional analysis of 419562 UK Biobank adult participants. *Lancet Planet Health, 1*, e277–e288.

Shindell, D., Borgford-Parnell, N., Brauer, M., Haines, A., Kuylenstierna, J. C. I., Leonard, S. A., et al. (2017). A climate policy pathway for near- and long-term benefits. *Science, 356*, 493–494.

Shindell, D., Kuylenstierna, J. C. I., Vignati, E., van Dingenen, R., Amann, M., Klimont, Z., et al. (2012). Simultaneously mitigating near-term climate change and improving human health and food security. *Science, 335*, 183–189.

Shindell, D. T., Lee, Y., & Faluvegi, G. (2016). Climate and health impacts of US emissions reductions consistent with 2°C. *Nature Climate Change, 6*, 503–507.

Springmann, M., Godfray, C. J., Rayner, M., & Scarborough, P. (2016). Analysis and valuation of the health and climate change co-benefits of dietary change. *Proceedings of the National Academy of Sciences of the United States of America, 113*, 4146–4151.

Stone, B., Hess, J. J., & Frumkin, H. (2010). Urban form and extreme heat events: Are sprawling cities more vulnerable to climate change than compact cities? *Environmental Health Perspectives, 118*, 1425–1428. https://doi.org/10.1289/ehp.0901879

Tainio, M., de Nazelle, A. J., Gotschi, T., Kahlmeier, S., Rojas-Rueda, D., Nieuwenhuijsen, M. J., et al. (2016). Can air pollution negate the health benefits of cycling and walking? *Preventive Medicine, 87*, 233–236.

Thompson, T. M., Rausch, S., Saari, R. K., & Selin, N. E. (2014). A systems approach to evaluating the air quality co-benefits of US carbon policies. *Nature Climate Change, 4*, 917–923.

Tilman, D., & Clark, M. (2014). Global diets link environmental sustainability and human health. *Nature, 515*, 518–522.

UNEP Emissions Gap Report (2018). https://www.unenvironment.org/resources/emissions-gap-report-2018 (accessed October 1st 2019)

United Nations Environment Programme (2013). *City-level decoupling: Urban resource flows and the governance of infrastructure transitions.* A Report of the Working Group on Cities of the International Resource Panel. Swilling M., Robinson B., Marvin S. and Hodson M. ISBN: 978-92-807-3298-6

West, J. J., Smith, S. J., Silva, R. A., Naik, V., Zhang, Y., Adelman, Z., et al. (2013). Co-benefits of mitigating global greenhouse gas emissions for future air quality and human health. *Nature Climate Change, 3*, 885–889.

Whitmee, S., Haines, A., Beyrer, C., Boltz, F., Capon, A. G., et al. (2015). Safeguarding human health in the Anthropocene epoch: Report of The Rockefeller Foundation-Lancet Commission on planetary health. *Lancet, 386*, 1973–2028.

WHO (2016). *Urban green spaces and health.* Copenhagen: WHO Regional Office for Europe. http://www.euro.who.int/__data/assets/pdf_file/0005/321971/Urban-green-spaces-and-health-review-evidence.pdf?ua=1

WHO/CCAC (2016). *Household air pollution and health.* Retrieved on February 16, 2020 from http://www.who.int/sustainable-development/LR-HAP-27May2016.pdf?ua=1

Wilkinson, P., & Haines, A. (2015). Diesel in the dock. *BMJ, 351*, h5415.

Wiser, R., Barbose, G., Heeter, J., Mai, T., Bird, L., Bolinger, M., et al. (2016). *A retrospective analysis of the benefits and impacts of U.S. renewable portfolio standards* (NREL/TP-6A20-65005). Lawrence Berkeley National Laboratory and National Renewable Energy Laboratory. Retrieved on February 16, 2020 from http://www.nrel.gov/docs/fy16osti/65005.pdf

Woodcock, J., Edwards, P., Tonne, C., Armstrong, B. G., Ashiru, O., Banister, D., et al. (2009). Public health benefits of strategies to reduce greenhouse-gas emissions: Urban land transport. *Lancet, 374*, 1930–1943.

CHAPTER 25
Well Under 2 °C: Ten Solutions for Carbon Neutrality and Climate Stability

V. Ramanathan, M. L. Molina, D. Zaelke, and N. Borgford-Parnell

Summary Climate change is becoming an existential threat with warming in excess of 2 °C within the next three decades and 4–6 °C within the next several decades. Warming of such magnitudes will expose as many as 75% of the world's population to deadly heat stress in addition to disrupting the climate and weather worldwide. Climate change is an urgent problem requiring urgent solutions. This chapter lays out urgent and practical solutions that are ready for implementation now, will deliver benefits in the next few critical decades, and place the world on a path to achieving the long-term targets of the Paris Agreement. The approach consists of four building blocks and three levers to implement ten scalable solutions described in this chapter. These solutions will enable society to decarbonize the global energy system by 2050 through efficiency and renewables, drastically reduce short-lived climate pollutants, and stabilize the warming well below 2 °C both in the near term (before 2050) and in the long term (after 2050). The solutions include an atmospheric carbon extraction lever to remove CO_2 from the air. The amount of CO_2 that must be removed ranges from negligible (if the emissions of CO_2 from the energy system and short-lived climate pollutants have started to decrease by 2020 and carbon neutrality is achieved by 2050) to a staggering one trillion tons (if the carbon lever is not pulled and emissions of climate pollutants continue to increase until 2030).

V. Ramanathan (✉)
Scripps Institution of Oceanography, University of California, San Diego,
San Diego, CA, USA
e-mail: vramanathan@ucsd.edu

M. L. Molina
University of California, San Diego, San Diego, CA, USA

D. Zaelke
Institute for Governance and Sustainable Development, Washington, DC, USA
Institute for Governance and Sustainable Development, Paris, France

N. Borgford-Parnell
Climate and Clean Air Coalition, Paris, France

© The Author(s) 2020
W. K. Al-Delaimy, V. Ramanathan, M. Sánchez Sorondo (eds.), *Health of People, Health of Planet and Our Responsibility*, https://doi.org/10.1007/978-3-030-31125-4_25

Bending the Curve: Four Building Blocks, Three Levers, and Ten Solutions

The Paris Agreement is an historic achievement. For the first time, effectively all nations have committed to limiting their greenhouse gas emissions and taking other actions to limit global temperature change (Fig. 25.1). Specifically, 197 nations have agreed to hold "the increase in the global average temperature to well below 2 °C above pre-industrial levels and pursue efforts to limit the temperature increase to 1.5 °C above pre-industrial levels" and achieve carbon neutrality in the second half of this century (UNFCCC, 2015).

The climate has already warmed by 1 °C (IPCC, 2013). The problem is running ahead of us, and under current trends we will likely reach 1.5 °C in less than 15 years (Xu and Ramanathan 2017) and surpass the 2 °C guardrail by midcentury with a 50% probability of reaching 4 °C by the end of the century (IPCC, 2014b; Ramanathan & Feng, 2008; World Bank, 2013). Warming in excess of 3 °C is likely to be a global catastrophe for three major reasons:

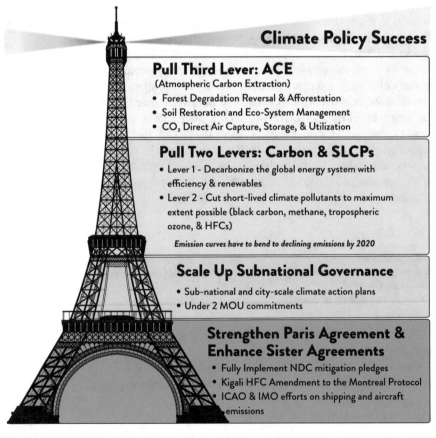

Climate Policy Success

Pull Third Lever: ACE
(Atmospheric Carbon Extraction)
• Forest Degradation Reversal & Afforestation
• Soil Restoration and Eco-System Management
• CO₂ Direct Air Capture, Storage, & Utilization

Pull Two Levers: Carbon & SLCPs
• Lever 1 - Decarbonize the global energy system with efficiency & renewables
• Lever 2 - Cut short-lived climate pollutants to maximum extent possible (black carbon, methane, tropospheric ozone, & HFCs)
Emission curves have to bend to declining emissions by 2020

Scale Up Subnational Governance
• Sub-national and city-scale climate action plans
• Under 2 MOU commitments

Strengthen Paris Agreement & Enhance Sister Agreements
• Fully Implement NDC mitigation pledges
• Kigali HFC Amendment to the Montreal Protocol
• ICAO & IMO efforts on shipping and aircraft emissions

Fig. 25.1 Four building blocks to achieve climate policy success. *HFC* hydrofluorocarbon, *MOU* memorandum of understanding, *NDC* nationally determined contribution, *SLCPs* short-lived climate pollutants

- Warming in the range of 3–5 °C is suggested as the threshold for several tipping points in the physical and geochemical systems; warming of about 3 °C carries a >40% probability of crossing over multiple tipping points, while warming close to 5 °C increases the probability to nearly 90% in comparison with baseline warming of less than 1.5 °C, which carries just over a 10% probability of exceeding any tipping point (Ramanathan & Feng, 2008).
- Health effects of such warming are emerging as a major if not dominant source of concern. Warming of 4 °C or more will expose more than 70% of the population—that is, about seven billion people by the end of the century—to deadly heat stress (Mora et al., 2017) and expose about 2.4 billion to vector-borne diseases such as dengue, chikungunya, and Zika virus, among others (Proestos et al., 2015; Ramanathan et al., 2017; WHO, 2016; Watts et al., 2015).
- Ecologists and paleontologists have postulated that warming in excess of 3 °C, accompanied by increased acidity of the oceans with the buildup of CO_2, could become a major causal factor for exposing more than 50% of all species to extinction. Twenty percent of species are in danger of extinction now due to population increase, habitat destruction, and climate change (Dasgupta et al., 2015).

The good news is that there may still be time to avert such catastrophic changes. The Paris Agreement and supporting climate policies must be strengthened substantially within the next 5 years to bend the emissions curve down faster, stabilize the climate, and prevent catastrophic warming. To the extent those efforts fall short, societies and ecosystems will be forced to contend with substantial needs for adaptation—a burden that will fall disproportionately on the poorest three billion people, who are least responsible for causing the climate change problem (Pope Francis, 2015).

Here we propose an emissions pathway and a policy roadmap with a realistic and reasonable chance of limiting global temperatures to safe levels and preventing unmanageable climate change—an outline of specific science-based policy pathways that serve as the building blocks for a three-lever strategy that could limit warming to well under 2 °C. The projections and the emission pathways proposed in this summary are based on a combination of published recommendations and new model simulations conducted by the authors of this study (see Fig. 25.2).

We have framed the plan in terms of four building blocks and three levers (Fig. 25.1), which would be implemented through ten solutions. The first building block would be full implementation of the nationally determined mitigation pledges under the Paris Agreement of the United Nations (UN) Framework Convention on Climate Change (UNFCCC). In addition, several sister agreements that provide targeted and efficient mitigation must be strengthened. Sister agreements include the Kigali Amendment to the Montreal Protocol to phase down hydrofluorocarbons (HFCs), efforts to address aviation emissions through the International Civil Aviation Organization (ICAO) and maritime black carbon emissions through the International Maritime Organization (IMO), and the commitment by the eight countries of the Arctic Council to reduce black carbon emissions by up to 33% (Arctic Council, 2017a, 2017b; ICAO, 2016; IMO, 2015; UNEP, 2016a). There are many other complementary processes that have drawn attention to specific actions on

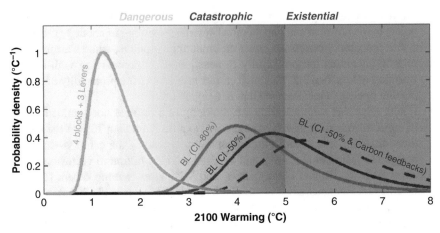

Fig. 25.2 Projected warming in four different scenarios from the preindustrial era to 2100, adopted from Xu and Ramanathan (PNAS, Vol. 114. No. 9, PP 10315–10323; Xu & Ramanathan, 2017). The warming is shown in terms of probability distribution instead of a single value because of uncertainties in climate feedbacks, which could make the warming greater or lesser than the central value shown by the peak probability density value. The *three curves on the right side* indicated by BL (for baseline) denote projected warming in the absence of climate policies. The baseline with a confidence interval of 80% (BL (CI-80%)) is for the scenario in which the energy intensity (the ratio of energy use to economic output) of the economy decreases by 80% in comparison with its value in 2010. For BL (CI-50%), the energy intensity decreases by only 50%. These scenarios bound the energy growth scenarios considered by Intergovernmental Panel on Climate Change Working Group III (IPCC-WGIII) (2014). The *extreme right curve*, BL (CI-50% and carbon feedbacks), includes the carbon cycle feedback due to the warming caused by the BL (CI-50%) case. The carbon cycle feedback adopts IPCC-recommended values for the reduction in CO_2 uptake by the oceans as a result of the warming, the release of CO_2 by melting permafrost, and the release of methane by wetlands. The *green curve* adopts the four building blocks and the three levers proposed in this chapter. There are four mitigation steps: (1) Improve the energy efficiency and decrease the energy intensity of the economy by as much as 80% from its 2010 value. This step alone will decrease the warming by 0.9 °C (1.6 °F) by 2100. (2) Bend the carbon emission curve further by switching to renewables before 2030 and achieving carbon neutrality within three decades. This step will decrease the warming by 1.5 °C (2.7 °F) by 2100. (3) Bend the short-lived climate pollutants curve, beginning in 2020, following the actions California has demonstrated. This step will decrease the warming by as much as 1.2 °C (2.2 °F) by 2100. (4) In addition, extract as much as 1 trillion tons (about half of what we have emitted so far) from the atmosphere by 2100. This step will decrease the warming by as much as 0.3 °C to 0.6 °C (0.5 °F to 1 °F). The 50% probable warming values in the four scenarios are (*from left to right*) 1.4 °C (2.5 °F), 4.1 °C (7.4 °F), 5 °C (9 °F), and 5.8 °C (10.4 °F), respectively. There is a 5% probability that the warming in the four scenarios could exceed (*from left to right*) 2.2 °C (4 °F), 5.9 °C (10.6 °F), 6.8 °C (12.2 °F), and 7.7 °C (14 °F), respectively. The *risk categories shown at the top* largely follow Xu and Ramanathan (2017), with slight modifications. Following the IPCC and Xu and Ramanathan (2017), we denote warming in excess of 1.5 °C as dangerous. Following the burning embers diagram from the IPCC as updated by O'Neill et al. (2017), warming in excess of 3 °C is denoted as catastrophic. We invoke recent literature on the health effects >4 °C warming, impacts including mass extinction with >5 °C warming, and a projected collapse of natural systems with warming in excess of 3 °C, to denote >5 °C warming as exposing the global population to existential threats

climate change, such as the Group of Twenty (G20), which has emphasized reform of fossil fuel subsidies, and the Climate and Clean Air Coalition (CCAC). HFC measures, for example, can avoid as much as 0.5 °C of warming by 2100 through the mandatory global phasedown of HFC refrigerants within the next few decades, and substantially more through parallel efforts to improve the energy efficiency of air conditioners and other cooling equipment, potentially doubling this climate benefit (Shah, Wei, Letschert, & Phadke, 2015; Xu et al. 2013; Zaelke, Andersen, & Borgford-Parnell, 2012).

For the second building block, numerous subnational and city-scale climate action plans have to be scaled up. One prominent example is California's Under 2 Coalition, signed by over 177 jurisdictions from 37 countries in six continents, covering a third of the world economy. The goal of this memorandum of understanding is to catalyze efforts in many jurisdictions that are comparable with California's target of 40% reductions in CO_2 emissions by 2030 and 80% reductions by 2050—emission cuts that, if achieved globally, would be consistent with stopping warming at about 2 °C above preindustrial levels (Under2MOU, 2017). Another prominent example is the climate action plans devised by over 50 cities and 65 businesses around the world, aiming to cut emissions by 30% by 2030 and 80–100% by 2050. There are concerns that the carbon-neutral goal will hinder economic progress; however, real-world examples from California and Sweden since 2005 offer evidence that economic growth can be decoupled from carbon emissions, and the data for CO_2 emissions and gross domestic product (GDP) reveal that growth in fact prospers with a green economy (Saha & Muro, 2016).

The third building block consists of two levers that we need to pull as hard as we can: one for drastically reducing emissions of short-lived climate pollutants (SLCPs), beginning now and completing this by 2030, and the other for decarbonizing the global energy system by 2050 through efficiency and renewables. Pulling both levers simultaneously could keep the global temperature rise below 2 °C through the end of the century (Shindell et al., 2012, 2017; Shoemaker, Schrag, Molina, & Ramanathan, 2013). If we bend the CO_2 emissions curve through decarbonization of the energy system such that global emissions have peaked in 2020, decrease steadily thereafter, and reach zero in 2050, there is less than a 20% probability of exceeding 2 °C warming (IPCC, 2013; Xu & Ramanathan, 2017). This call for bending the CO_2 curve by 2020 is one key way in which this chapter's proposal differs from the Paris Agreement, and it is perhaps the most difficult task of all those envisioned here. Many cities and jurisdictions are already on this pathway, thus demonstrating its scalability. Achieving carbon neutrality and reducing emissions of SLCPs would also drastically reduce air pollution globally, including air pollution in all major cities, thus saving millions of lives and over 100 million tons of crops lost to air pollution each year (Shindell et al., 2012). In addition, these steps would provide clean energy access for the world's poorest three billion people, who are still forced to resort to nineteenth-century technologies to meet basic needs such as cooking (Dasgupta et al., 2015; IPCC, 2014a).

For the fourth and final building block, we are adding a third lever, ACE (atmospheric carbon extraction, also known as carbon dioxide removal (CDR)). This lever is added as insurance against unwanted surprises (due to policy lapses, mitigation delays, or nonlinear climate changes) and would require development of scalable measures for removing the CO_2 already in the atmosphere. The amount of CO_2 that must be removed will range from negligible (if the emissions of CO_2 from the energy system and SLCPs have started to decrease by 2020 and carbon neutrality is achieved by 2050) to a staggering one trillion tons (if CO_2 emissions continue to increase until 2030 and the carbon lever is not pulled until after 2030) (Xu & Ramanathan, 2017). This issue is raised because the NDCs (nationally determined contributions) accompanying the Paris Agreement would allow CO_2 emissions to increase until 2030 (UNEP, 2016b). We call on economists and experts in political and administrative systems to assess the feasibility and cost effectiveness of reducing carbon and SLCP emissions beginning in 2020, in comparison with delaying it by 10 years and then being forced to pull the third lever to extract one trillion tons of CO_2 from the atmosphere.

The fast mitigation plan of requiring emissions reductions to begin by 2020—which means that many countries already need to be cutting their emissions now—is urgently needed to limit the warming to well under 2 °C. Climate change is not a linear problem. Instead, we are facing nonlinear climate tipping points that can lead to self-reinforcing and cascading climate change impacts (Lenton et al., 2008). Tipping points and self-reinforcing feedbacks are wild cards that are more likely with increased temperatures, and many of the potential abrupt climate shifts could happen as warming goes from 1.5 °C in 15 years to 2 °C by 2050, with the potential to push us well beyond any possibility of achieving Paris Agreement goals (Drijfhout et al., 2015).

Ten Scalable Solutions to Bend the Curve

The four building blocks and the three levers require global mobilization of human, financial, and technical resources. For the global economy and society to achieve such rapid reductions in SLCPs by 2030 and carbon neutrality and climate stability by 2050, we will need multidimensional and multisectoral changes and modifications to bend the warming curve, which are grouped under the ten scalable solutions listed below. We have adapted these solutions with some modifications from a book chapter titled "Bending the Curve," written by 50 researchers from the University of California system (Ramanathan et al., 2015). The ten solutions are grouped under six clusters as listed below:

Science Pathways Cluster

1. Show that we can bend the warming curve immediately by reducing SLCPs and long-term by replacing current fossil fuel energy systems with carbon-neutral technologies.

Societal Transformation Cluster

2. Foster a global culture of climate action through coordinated public communication and education at local to global scales.
3. Build an alliance among science, religion, health care, and policy to change behavior and garner public support for drastic mitigation actions.

Governance Cluster

4. Build upon and strengthen the Paris Agreement, and strengthen sister agreements like the Montreal Protocol's Kigali Amendment to reduce HFCs.
5. Scale up subnational models of governance and collaboration around the world to embolden and energize national and international action. California's Under 2 Coalition and the climate action plans devised by over 50 cities are prime examples.

Market-Based and Regulation-Based Cluster

6. Adopt market-based instruments to create efficient incentives for businesses and individuals to reduce CO_2 emissions.
7. Target direct regulatory measures—such as rebates and efficiency and renewable energy portfolio standards—for high-emissions sectors not covered by market-based policies.

Technology Cluster

8. Promote immediate and widespread use of mature technologies such as photovoltaics, wind turbines, biogas, geothermal energy, batteries, hydrogen fuel cells, electric light-duty vehicles, and more efficient end-use devices, especially in lighting, air conditioning and other appliances, and industrial processes. Aggressively support and promote innovations to accelerate the complete electrification of energy and transportation systems and improve building efficiency.
9. Immediately make maximum use of available technologies combined with regulations to reduce methane emissions by 50%, reduce black carbon emissions by 90%, and eliminate hydrofluorocarbons with high global-warming potential (high-GWP HFCs) ahead of the schedule in the Kigali Amendment while fostering energy efficiency.

Atmospheric Carbon Extraction Cluster

10. Regenerate damaged natural ecosystems and restore soil organic carbon. Urgently expand research and development of atmospheric carbon extraction, along with carbon capture, utilization, and storage.

Concluding Remarks: Where Do We Go from Here?

A massive effort will be needed to stop warming at 2 °C, and time is of the essence. With unchecked business-as-usual emissions, global warming has a 50% likelihood of exceeding 4 °C and a 5% probability of exceeding 6 °C in this century, raising existential questions for most, but especially the poorest three billion people. Dangerous to catastrophic impacts on the health of people (including generations yet to be born), on the health of ecosystems, and on species extinction have emerged as major justifications for keeping climate change well below 2 °C, although we must recognize that the intrinsic uncertainties in climate and social systems make it hard to pin down exactly the level of warming that will trigger possibly catastrophic impacts. To avoid these consequences, we must act now, and we must act fast and effectively. This chapter sets out a specific plan for reducing climate change in both the near and long terms. With aggressive urgent actions, we can protect ourselves. Acting quickly to prevent catastrophic climate change by decarbonization will save millions of lives, trillions of dollars in economic costs, and massive suffering and dislocation of people around the world. This is a global security imperative, as it can avoid the migration and destabilization of entire societies and countries, and reduce the likelihood of environmentally driven civil wars and other conflicts.

We must address everything from our energy systems to our personal choices to reduce emissions to the greatest extent possible. We must redouble our efforts to invent, test, and perfect systems of governance so that the large measure of international cooperation needed to achieve these goals can be realized in practice. The health of people for generations to come and the health of ecosystems crucially depend on an energy revolution beginning now that will take us away from fossil fuels and toward the clean renewable energy sources of the future. It will be nearly impossible to achieve other critical social goals—for example, the UN's Agenda 2030 and Sustainable Development Goals—if we do not make immediate and profound progress in stabilizing the climate, as we are outlining here.

Fortunately, there is momentum behind climate mitigation. Twenty-four countries have already embarked on a carbon-neutral pathway, and there are numerous living laboratories, including 53 cities, many universities around the world, and the state of California. These laboratories have already created eight million jobs in the clean energy industry; they have also shown that emissions of greenhouse gases and air pollutants can be decoupled from economic growth. We have a long way to go and very little time to achieve carbon neutrality. We need institutions and enterprises that can accelerate this bending by scaling up the solutions that are being proven in the living laboratories. We have less than a decade to put these solutions in place around the world to preserve nature and our quality of life for generations to come. The time is now.

Acknowledgements This chapter is based on a full report, which can be accessed at: http://www. ramanathan.ucsd.edu/about/publications.php. The authors of the report include, in addition to the four authors of this chapter, the following: Ken Alex, Max Auffhammer, Paul Bledsoe, William Collins, Bart Croes, Fonna Forman, Örjan Gustafsson, Andy Haines, Reno Harnish, Mark Jacobson, Shichang Kang, Mark Lawrence, Damien Leloup, Tim Lenton, Tom Morehouse, Walter Munk, Romina Picolotti, Kimberly Prather, Graciela B. Raga, Eric Rignot, Drew Shindell, K. Singh, Achim Steiner, Mark Thiemens, David W. Titley, Mary Evelyn Tucker, Sachi Tripathi, David Victor, and Yangyang Xu.

References

Arctic Council (2017a). *Fairbanks declaration*. Tromsø, Norway: Artic Council.
Arctic Council (2017b). *Expert group on black and carbon methane: Summary of Progress and recommendations*. Tromsø, Norway: Artic Council.
Carbon Neutrality Coalition welcomes new members, pledges renewed ambition at UN Climate Action Summit (23 September 2019), https://www.carbon-neutrality.global/carbon-neutrality-coalition-welcomes-new-members-pledges-renewed-ambition-at-un-climate-action-summit/
Dasgupta, P., Ramanathan, V., Raven, P., Sorondo, Mgr M., Archer, M., Crutzen, P. J., et al. (2015). *Climate change and the common good: A statement of the problem and the demand for transformative solutions*. Vatican, Rome: The Pontifical Academy of Sciences and the Pontifical Academy of Social Sciences.
Drijfhout, S., Bathiany, S., Beaulieu, C., Brovkin, V., Claussen, M., Huntingford, C., et al. (2015). Catalogue of abrupt shifts in intergovernmental panel on climate change climate models. *Proceedings of the National Academy of Sciences of the United States of America, 112*, E5777–E5786.
Intergovernmental Panel on Climate Change (IPCC) (2013). *Climate change 2013: The physical science basis: Working group I contribution to the fifth assessment report of the intergovernmental panel on climate change*. Geneva, Switzerland: IPCC.
Intergovernmental Panel on Climate Change (IPCC) (2014a). *Climate change 2014: Impacts, adaptation, and vulnerability: Working group II contribution to the fifth assessment report of the intergovernmental panel on climate change*. Geneva: IPCC.
Intergovernmental Panel on Climate Change (IPCC) (2014b). *Climate change 2014: Synthesis report: Contribution of working groups I, II, and III to the fifth assessment report of the intergovernmental panel on climate change*. Geneva: IPCC.
International Civil Aviation Organization (ICAO) (2016). *Historic agreement reached to mitigate international aviation emissions*. Montreal: ICAO. Press Release.
International Maritime Organization (IMO) (2015). *Investigation of appropriate control measures (Abatement technologies) to reduce black carbon emissions from international shipping*. London: IMO.
Lenton, T. M., Held, H., Kriegler, E., Hall, J. W., Lucht, W., Rahmstorf, S., et al. (2008). Tipping elements in the Earth's climate system. *Proceedings of the National Academy of Sciences of the United States of America, 105*, 1786–1793.
Mora, C., Dousset, B., Caldwell, I. R., Powell, F. E., Geronimo, R. C., Bielecki, C. R., et al. (2017). Global risk of deadly heat. *Nature Climate Change, 7*, 501–506.
O'Neill, B. C., Oppenheimer, M., Warren, R., Hallegatte, S., Kopp, R. E., Pörtner, H. O., (2017). IPCC reasons for concern regarding climate change risks. *Nature Climate Change, 7*, 28–37.

Pope Francis. (2015). Praise Be To You *Laudato Si: On care for our common home*. Vatican City, Libreria Editrice Vaticana. ISBN 978-1-68149-677-2.

Proestos, Y., Christophides, G. K., Ergüler, K., Tanarhte, M., Walkdock, J., & Lelieveld, J. (2015). Present and future projections of habitat suitability of the Asian tiger mosquito, a vector of viral pathogens, from global climate simulation. *Philosophical Transactions of the Royal Society, B: Biological Sciences, 370*, 1–16.

Ramanathan, V., & Feng, Y. (2008). On avoiding dangerous anthropogenic interference with the climate system: Formidable challenges ahead. *Proceedings of the National Academy of Sciences of the United States of America, 105*(38), 14245–14250.

Ramanathan, V., Molina, M. J., Zaelke, D., Borgford-Parnell, N., Xu, Y., Alex, K., et al. (2017). *Full: Well under 2 degrees celsius: Fast action policies to protect people and the planet from extreme climate change*. Washington, DC: Institute of Governance and Sustainable Development.

Ramanathan, V., Allison, J. E., Auffhammer, M., Auston, D., Barnosky, A. D., Chiang, L., et al. (2015). *Bending the curve: Executive summary*. Oakland, CA: University of California. Retrieved on February 16, 2020 from https://uc-carbonneutralitysummit2015.ucsd.edu/_files/Bending-the-Curve.pdf

Saha D. & Muro M. (2016). *Growth, carbon, and trump: State progress and drift on economic growth and emissions 'decoupling'*. Brookings report: Metropolitan Policy Program. Retrieved on February 16, 2020 from https://www.brookings.edu/research/growth-carbon-and-trump-state-progress-and-drift-on-economic-growth-and-emissions-decoupling/

Shah, N., Wei, M., Letschert, V., & Phadke, A. (2015). *Benefits of leapfrogging to Superefficiency and low global warming potential refrigerants in air conditioning*. Berkeley, CA: Lawrence Berkeley National Laboratory.

Shindell, D., Kuylenstierna, J. C. I., Vignati, E., van Dingenen, R., Amann, M., Klimont, Z., et al. (2012). Simultaneously mitigating near-term climate change and improving human health and food security. *Science, 335*, 183–189.

Shindell, D., Borgford-Parnell, N., Brauer, M., Haines, A., Kuylenstierna, J. C. I., Leonard, S. A., et al. (2017). A climate policy path- way for near- and long-term benefits. *Science, 356*, 493–494.

Shoemaker, J. K., Schrag, D. P., Molina, M. J., & Ramanathan, V. (2013). What role for short-lived climate pollutants in mitigation policy? *Science, 42*, 1323–1324.

UN Framework Convention on Climate Change (UNFCCC) (2015). *Paris agreement*. Bonn: UNFCCC.

Under2MOU (2017). *The memorandum of understanding (MOU) on subnational global climate leadership*. Subnational Global Climate Leadership Memorandum of Understanding. Retrieved on February 16, 2020 from https://www.theclimategroup.org/sites/default/files/under2-mou-with-addendum-english-a4.pdf

United Nations Environment Programme (UNEP) (2016a). *Decision XXVIII/1: Further amendment of the Montreal protocol*. Nairobi: UNEP.

United Nations Environment Programme (UNEP) (2016b). *The emissions gap report 2016: A UNEP synthesis report*. Nairobi: UNEP.

Watts, N., et al. (2015). Health and climate change: Policy responses to protect public health. *The Lancet, 386*, 1861–1914.

World Bank. (2013). *Turn down the heat: Why a 4°C warmer world must be avoided*. Washington, DC: World Bank.

World Health Organization (WHO) (2016). *Second global conference: Health and climate, conference conclusions and action agenda*. Geneva: WHO.

Xu, Y., & Ramanathan, V. (2017). Well below 2°C: Mitigation strategies for avoiding dangerous to catastrophic climate changes. *Proceedings of the National Academy of Sciences of the United States of America, 114*, 10315–10323.

Xu, Y., Zaelke, D., Velders, G. J. M., & Ramanathan, V. (2013). The role of HFCs in mitigating 21st century climate change. *Atmospheric Chemistry and Physics, 13*, 6083–6089.

Zaelke, D., Andersen, S. O., & Borgford-Parnell, N. (2012). Strengthening ambition for climate mitigation: The role of the Montreal protocol in reducing short-lived climate pollutants. *Review of European Community and International Environmental Law, 21*, 231–242.

CHAPTER 26
Defeating Energy Poverty: Invest in Scalable Solutions for the Poor

Daniel M. Kammen ⓘ

Energy and persistence conquer all things.

—Benjamin Franklin

Summary Energy poverty is arguably the most pervasive and crippling threat society faces today. Lack of access impacts several billion people, with immediate health, educational, economic, and social damage. Furthermore, how this problem is addressed will result in the largest accelerant of global pollution or the largest opportunity to pivot away from fossil fuels onto the needed clean energy path. In a clear example of the power of systems thinking, energy poverty and climate change together present a dual crisis of energy injustice along gender, ethnic, and socio-economic grounds, which has been exacerbated if not outright caused by a failure of the wealthy to see how tightly coupled our global collective fate is if addressing climate change fairly and inclusively does not become an immediate, actionable priority.

While debate exists on the optimal path or paths to wean our economy from fossil fuels, there is no question that technically we now have sufficient knowledge and a sufficient technological foundation to launch and to even complete decarbonization. What is critically needed is an equally powerful social narrative to accelerate the clean energy transition. *Laudato Si'* provides a compelling foundation built on the narrative around the health, climate, and social benefits of a global green energy transition. The Green New Deal in the USA is a political movement that grows from this new understanding of sound stewardship and respect for the planet and its inhabitants.

This chapter presents examples and formulation of an action agenda to defeat energy poverty and energy injustice.

D. M. Kammen (✉)
Energy and Resources Group & Goldman School of Public Policy,
University of California, Berkeley, Berkeley, CA, USA
e-mail: kammen@berkeley.edu

© The Author(s) 2020
W. K. Al-Delaimy, V. Ramanathan, M. Sánchez Sorondo (eds.), *Health of People, Health of Planet and Our Responsibility*, https://doi.org/10.1007/978-3-030-31125-4_26

The Multifaceted Nature of Energy Poverty and Injustice

In what has become one of the most important and widely quoted assessments of the current state of energy poverty, we now hear daily that "over one billion people today lack access to modern energy services."

This striking statement has become the rallying cry of the United Nations, the World Bank, and a myriad of nongovernmental groups. What is less widely discussed is how central energy poverty is related to lack of food quality, high levels of indoor air pollution, and lack of quality water resources. Energy poverty is health poverty.

A Framework for Learning and Innovation

A variety of factors—all shameful in hindsight—have retarded the progress of a pro-poor, pro-environment agenda for research, testing, and action. The problems of household particulate pollution—black carbon—and trace gas pollution from indoor cooking and heating have been known and studied since the 1950s, and yet only recently has there been any significant scale-up of the solutions.

These are many examples of neglected yet critically needed science for these seemingly commonplace or mundane issues. Cookstoves, clean energy, climate change, and social, ethnic, and other injustices all fall squarely into this area and have arguably been under-researched. One useful theoretical framework in which to assess the impact of innovations is to distinguish between "fundamental science" and "use-inspired" or applications-motivated science. In his seminal book *Pasteur's Quadrant*, the political scientist Donald Stokes summarized this value in a classic 2 × 2 matrix of opportunities (Fig. 26.1).

It is clear that raising the mundane to an intellectual, political, and commercial level comparable to other texts and tenets of society is vitally needed for the natural, social, psychological, and policy tools required for energy access sustainability. If we utilize it effectively, *Laudato Si'* can be one of these grand ethical and intellectual levers.

Energy poverty exists at many scales: from the classic "three-stone fires" of rural homes in the poorest nations to rapidly growing urban communities, to rich nations that deny their own citizens even the most basic forms of energy infrastructure. In the next three sections, we illustrate the power of an approach that values the Pasteur's quadrant perspective for (a) rural cookstoves, (b) mini-grids for community energy services, and (c) the need and opportunity to transform the dirtiest large-scale grids to meet both household and industrial energy needs. Sadly, while attention is often focused on the lack of energy for the poor, we recently found that even where rooftop solar power is expanding rapidly, such as in parts of the USA, access is often socially uneven and unjust. In a survey across the USA, we recently found 40–70% lower rates of photovoltaics (PV) in deployment in Latino and African American neighborhoods than the average, and 20–30% higher rates of deployment in white majority areas (Sunter, Castellanos, & Kammen, 2019).

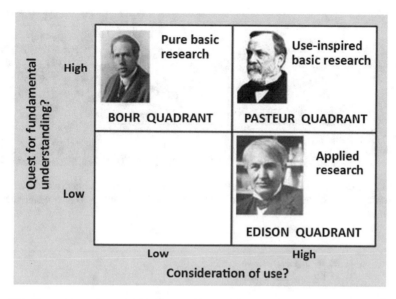

Fig. 26.1 Pasteur's quadrant and the power of formulating formerly applied research as fundamental (Stokes, 1997). The empty "Low–Low" quadrant has been variously and humorously described, generally with self-deprecation as the zone of "university presidents" or of "social media gadflies"

The Intra-household Scale: Cookstoves, Women, and Health

Women—who gather or purchase fuel, cook, and handle fire considerably more frequently than men—also have much higher exposure to the health impacts of energy use, including respiratory or eye diseases due to indoor smoke, burns, or back pain, and injuries from carrying heavy loads. While these increased health risks have long been recognized (Kammen & Lankford, 1990; Kammen, 1995), detailed analysis illustrates that even the scientific tools used for quantifying exposure and disease relationships need to pay attention to details of women's work to properly assess the magnitude of the health risks associated with household energy tasks.

A rich literature exists on the exposure to indoor smoke under actual conditions of use and has shown that stove emissions are highly episodic and that peaks in emissions commonly occur when fuel is added or moved, the stove is lit, the cooking pot is placed on or removed from the fire, or food is stirred (Fig. 26.2; Ezzati & Kammen, 2001, Bailis, Ezzati, & Kammen, 2005). Quantitative and qualitative data on time–activity budgets also indicate that female household members are consistently closest to the fire when the pollution level is the highest. Other household members may be outside or away from the house at such times, especially during the hours when the fire is lit or extinguished.

In a long-term study of 500 villagers in central Kenya, we explored both indoor air pollution and health impacts on families using a range of stove and fuel combinations (Bailis et al., 2005; Ezzati & Kammen 2001).

Fig. 26.2 Women collecting firewood in Kitui District, Kenya. Ethiopian injera stoves showing a common open access design with ease of use but also high energy loss. (Photo credits: Daniel M. Kammen)

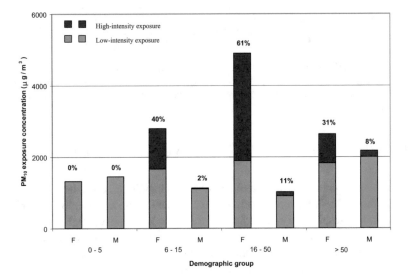

Fig. 26.3 Total daily exposure to particulate matter 10 μm or less in diameter (PM$_{10}$), subdivided into high-intensity (*darker shade*) and low-intensity (*lighter shade*) exposure of each demographic subgroup (Females and Males, by age), where the total height of the column is the group average exposure concentration divided into the average for high- and low-intensity components. The percentages indicate the share of total exposure that is high-intensity exposure. The high-intensity component of exposure occurs in less than 1 h, emphasizing the intensity of exposure in these episodes. For details, see Ezzati and Kammen (2001)

Figure 26.3 shows the exposure data obtained using an approach that considers the full exposure patterns and profile of individuals, divided into exposure during high-intensity and low-intensity episodes, respectively.

Increased access to clean energy sources can improve the day-to-day and long-term welfare of female household members. In fact, it has been argued that the increased prominence of biomass as an economic commodity (e.g., as a source of energy for small-scale manufacturing) has attracted local entrepreneurs and business actors—mostly men—driving women to assume more marginal social roles and depend on inferior sources of energy (Kammen, 1995). In our study, we found that the transition from traditional wood-burning stoves to the cleanest charcoal-burning stoves could cut pollution levels so significantly—largely for women and children—that there was literally no more cost-effective intervention (Ezzati & Kammen, 2001) on a DALY (disability-adjusted life year) than childhood immunizations and improved cookstoves.

Cookstoves and cookstove programs have also been the subject of repeated criticisms that the stoves do not reduce pollution in practice in the field and thus the claimed benefits are illusory (see, for example, Gunther, 2015). As comes as no surprise to those who have studied not only stoves but the full life-cycle of maintenance, deployment, and education, stove/fuel/community partnerships that use the best stoves and focus on human economic and health benefits, and not just hardware, have proven successful (Goodman, 2018). Here again, we see a need to look beyond a technical fix and embrace the needed evolution from mundane science (Dove & Kammen, 2015; Kammen & Dove, 1997) to fulfil the full narrative of use-inspired basic research—namely, that of Pasteur's quadrant.

The Community Scale: Mini-grids

The International Energy Agency (IEA) projects that over 900 million people in rural areas will remain without electricity by 2030, in contrast to only about 100 million in urban areas, with the vast majority being in sub-Saharan Africa (Alstone, Gershenson, & Kammen, 2015). Sustainable Energy for All (SE4ALL, 2013), using data from the IEA, expects that achieving universal access will require grid extension for all new urban connections and 30% of rural populations, with the remaining 70% of rural people gaining access through decentralized solutions (via mini-grids, solar home systems (SHS), and intra-household (or "pico-") products, known widely as "pay-as-you-go" [PAYG] energy access technologies). In fact, a legitimate argument exists that despite "utility apologist" arguments, mini-grids will in time be cheaper, more reliable, and cleaner than large-grid systems (Casillas & Kammen, 2010).

The economic limitations of the rural poor are reflected in their low energy consumption, struggle to pay connection fees, and challenges in procuring household wiring and appliances. In fact, many households and businesses in "electrified" areas lack access, even directly beneath power lines. People and communities without property rights may lack the stability to justify investments in fixed infrastructure, or may not have permission from central authority to do so. Figure 26.4 presents a representative mini-grid in Sabah, Malaysian Borneo, showing solar power (and micro-hydropower) powering telecommunications hardware and village lighting.

Figure 26.5 provides an overview of energy service provision from PAYG, mini-grid, and large-grid systems. The battle for customers is one that distributed PAYG and mini-grid systems are increasingly winning—one home and 1 kW at a time.

Fig. 26.4 A village micro-grid energy and telecommunications system in the Crocker Highlands of Sabah, Malaysian Borneo. The system serves a community of 200 and provides household energy services, telecommunications, and a satellite communications (dish shown), water pumping for fish ponds (seen in the *center*), and refrigeration. The supply includes micro-hydropower and solar generation (one small panel is shown here; others are distributed on building rooftops). (Photo credit: Daniel M. Kammen)

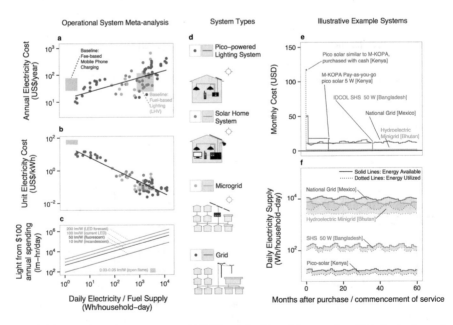

Fig. 26.5 Mapping of pay-as-you-go (PAYG), mini-grid, and large-scale grid technologies and levels of service provision (Alstone, Gershenson, and Kammen, 2015). *IDCOL* Infrastructure Development Company Limited, *LED* light-emitting diode, *SHS* solar home systems

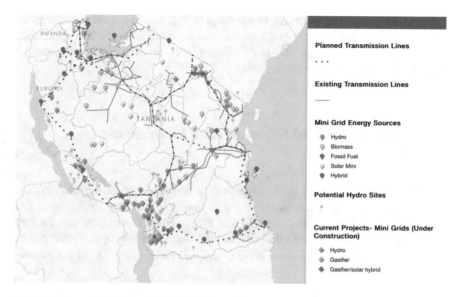

Fig. 26.6 Map of existing and planned utility grids and the proliferation of mini-grid systems across Tanzania. (Source: http://www.wri.org/resources/maps/tanzania-energy-access-maps). For additional mini-grid assessments, see Schnitzer et al., 2014. Similar maps are emerging worldwide

Globally, utility-scale grid expansion has roughly kept pace with the increase in the global population. In 2016, about 1.2 billion people were completely off-grid and many ostensibly connected people in the developing world experienced significant outages that ranged from 20 to 200+ days a year. Current forecasts are that this number will remain roughly unchanged until 2030, which would relegate a significant portion of the population and the economies of many of the neediest countries on earth to fragile, under productive lives, with fewer options than they could otherwise have (IPCC, 2011) (Fig. 26.6).

Scientific Advances in Off-Grid Systems

Mini-grids and products for individual user end-use, such as solar home systems, have benefited from dramatic price reductions and performance advances in solid-state electronics, cellular communications technologies, and electronic banking, and from the dramatic decrease in solar energy costs and more recently in storage costs (Kittner, Lil, & Kammen, 2017). This mix of technological and market innovation has contributed to a vibrant new energy services sector, which in many nations has outpaced traditional grid expansion.

High-performance, low-cost photovoltaic generation paired with advanced batteries and controllers provide scalable systems across much larger power ranges than central generation, from megawatts down to fractions of a watt. As these mini-grid and reliable utility-scale grid systems become more common and affordable, electrified cooking and thus very-low-pollution home environments become possible—further improving health outcomes.

The Utility Scale: Grid Expansion and Its Dramatic Alternatives

In 2008, I was approached by a consortium of nongovernmental organizations in Sabah, Malaysia, to offer alternatives to a coal-fired power plant. Sabah, a small and generally well-managed state, is the former British North Borneo, famous for rare megafauna and Mount Kinabalu. Our report *Clean Energy Options for Sabah* (McNish, Kammen, & Gutierrez, 2010) (Fig. 26.7) and a great deal of community

Fig. 26.7 *Top left*: The Renewable and Appropriate Technology Laboratory (RAEL) report, *Clean Energy Options for Sabah*. *Top right*: Presenting the Report to the environment minister. *Bottom left*: Coverage in *The Borneo Post*: "Anti-coal stand has support." *Bottom right*: Large public turnouts at events to discuss and debate the report's findings

education and outreach helped to overturn a decision to purchase a 300-megawatt (MW) used coal-fired power plant from China in favor of investment in expanded distributed renewables, an improved grid, and natural gas. This victory helped to connect our team to a more challenging case, that of the larger and more ecologically and culturally brutalized state of Sarawak.

The victory against big coal in Sabah inspired a new kind of discussion around energy to happen across the region. Activists from throughout Asia working on campaigns against "dirty" energy projects like coal and dams were talking to clean and appropriate energy practitioners.

In 2002, as Sarawak was completing the Bakun Dam, one indigenous group— the Kenyah-Badang community—that would have been displaced refused to accept the resettlement package and instead relocated their settlement just outside the dam's reservoir. Gara Jalong, the tenacious and charismatic leader of Long Lawen, had led the village against the many timber companies encroaching on indigenous lands in the 1980s, then against dam developers in the 1990s. Gara also sought out professionals with experience in rural electrification via micro-hydropower, to provide a sustainable source of energy for his people.

A Clever Deception: The Sarawak Corridor of Renewable Energy

Malaysian Borneo, like much of Southeast Asia, has experienced not only a brutal wave of deforestation but also a surge in mega-hydropower projects, ostensibly to facilitate industrial productivity and consumption. The Three Gorges Dam in China was completed in 2006, while the Nam Theun Dam (completed in 2010) and the Xayaburi Dam (under construction) in Laos were the first of a series of dams to be built in the trans-boundary Lower Mekong Basin. The island of Borneo, too, has abundant natural resources, immense global ecological importance, exotic megafauna (orangutans, dwarf rhino, and pygmy elephants) and hornbills, a largely rural human population (some of them former headhunters), and an agrarian economy, which is also on the cusp of major industrial dislocation due to hydropower development.

The Sarawak Corridor of Renewable Energy (SCORE) is a corridor in central Sarawak, an East Malaysian state on the island of Borneo. SCORE differs fundamentally from the other Malaysian economic corridor projects in its predominant emphasis on hydropower. At least 12 large hydroelectric dams and two coal power plants, together constituting 9380 MW of capacity, were scheduled to be built before 2030 (Sovacool & Bulan, 2012). Six dams were scheduled to be completed by 2020, with three major dams already at different stages of development.

In 2012, the 2400 MW Bakun dam became operational. At 205 meters high, it is Asia's largest dam outside China. The dam's reservoir submerged 700 km^2 of land and displaced about 10,000 people. In 2013, the 944 MW Murum Dam was completed and the 1200 MW Baram Dam was scheduled for construction.

These projects—SCORE—summed to a targeted ninefold increase in energy output between 2010 and 2020, or from 5921 to 54,947 GWh, representing a 16% annual growth rate. In terms of installed capacity, this would translate into an expansion from 1300 MW in 2010 to between 7000 and 8500 MW in 2020. No extant demand existed for this power, yet the ecological and community damage would be extreme.

The story of SCORE was further complicated because with so many proposed dams but little quantitative analysis of the energy options or cost and benefit trade-offs in the literature or the public discussion over this conflict, it was very hard for opposition to develop a clear narrative.

In response, we developed a modeling tool to examine the energy costs and land impacts of different energy choices. While that model is not described in technical detail here, the interested reader is directed to our papers that detail the analytic tool itself, the data sets, and many of the specific results (Shirley & Kammen, 2015). Our results appeared locally in late 2015, and we immediately began to hold open meetings to present these models and our findings (Fig. 26.8).

Our findings were stark and robust: Sarawak's already installed capacity including Bakun already exceeded expected demand in 2030 under the business-as-usual (BAU) growth assumption, so there was no additional need to build more capacity under the government's own BAU growth forecast. The Bakun Dam itself could provide more than 10,000 GWh per annum. Under a 7% electricity demand growth assumption, this represented half of the expected demand in 2030. Even under the more aggressive 10% growth assumption, Bakun alone would satisfy a third of the demand in 2030. Completion of the two additional dams already under construction (Murum and Baram) would oversupply the 2030 demand under 7% growth, leading to a large excess capacity, and would require only a marginal amount of additional generation under 10% growth. Community energy—mini-grids—would preserve forests and provide local energy without household pollution.

Fig. 26.8 *Left*: Activist team dinner with Dr. Rebekah Shirley (*second from left*) and Prof. Daniel Kammen (*middle*), flanked by community activist leaders from across Sabah and Sarawak. *Right*: Peter Kallang, the chairman of SAVE Rivers (*far left*), Gabriel Sundoro Wynn, the Asia regional director for Green Empowerment (*second from left*), Prof. Daniel Kammen (*second from right*), and See Hee Chow, a lawyer and Sarawak assemblyman (*far right*)

Fig. 26.9 Protesters citing (and holding) copies of the Shirley and Kammen study add ammunition to the wider social dialogue around the need for mega-dams and the dubious possibility of billions in foreign investment (from the article in *The Borneo Post*)

In response to the new study, press coverage (Fig. 26.9), indigenous protests, and a meeting I had with the chief minister, something remarkable and unexpected happened. On July 30, 2015, a moratorium on further work on the Baram Dam was announced by the chief minister.

Then, on March 21, 2016, a legal decision to solidify this position was announced: the Government of Sarawak revoked the classification of the land that the dam developers had used to extinguish the native rights of ownership of the contested dam site. This remarkable turnaround victory was a demonstration of local community activism and the importance of science communication, as research published by the Renewable and Appropriate Energy Laboratory, identifying financially viable commercial power production alternatives for the state, played a major role in the developments.

No environmental victory is ever final, of course. One moment's victory then demands constant vigilance, as anything can be overturned by a change in a political party, a shocking election result, or illegal land grabs and invasions. In the case of the Baram Dam, the paper was clear, but shortly thereafter the chief minister passed away. While the law stands, what was once final is never quite so final with the inevitable change of officials. For now, however, things are in a good place.

Assessment: The Benefits of Diverse Technology Options to Provide Energy Services

With these technological cornerstones at the household to community to large scales, aid organizations, governments, academia, and the private sector are developing and supporting a wide range of approaches to serve the needs of the poor, including pico-lighting devices (often very small 1- to 2-watt solar panels charging lithium ion batteries, which in turn power low-cost, high-efficiency light-emitting diode [LED] lights), solar home systems, and community-scale micro- and mini-grids. Decentralized systems are clearly not complete substitutes for a reliable grid connection, but they represent an important level of access until a reliable grid is available and feasible. They provide an important platform from which to develop more distributed energy services. By overcoming access barriers, often through market-based structures, these systems provide entirely new ways to bring energy services to the poor and formerly unconnected people.

Meeting peoples' basic lighting and communication needs is an important first step on the "modern electricity service ladder" (Masera, Saatkamp, & Kammen, 2000; Saatkamp, Masera, & Kammen, 2000). Eliminating kerosene lighting from a household improves household health and safety while providing significantly higher quality and greater quantities of light. Fuel-based lighting is a $20 billion industry in Africa, and tremendous opportunities exist to both reduce energy costs for the poor and improve the quality of service. Charging a rural or village cell phone can cost $5–10/kWh at a pay-for-service station but less than $0.50/kWh via an off-grid product or on a mini-grid.

This investment frees up income and also tends to lead to higher rates of utilization of cell phones and other small devices. Overall, the first few watts of power mediated through efficient end-uses lead to benefits in household health, education, and poverty reduction. Beyond basic needs, there can be a wide range of important and highly valued services from decentralized power (e.g., television, refrigeration, fans, heating, ventilation and air conditioning, and motor-driven applications), depending on the power level and its quality along with demand-side efficiency.

Testing laboratories that rate the quality of lighting products and disseminate the results are an invaluable step in increasing the quality and competitiveness of new entrants into the off-grid and mini-grid energy services space. The *Lighting Global* (https://www.lightingglobal.org) program is one example of an effort that began as an industry watchdog but has now become an important platform that provides market insights; steers quality assurance frameworks for modern, off-grid lighting devices and systems; and promotes sustainability through a partnership with industry.

As we explore the human and environmental impacts of pollution, it comes as no surprise that the range of costs only rises. In a recent study of the legacy of coal in Kosovo, the trace metals recorded in both the coal and coal dust that are deposited in the local area (which contains Pristina, the capital city, only 10 km from the mine and power plant) are at health-damaging levels, and have been for decades (Kittner, Fadadu, Buckley, Schwarzmann, & Kammen, 2018).

An Action Agenda for Clean Energy for Human and Environmental Health and Energy Access

Exploit Information Technology

The information technology revolution has dramatically closed the gap between off-grid, mini-grid, and utility-grid-scale management solutions and costs. In fact, the diversity of new energy service products available and the rapidly increasing demand for information and communication services, water, health, and entertainment in villages worldwide have built a very large demand for reliable and low-cost energy. Combining this demand with the drive for clean energy brings two important objectives, which were for many years seen as being in direct competition with alignment around the suite of new clean energy products that can power village energy services.

To enable and expand this process, a range of design principles that can form a roadmap to clean energy economies emerges:

Establish clear goals at the local level: Universal energy access is the global goal for 2030, but establishing more near-term goals that embody meaningful steps from the present situation will show what is possible and at what level of effort. Cities and villages have begun with audits of energy services, costs, and environmental impacts.

Empower villages as both designers and consumers of localized power: Village solutions necessarily vary greatly, but clean energy resource assessments, evaluation of the needs of the community, planning infrastructure, and training are all key to making the village energy system a success. In a pilot project in rural Nicaragua, once the assessment was complete, movement from evaluation to implementation quickly became a goal of both the community and a local commercial plant.

Make equity a central design consideration: Community energy solutions have the potential to liberate women entrepreneurs and disadvantaged ethnic minorities by tailoring user materials and energy plans to meet the cultural and linguistic needs of these communities. National programs often ignore business specialties, culturally appropriate cooking, and other home energy needs. Thinking explicitly about this is good business and also makes the solutions much more likely to be adopted.

Each of these steps warrants a further narrative, one that relies not only on science and technology but also centrally on the narrative. *Laudato Si'* is such a text: a compelling argument that blends science, technology, and economics in a larger *integral ecology* that is as respectful of culture as it is of community and social justice, but one that at the same time presents a vision of the positive nexus of science, technology, policy, and, above all, culture and community.

Acknowledgements This research was conducted in partnership with Dr. Rebekah Shirley, Green Empowerment, and the community nongovernmental organization TONIBUNG. The Save Sarawak Rivers Network (SAVE Rivers) was the central coordinating group that brought together community leaders, researchers, and public officials, and ensured continued attention and efforts over the years of work, and blockades, of the dams. SAVE Rivers was the group that proposed that the study be carried out and then involved the Bruno Manser Fonds. This research was funded by the Bruno Manser Fonds, the Rainforest Foundation Norway, the Karsten Family Foundation, and the Zaffaroni Family. Gabriel Sundoro Wynn and Peter Kallang provided much-needed corrections, added details, and have been inspiring colleagues throughout this work.

References

Alstone, P., Gershenson, D., & Kammen, D. M. (2015). Decentralized energy systems for clean electricity access. *Nature Climate Change, 5,* 305–314.

Bailis, R. R., Ezzati, M., & Kammen, D. M. (2005). Mortality and greenhouse gas impacts of biomass and petroleum energy futures in Africa. *Science, 308,* 98–103.

Casillas, C., & Kammen, D. (2010). The energy-poverty-climate nexus. *Science, 330,* 1182.

Dove, M., & Kammen, D. (2015). *Science, society and environment: Applying physics and anthropology to sustainability.* Taylor and Francis: London.

Ezzati, M., & Kammen, D. M. (2001). Indoor air pollution from biomass combustion and acute respiratory infections in Kenya: An exposure-response study. *The Lancet, 358,* 619–624.

Goodman, J. (2018). Toxic smoke is Africa's quiet killer. An entrepreneur says his fix can make a fortune, *The New York Times,* December 6.

Gunther, M. (2015). These cheap, clean stoves were supposed to save millions of lives. What happened?, *The Washington Post,* October 29.

Intergovernmental Panel on Climate Change (IPCC) (2011). *Special report on renewable energy sources and climate change mitigation.* Cambridge: Cambridge University Press.

Kammen, D. M., & Lankford, W. F. (1990). Cooking in the sunshine. *Nature, 348,* 385–386.

Kammen, D. M. (1995). Cookstoves for the developing world. *Scientific American, 273,* 72–75.

Kammen, D. M., & Dove, M. R. (1997). The virtues of Mundane science. *Environment, 39,* 10–15. 38–41.

Kittner, N., Fadadu, R., Buckley, H. L., Schwarzmann, M., & Kammen, D. M. (2018). Trace metal presence in lignite coal from Kosovo and associated air pollution-related risk. *Environmental Science & Technology, 52,* 2359–2367.

Kittner, N., Lil, F., & Kammen, D. M. (2017). Energy storage deployment and innovation: A multi-technology model for the clean energy transition. *Nature Energy, 2,* 17125.

Masera, O., Saatkamp, B., & Kammen, D. M. (2000). From fuel switching to multiple cooking fuels: A critique of the energy ladder model in rural households. *World Development, 28,* 2083–2103.

McNish, T., Kammen, D. M., & Gutierrez, B. (2010). *Clean energy options for Sabah.* Sabah, Malaysia: Sabah Unite to Re-power the Future & World Wide Fund for Nature, WWF.

Saatkamp, B., Masera, O., & Kammen, D. M. (2000). Energy and health transitions in development science and planning: Fuel use, stove technology, and morbidity in Jarácuaro, México. *Energy for Sustainable Development, 4,* 5–14.

Schnitzer, D., Lounsbury, D. S., Carvallo, J. P., Deshmukh, R., Apt, J., & Kammen, D. M. (2014). *Microgrids for rural electrification: A critical review of best practices based on seven case studies.* New York, NY: United National Foundation. Retrieved on February 16, 2020 from http://energyaccess.org/wp-content/uploads/2015/07/MicrogridsReportFINAL_high.pdf

SE4ALL (2013). *Global tracking framework.* New York, NY: United Nations Sustainable Energy For All.

Shirley, R., & Kammen, D. M. (2015). Energy planning and development in Malaysian Borneo: Assessing the benefits of distributed technologies versus large scale energy mega-projects. *Energy Strategy Reviews, 8,* 15–29.

Sovacool, B., & Bulan, L. C. (2012). Energy security and hydropower development in Malaysia: The drivers and challenges facing the Sarawak Corridor of Renewable Energy (SCORE). *Renewable Energy, 40,* 113–129.

Stokes, D. (1997). *Pasteur's quadrant: Basic science and technological innovation.* Washington, DC: Brookings Institution.

Sunter, D., Castellanos, S., & Kammen, D. M. (2019). Disparities in rooftop photovoltaics deployment in the United States by race and ethnicity. *Nature Sustainability, 2,* 71–76.

CHAPTER 27
Sensor-Enabled Climate Financing for Clean Cooking

Nithya Ramanathan, Erin Ross, Tara Ramanathan, Jesse Ross, Martin Lukac, and Denisse Ruiz

Traditional mud stove (left) and improved cookstove with sensor (right). *Source*: Nexleaf Analytics

Summary Household indoor air pollution kills approximately four million people every year. **Affordable** access to **usable** clean energy technologies is essential for the poorest three billion, who rely on solid fuels for cooking, heating, and lighting, to protect themselves from air pollution and to adapt to devastating climate changes. Project Surya has developed an innovative, collaborative approach, called Sensor-enabled Climate Financing (SCF) (Ramanathan et al., 2017), that aligns the objectives of clean energy implementers, stove manufacturers, governments, and multinationals with the needs of the individual women and families they strive to impact. SCF is a verified solution to the sector-wide problems of clean cookstove *affordability* and *adoption*.

N. Ramanathan (✉) · E. Ross · T. Ramanathan · J. Ross · M. Lukac · D. Ruiz
Nexleaf Analytics, Los Angeles, CA, USA
e-mail: nithya@nexleaf.org; erin@nexleaf.org; tara@nexleaf.org; jesse@nexleaf.org;
martin@nexleaf.org; denisse@nexleaf.org

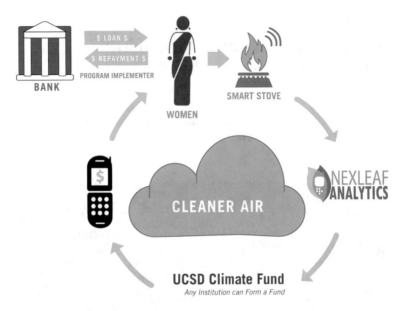

Fig. 27.1 SCF model. *Source*: Nexleaf Analytics

In the SCF implementation model (Fig. 27.1), a woman secures a microloan from a local bank to purchase an improved biomass cookstove equipped with a sensor that transmits objective cooking data in near real time to a Nexleaf server. Small payments to women are made through a donor-based climate fund established at the University of California at San Diego.

Women receive payments through a mobile money app, and clean cooking payments are deposited electronically. This approach funnels results-based financing directly to the individual women who cook, financing the purchase of the cleaner-burning cookstove.

Sustained adoption of clean cooking technologies is required to bring about any of the promised health- or climate-related impacts of these innovations. With SCF, we are seeing sustained improved cookstove adoption in multiple villages (over 89% for over 18 months).

Background

Reducing harmful emissions from traditional solid biofuel (firewood, dung, and crop residues) cookstoves is at the forefront of global efforts to improve the lives of the approximately three billion people worldwide who depend on biofuels for their basic needs. Close to four million people die every year from health complications attributable to household smoke (World Health Organization, 2018). Women and children are particularly impacted by polluting stoves inside the home.

What's More, Traditional Cooking Is a Major Contributor to Global Climate Change

Burning biomass in inefficient handmade stoves, a common practice in more than 600 million households worldwide, contributes to the carbon crisis through carbon dioxide emissions and deforestation. Even more damaging are the short-lived climate pollutants (SLCPs), including soot, also known as black carbon (BC), emitted by traditional cooking. These pollutants have a warming impact many thousands of times greater per unit than carbon dioxide (Ramanathan & Carmichael, 2008).

Since the 1970s, scientists have been working to study the scope of the household smoke crisis, and stove designers have been designing cleaner-burning biomass cookstoves for scale.

Project Surya researchers, through our field work deploying advanced stoves and advanced wireless cookstove usage and soot monitoring systems (Graham et al., 2014; Ramanathan et al., 2011), have published several peer-reviewed papers that document the extent of pollution from stoves (Kar et al., 2012), their impacts on indoor human exposure and ambient levels of BC (Patange et al., 2015; Praveen, Ahmed, Kar, Rehman, & Ramanathan, 2012; Rehman, Ahmed, Praveen, Kar, & Ramanathan, 2011), and the cooking technology that can drastically reduce BC emissions. Our data indicate that among available clean cookstoves, forced-draft stoves, also known as forced-air or fan stoves, burn most efficiently and reduce climate-warming particulates the most. We found that forced-draft stoves, which burn locally available biomass, can reduce cooking-related emissions of BC by approximately 90% (Kar et al., 2012) and, when completely displacing traditional cooking, can reduce the overall concentration of BC in the kitchen by approximately 40% (Patange et al., 2015) (with kerosene lamps and outdoor emissions continuing to contribute to overall kitchen BC concentrations).

Alternatives to forced-draft stoves include gas (or LPG) stoves and electric induction cookstoves, which are among the cleanest technologies. However, the associated fuel costs are prohibitive, and there is no reliable fuel supply or supply chain to deliver the required energy to all. For the many off-grid communities and villages around the world, waiting for grid power or an LPG distribution channel is simply not possible. Forced-draft stoves are relatively cheaper, at a cost of approximately US$80 (which includes the cost of delivery), but still unaffordable for a majority of the underserved, many of whom earn less than US$2 a day. However, these stoves will likely remain the best option until a more cost-effective and feasible solution becomes available for large-scale adoption.

A Data-Driven Approach to Adoption and Impact

Project Surya, launched in 2009, is an international collaboration between NGOs, academic institutions, and the private sector. Given the potential impact of forced-draft stoves, Project Surya set out to understand women's adoption of these stoves and to develop a methodology to tie adoption to impact. Project Surya was also convinced that getting climate financing directly into the hands of low-income women was a moral imperative. Because the world's poorest three billion will bear the brunt of dislocation and disruption caused by global climate change, it is especially important to ensure that the financial benefits of their climate stewardship reach them directly.

Clean-energy implementers often use surveys to verify the adoption of newly introduced clean technologies in remote places. Project Surya's early research included an investigation into the accuracy of using self-reported stove usage data. StoveTrace sensors—wireless thermal sensors attached to stoves in household kitchens that transmit, using local cellular networks, in near real time information about stove use —were deployed in households with traditional cookstoves. Women in those same households were then asked to report how much time they spent cooking each day (Ramanathan et al., 2017). The data (reproduced in the plot in Fig. 27.2) show the lack of correlation between individual stove users' self-reported

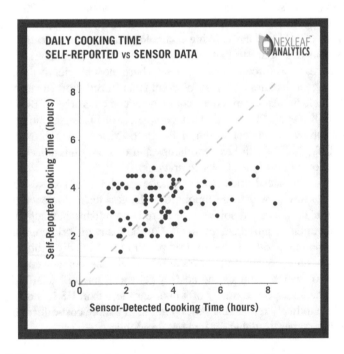

Fig. 27.2 Self-reported cooking data vs. sensor data. *Source*: Nexleaf Analytics

cooking (y-axis) and sensor-verified measurements of actual cooking (x-axis) in households.

These early data demonstrated the importance of employing objective sensors for measuring usage of the improved cookstoves. The objective sensor data on cookstove usage are crucial not only to verify the use of improved cookstove interventions, but also to validate feedback from women to better understand what is working in terms of cookstove technologies and last-mile service and to address barriers to adoption.

Surya researchers developed the SCF methodology to use objective sensor data from both improved and traditional stoves to more accurately quantify the reductions in emissions generated when women switch from a traditional mud stove to an improved biomass stove. Surya researchers published the SCF methodology, which translated emissions reductions into tons of carbon dioxide equivalents, or CO_2e (Ramanathan et al., 2017), based on objective data on the usage of the stove. CO_2e includes reductions of SLCPs, including BC, which contribute to global climate change, and cooling agents such as organic carbon. The methodology integrates available knowledge from peer-reviewed publications with data collected by the Surya team on air pollution levels in the field, compliance on the use of clean technologies, and types and amount of fuel consumed to quantify the emissions reductions achieved. Because clean energy implementers seek to measure impact and hit emissions reduction targets, quantifying baseline emissions for existing traditional cookstove technologies is crucial.

Climate Credit Pilot Project: Findings from a Woman-Centered Approach

Using the SCF implementation model, real-time sensor technologies, and methodology described earlier, Project Surya launched the Climate Credit Pilot Project, or C2P2, in 2014 in Odisha, India.

Project Surya implementers distributed over 4000 forced-draft cookstoves—the cleanest available biomass-burning cookstoves—to households in Odisha, India. Implementers equipped over 450 of the forced-draft cookstoves with Nexleaf's StoveTrace remote wireless sensors. These sensors uploaded cooking event data automatically to Nexleaf's servers in real time, and the data were immediately available on display on the web-based StoveTrace data dashboard.

In 2016–2017, we published findings from the C2P2 stove distribution pilot in India in *Nature Climate Change* (Ramanathan et al., 2017). Analysis of pilot data verified that our financing mechanism works. However, the sensor measurements of cookstove usage by women (y-axis) showed low usage of the forced-draft cookstove over time (Fig. 27.3).

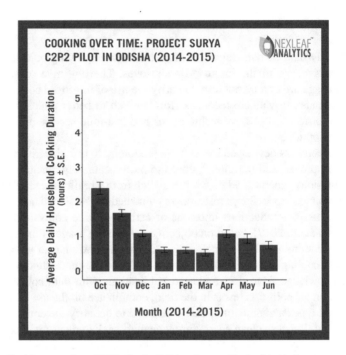

Fig. 27.3 Cooking over time, C2P2 pilot in Odisha. *Source*: Nexleaf Analytics

Quantitative and qualitative follow-up with stove users identified key challenges that likely plague most clean cooking interventions.

First, there is a need for last-mile distribution and service. Like any household appliance, clean cookstoves sometimes break. Technicians need to be trained, resourced with tools and spare parts, and paid to ensure that stoves are repaired and continue to operate.

Second, distributing results-based financing through rural banks alone is ineffective. The individual women enrolled in C2P2 rarely if ever visited their banks to check their balance or verify their receipt of the micropayments they were receiving for their clean cooking.

Third, stoves must be usable, and stove users want consumer choice. Too many of the clean cookstoves available on the market do not adequately reduce emissions, and those that do are often difficult or inconvenient for women to use. Stove users want the opportunity to select a stove among multiple options that best fits their lifestyle. Therefore, a variety of clean, usable stove models must be available to women if they are to switch away from traditional cooking.

Fortunately, an analysis of these data revealed the path to sustained adoption.

Improved Sensor-Enabled Climate Financing Model: 2016–2017 Implementations

Project Surya improved the SCF model to address the gaps identified. Implementations in Odisha beginning in 2016 demonstrated the effectiveness of SCF in India in promoting sustained usage of improved cookstoves. Initial data on adoption (Fig. 27.4) suggest that the improved model leads to sustained usage of the clean cookstoves at levels commensurate with scientific data on cooking behavior.

The aforementioned data set is small, but success has continued beyond these initial pilot homes. By the time of the Vatican meeting in November 2017, program households were using their improved cookstoves consistently, achieving 93% adoption. A full report can be found at nexleaf.org/impact/iot-for-development-stovetrace/. The following modifications made this quantifiable improvement possible.

Introducing Mobile Money Instead of obliging women to visit distant banks to see the rewards of their clean cooking, we worked with Vodafone M-Pesa to register the women on the M-Pesa mobile money application. The women then received payments via their cell phones. For women without a cell phone, a local M-Pesa cash-out agent provides payments in cash.

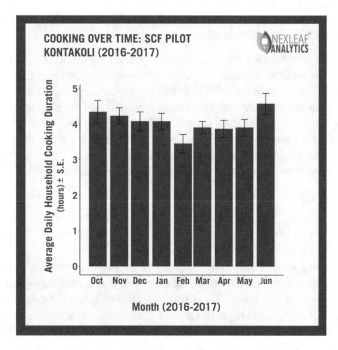

Fig. 27.4 Cooking over time, SCF pilot in Kontakoli. *Source*: Nexleaf Analytics

Strengthening Supply Chains by Mobilizing Clean Energy Entrepreneurs
Drawing upon existing networks of rural women entrepreneurs, we developed a distribution and after-sales service model for cookstoves that employs women entrepreneurs. Women energy entrepreneurs want the chance to work with their community to increase access to great products.

Independent Stove Testing We designed a laboratory testing protocol to evaluate improved cookstoves for BC reductions and emissions.

Achieving adoption of new technologies such as improved cookstoves requires multidimensional interventions. However, the dearth of clean, usable biomass stoves in the marketplace continues to pose a critical and persistent challenge.

In recent months, Nexleaf has developed a mechanism for in-home usability testing using StoveTrace sensors and detailed, data-enabled user feedback to continue to identify promising technologies for potential scale. This new standard for clean cooking evaluation helps ensure that manufactured stoves meet women's needs before deployment at scale:

1. Rigorous independent lab testing of stove emissions, including BC;
2. In-home usability testing conducted by women in their homes for at least 3 months to identify the most user-friendly stoves;
3. Due diligence with stove manufacturers to determine whether they can meet last-mile distribution demands, including fulfilling warrantees and providing training and capacity building to last-mile technicians;
4. Long-term cost of ownership, including purchase, financing, maintenance and fuel, to evaluate true affordability.

These four measures inform our evaluation process for clean cooking implementations, which focuses on an incremental scale-up of cooking technologies. We begin with pilots of ten households and deliberately increase the program size only after evaluating the performance of the cookstove and proving 80% adoption.

Recent data confirm sustained adoption (above 89%) of improved cookstoves in SCF households for more than 18 months. A 2018 report including data from SCF households and explaining Nexleaf's adoption metric can be found at nexleaf.org/reports/joint-learning-series/beyond-monitoring-and-evaluation.pdf.

The Future of SCF

Moving forward, Project Surya aims to preserve the environment, empower women, improve health, and bolster the market for clean and usable cooking solutions, organizing our actions around operationalizing objective data. In the future, we will test the SCF model's resilience at greater scale and continue using data and analytics to reveal and help us resolve any challenges that arise. We will expand SCF by scaling the best stove models we have identified, and we will continue lab and usability testing to identify more SCF-eligible stoves to address the broader market failure and increase women's choice.

Our SCF method has sparked interest across multiple sectors because it addresses 10 of the 20 United Nations Sustainable Development Goals.

Conclusion: Transparency in Clean Energy Interventions Is Key to Climate Justice

One key to our success has been an unflinching dedication to the power of data. Even—and perhaps especially—when the data contradict our assumptions, we must continue to adjust and iterate based on objective and transparent measures of impact. In the clean energy sector as in other sectors Nexleaf works in, we have seen how introducing sensor data can bring about anxiety. Too many development initiatives have faced punitive measures or encountered censure for falling short of their objectives. Changing the world for the better is difficult, complicated work, and we assume good will on the part of all global stakeholders, from multinationals to governments to stove manufacturers. That's why, for example, Nexleaf provides feedback to stove manufacturers whose stoves do not pass our evaluation standards.

Clean energy stakeholders can embrace the power of data to reveal both unforeseen problems and their potential solutions. Additionally, small and well-measured pilots must be scaled at a pace that aligns with the capacity of the data and the implementation team to reveal and react to the objective truth. Every village, district, province, state, and country has its own context, and data-driven implementations provide the flexibility necessary to serve cookstove users with many different preference and behavior profiles.

Data not only help steer clean cooking implementations in the right direction; they also enable all stakeholders to tap into new and exciting avenues of financing. From corporate social responsibility campaigns to individual buyers who wish to offset their carbon expenditures, an objective feedback portal will serve eager donors and climate credit buyers.

SCF has the power to accomplish three primary objectives critical to successful climate change interventions, climate justice, global emissions reductions, and economic inclusion for the world's poorest:

1. Align incentives and provide feedback loops necessary to bring about sustained adoption of clean cooking technologies.
2. Provide verification of climate impact of interventions against agreed-upon emissions reductions targets (i.e., Paris Accords).
3. Revolutionize access to results-based financing driven by objective data. This funding can be used to reward individual rural women and women entrepreneurs with results-based financing for their roles as climate stewards.

Ultimately, we believe in unlocking this financing for the world's poorest and using cutting-edge technology both to verify and to disburse these funding streams.

Acknowledgements Project Surya is an international collaboration between NGOs, academic institutions, and the private sector, initiated by the Scripps Institution of Oceanography, UCSD, the Energy Resources Institute, and Nexleaf Analytics, and now led by Nexleaf Analytics. The authors would also like to thank Drs. Pattanayak and Jeuland for leading the survey data collection for the women's self-report in the early investigation of the accuracy of stove usage sensors.

References

Graham, E. A., Patange, O., Lukac, M., Singh, L., Kar, A., Rehman, I. H., et al. (2014). Laboratory demonstration and field verification of a Wireless Cookstove Sensing System (WiCS) for determining cooking duration and fuel consumption. *Energy for Sustainable Development, 23,* 59–67.

Kar, A., Rehman, I. H., Burney, J., Puppala, S. P., Suresh, R., Singh, L., et al. (2012). Real-time assessment of black carbon pollution in Indian households due to traditional and improved biomass cookstoves. *Environmental Science and Technology, 46,* 2993–3000.

Patange, O. S., Ramanathan, N., Rehman, I. H., Tripathi, S. N., Misra, A., Kar, A., et al. (2015). Reductions in indoor black carbon concentrations from improved biomass stoves in rural India. *Environmental Science and Technology, 49,* 4749–4756.

Praveen, P. S., Ahmed, T., Kar, A., Rehman, I. H., & Ramanathan, V. (2012). Link between local scale BC emissions in the Indo-Gangetic Plains and large scale atmospheric solar absorption. *Atmospheric Chemistry and Physics, 12,* 1173.

Ramanathan, N., Lukac, M., Ahmed, T., Kar, A., Praveen, P. S., Honles, T., et al. (2011). A cell-phone based system for large-scale monitoring of black carbon. *Atmospheric Environment, 45,* 4481–4487.

Ramanathan, T., Ramanathan, N., Mohanty, J., Rehman, I. H., Graham, E., & Ramanathan, V. (2017). Wireless sensors linked to climate financing for globally affordable clean cooking. *Nature Climate Change, 7,* 44.

Ramanathan, V., & Carmichael, G. (2008). Global and regional climate changes due to black carbon. *Nature Geoscience, 1,* 221.

Rehman, I. H., Ahmed, T., Praveen, P. S., Kar, A., & Ramanathan, V. (2011). Black carbon emissions from biomass and fossil fuels in rural India. *Atmospheric Chemistry and Physics, 11,* 7289–7299.

World Health Organization (2018). *Household air pollution and health. Fact Sheet 292.* Geneva: World Health Organization.

CHAPTER 28
Research Is Vital to Tackling Climate Change, But It Cannot Succeed Alone

Jeremy Farrar

Summary Our Planet, Our Health is Wellcome's contribution to the growing field of planetary health. The world needs focused research to better understand how health is affected by global environmental changes, including climate change, and how we can mitigate these health impacts and the environmental changes. We call on researchers to work with us, and on civic society, industry and governments to support researchers, apply their findings and build an equitable movement of research into practice to improve our health. We call on governments, industry and civic society to support these researchers—commit to acknowledging and employing their findings and recommendations. Through policies and legislation, governments can do what no one else can, while foundations like Wellcome and others complement their work, taking risks and doing what governments cannot, and together we can help industry play its part, secure in the knowledge that there is a pipeline of great people and ideas coming through. But if governments step back from progress in planetary health, the gaps cannot be filled by anyone else. There is reason for optimism, if our optimism drives us all to act now for the sake of current and future generations.

Background

In Bangladesh, the dry season runs from November to early May. Less rainfall means shrinking rivers, and what water there is becomes more saline as a result. When the ponds have dried up completely, people can pump groundwater to use for drinking and cooking. But the groundwater, especially in coastal areas, is contami-

J. Farrar (✉)
Wellcome Trust, London, UK
e-mail: J.Farrar@Wellcome.ac.uk

© The Author(s) 2020
W. K. Al-Delaimy, V. Ramanathan, M. Sánchez Sorondo (eds.), *Health of People, Health of Planet and Our Responsibility*, https://doi.org/10.1007/978-3-030-31125-4_28

nated with salt, too. Seawater has infiltrated the natural aquifers, a problem that is likely to get worse as climate change threatens to bring even higher sea levels, more floods and tidal surges, and more cyclones (Naser et al., 2017a).

The high salinity of available water in the dry season means people consuming anything from double to eight times the World Health Organization's (WHO) recommended daily intake of sodium. This has important health implications: research has revealed associations between salinity and high blood pressure in adults and preeclampsia in pregnant women. On the other hand, saving and storing rainwater from the wet season may also pose a health risk, as rainwater lacks minerals such as calcium and magnesium, which help protect against heart disease but are often deficient in Bangladeshi diets.

In the 1970s, millions of tube wells were dug to provide new sources of fresh water in Bangladesh. While this reduced gastrointestinal infections and cases of diarrhoea from contaminated surface water, it inadvertently and disastrously exposed the population to naturally occurring arsenic deeper underground (Siddiqi, 2017); arsenic is a major risk factor in heart disease, stroke, cancer and other serious health conditions.

So what solutions are available today? As with all water, sanitation and hygiene (WASH) challenges, any response will require many different sectors of society and will rely on the active engagement and participation of individuals, governments and non-governmental organisations (NGOs). People will need to change the way they obtain, treat and store water, as will farmers and factories, just as governments will have to take on responsibility for implementing and maintaining any new infrastructure, physical or political.

One new idea for securing healthy drinking water in Bangladesh is to recharge the natural underground aquifers with rainwater or surface water—effectively putting fresh water into the ground to dilute the high-salt water already there. Before this can be rolled out everywhere, however, we need to know whether it works. Wellcome has funded a research study to find out (Naser et al., 2017b). It is the first study to assess the health impact of an environmental intervention to reduce the salinity of groundwater in Bangladesh. The randomised controlled trial design being used by Mahbubur Rahman and colleagues at the International Centre for Diarrhoeal Disease Research, Bangladesh, will determine whether managed aquifer recharge lowers the blood pressure of people in communities adopting the new approach. If the results are positive, it can be adopted elsewhere with confidence, supporting more people as the effects of climate change intensify.

Humanity's response to climate change will define the twenty-first century. Pressures on water and sanitation are far from the only issues we face. We expect increased heat stress and droughts as well as floods and more frequent, more powerful storms. Land use and ecological systems will change as the environment changes. Food and nutrition will be harder to provide for all, and more people will migrate or be displaced, experiencing more mental health conditions as a result. Insects will spread to new territories, bringing infectious diseases with them. Air pollution will rise, especially as more people move to cities and other urban locations (Watts et al., 2015, 2016).

People in every part of the world are now coping with increasing challenges. Those in communities that are already vulnerable will be hit hardest, and in recent months and years, we have seen that many people are already suffering and dying from the effects of climate change. That includes severe storms on the southern coast of the USA as well as saline drinking water on the southern coast of Bangladesh. It includes islanders dealing with hurricanes and Europeans responding to intense heat waves (Christidis et al., 2015). The WHO predicts that by 2050 there will be an extra 250,000 deaths a year due to increases in malnutrition, heat stress, diarrhoeal diseases and malaria as a direct result of climate change (WHO, 2017).

To save those lives and avoid further risks to our health, we must act now to protect and sustain the planet. We have to come together to collaborate on devising and delivering solutions to everyone's benefit. We have to change our own behaviour and persuade others to do the same.

Our Planet, Our Health

In 2015, following an earlier exploratory programme, Wellcome launched Our Planet, Our Health as a priority project, with the aim of developing and supporting focused research in the growing field of planetary health (Wellcome Trust, 2017b). The world needs more and higher-quality research, combining different disciplines and spanning all geographies, in order to understand the effects of global environmental change and the actions we might take in response.

The climate change community has done a remarkable job providing research evidence to inform policy. But it has perhaps too often presented the impact of climate change as something that will affect our grandchildren in decades to come. We want to support and spearhead research to show the impact today of climate change on human and animal health and through research help provide answers that we hope will inform policy and behaviour.

Wellcome is by no means the only organisation committed to this area. But we are in a unique position to make a significant contribution. Our experience as an international research funder with global networks across science, innovation and society mean we can bring people together and broker innovative activities.

Our Planet, Our Health has three inter-related pillars. The first pillar is understanding the impact of climate change on human health now and in the future. In partnership with *The Lancet*, we have set up the Lancet Countdown (Watts et al. 2016) an international research collaboration that tracks global progress on climate change (Costello et al., 2017; Watts et al., 2016). This is a first step towards lessening the effects of climate change on health and adapting to its impact.

The second pillar of Our Planet, Our Health is the role of global urbanisation as a major driver of health. More than half the people in the world now live in urban areas—by 2050, it will be close to 70% (United Nations Department of Economic and Social Affairs, 2014). The intersection of social, natural and built environments in cities brings many benefits but also challenges for human health. We have invested £17.8 million in research looking at how urban design and policy can work to improve health and make cities environmentally friendlier.

The third pillar of Our Planet, Our Health is feeding the world healthy diets without destroying the planet. For example, people tend to switch to eating more meat as they get wealthier, but too much meat consumption is far more damaging than a plant-based diet for both humans and the environment. Food and land use already account for more than one-quarter of greenhouse gas emissions and 70% of water use. We co-founded the EAT Foundation, bringing together scientists, businesses and policy-makers to transform food systems for the future.

We have funded focused research across all of these areas, generating new knowledge, data and models for understanding current and future effects of environmental changes (Wellcome Trust, 2017a). Fifteen pilot projects galvanised collaboration across the globe and across disciplines, from economics and social sciences to epidemiology and ecology. To complement these open funding calls, four major themed programmes were established to nucleate focused centres of excellence and provide a thriving, supportive environment for research, mentorship, policy and training.

For example, one project explored innovative designs for healthy and sustainable low-income housing in the UK, India, Mexico and South Africa. Another looked at the health impacts of poor-quality urban development, from fast-emerging effects like air pollution and heat waves to slower ones such as changing weather patterns and the rise of non-communicable diseases.

A recent paper from another group we funded explored the influence of climate change on infectious disease dynamics (Metcalf et al., 2017). Infections are sensitive to climate change for a variety of reasons—environmental contamination can expose more people to water-borne diseases, while the survival patterns of insects and other disease vectors may change, putting different populations at risk (Altizer et al., 2013). The research led to the conclusion that we need to move beyond simple statistical correlation and association between environmental factors and disease and must develop more mechanistic models to predict trends in incidence.

This year, Our Planet, Our Health announced four major new partnerships, all focusing on providing policy-makers with innovative and interdisciplinary research evidence. Two are in the area of global food systems, looking at everything from farm production to consumer choices and testing how changes could be made. The other two concern urban environments, learning from South America what makes cities healthy or not and looking at greener ways to provide clean water to informal settlements in Indonesia and Fiji.

This last partnership is a good example of how researchers and funders will have to work together if we are to adapt to our changing world. More than a billion people live in slums and other informal settlements around the world. They face serious public health challenges relating to WASH issues. The "big pipes" traditionally used to provide water and sanitation to cities are extremely costly in financial, environmental and social terms and especially impractical in informal settlements.

Revitalising Informal Settlements and their Environments (RISE), a programme led by Rebekah Brown at Monash University in Melbourne, Australia, is testing ecologically sustainable water infrastructure—such as grass channels, wetlands and natural filters—in 24 locations in Indonesia and Fiji (Monash University, 2017). In addition to research funding from Wellcome, RISE has received support from the Asian Development Bank and from partners in government, industry and academia. They will determine whether these alternative solutions improve environmental and human health outcomes, such as gastrointestinal infections in children under five. If they do, the challenge will quickly become one of scaling up the solutions and employing them elsewhere without delay.

We see this combination of focused centres of excellence and open research funding calls linked with communication, public engagement, advocacy and policy and framing this as an equitable, civil rights issue as the best way to ensure research will be supported in the best environments, open to a diverse range of new ideas and linked by a commitment to impacting and enhancing all our lives.

Together, Science Can

Our challenge now is to support and encourage more research like this—much more, and of a consistently high quality such that it can be relied on and used at all levels of political decision-making. As we look forward, Our Planet, Our Health must grapple with questions of standards, knowledge sharing, interdisciplinarity and scale.

How do we achieve standards of excellence so that governments and communities can trust and apply research findings? How can we share methodologies and findings so that research can be combined to create a clear and useful picture of the situation and potential actions? How do we make sure research is fully interdisciplinary, in order to get the most complete and accurate view of the situation we face? How do we accelerate the translation of research findings into policies and actions that work at scale for communities, cities and countries?

For example, an intervention designed to reduce greenhouse gas emissions that focused on one area, such as food or transport, could have negative impacts on water or land use. Joining up our thinking will help create lasting solutions and avoid shifting potential unintended problems onto other sectors.

This extends across Wellcome's other priority areas, too. We know that issues around migration, water use and farming will affect the emergence of drug-resistant and emerging infections and impact people's mental health. Our priority area around research ecosystems in Africa and Asia is ensuring that the necessary skills, evidence and expertise are supported in regions where climate and environmental change will have the biggest impact. And we must make much more significant progress in the areas of diversity and inclusion in science and research, given that these challenges affect everyone, across every continent and all sectors of society.

Wellcome's Current Platforms
- Diversity and Inclusion
- Drug-Resistant Infections
- Education for Science
- Mental Health
- Our Planet, Our Health
- Research Ecosystems in Africa and Asia
- Vaccines and Epidemics

These are the focused challenges Wellcome is taking on, but we cannot do it alone. No one can.

We call on researchers to accept the challenge with us—apply your skills to the most fundamental health challenge of our time.

We call on governments, industry and civic society to support these researchers—commit to acknowledging and employing their findings and recommendations. Through policies and legislation, governments can do what no one else can, while foundations like Wellcome and others complement their work, taking risks and doing what governments cannot. Together, we can help industry to play its part, secure in the knowledge that there is a pipeline of great people and ideas coming through. But if governments step back from progress in planetary health, the gaps cannot be filled by anyone else.

And we call on communities to engage with research—participate, shape, even lead the development and direction of research. Tell us what you need to be studied, use the results and share your successes. None of the great challenges of our time will be solved by technology alone. We must break down barriers within science, barriers between nations, across industries and especially barriers between science and society. We need the greatest diversity of thought to solve these problems.

The world is changing at an unprecedented rate. Climate and environmental change, demographic shifts, the rise of non-communicable diseases, emerging and endemic infectious diseases, and drug-resistant infections, and mental health—these interconnected challenges are putting our health, and the health of the planet, at risk.

Solutions to the health effects of climate change will be, on the whole, "no regret" solutions (Watts et al., 2015). They will improve people's health no matter what the climate does. And what is good for us will generally be good for the planet as well and we can be both pro-growth and pro-planet; they are not mutually exclusive and indeed can be mutually reinforcing. Such hope, optimism and confidence have too often been missing from political debates that characterise the options as economic growth versus a sustainable future—it does not have to be so; we can ensure a sustainable future whilst also ensuring more equitable economic growth. We do not have to make people poorer to ensure a healthy planet and build a sustainable future.

Increasing the resilience of national healthcare systems and strengthening health infrastructure through the Sustainable Development Goals and Universal Health

Coverage; reducing poverty and offering support to vulnerable communities; maintaining surveillance of existing and potential health threats and changing cities into places that support and promote healthy lifestyles; creating sustainable food and water systems—such actions will improve the health of our generation but also those of our children, our grandchildren, and all those who will follow.

A commitment to preventing a potential global catastrophe has been made by 196 countries (Costello et al., 2017), and in many places cities have led the way with empowered local government, responsive to the needs of its citizens to be able to make locally driven solutions to enhance sustainability and improve health and well-being. It will be our collective failure if we cannot now use that commitment and goodwill. Science and research are integral to the world's combined response, and science can also help to bring people together. It can be a force for breaking down barriers, building social cohesion and reducing inequality. If we stand together, science can help us understand and define the challenges we all face and, together, solve them.

References

Altizer, S., Ostfeld, R. S., Johnson, P. T., Kutz, S., & Harvell, C. D. (2013). Climate change and infectious diseases: From evidence to a predictive framework. *Science, 341*, 514–519.

Christidis, N., Jones, G., & Stott, P. A. (2015). Dramatically increasing chance of extremely hot summers since the 2003 European heat wave. *Nature Climate Change, 5*, 46–50.

Costello, A., Gong, P., Montgomery, H., Watts, N., & Wheeler, N. (2017). Christiana Figueres joins the Lancet countdown—Delivering on the promise of Paris. *The Lancet, 389*, e16.

Metcalf, C. J. E., Walter, K. S., Wesolowski, A., Buckee, C. O., Shevliakova, E., Tatem, A. J., et al. (2017). Identifying climate drivers of infectious disease dynamics: Recent advances and challenges ahead. *Proceedings of the Royal Society B, 284*.

Monash University. (2017). *Monash awarded grant to lead global slum revitalisation research.* Retrieved October 16, 2017, from Monash University website https://www.monash.edu/news/articles/monash-awarded-grant-to-lead-global-slum-revitalisation-research

Naser, A. M., Martorell, R., KMV, N., & Clasen, T. F. (2017a). First do no harm: The need to explore potential adverse health implications of drinking rainwater. *Environmental Science and Technology, 51*, 5865–5866.

Naser, A. M., Unicomb, L., Doza, S., Ahmed, K. M., Rahman, M., Uddin, M. N., et al. (2017b). Stepped-wedge cluster-randomised controlled trial to assess the cardiovascular health effects of a managed aquifer recharge initiative to reduce drinking water salinity in southwest coastal Bangladesh: Study design and rationale. *BMJ Open, 7*, e015205.

Siddiqi, Y. (2017). Empty reservoirs, dry rivers, thirsty cities—and our water reserves are running out. *The Guardian.*.

United Nations Department of Economic and Social Affairs (2014). *2014 revision of the World Urbanization Prospects.* Retrieved October 16, 2017, from UN website http://www.un.org/en/development/desa/publications/2014-revision-world-urbanization-prospects.html

Watts, N., Adger, W. N., Agnolucci, P., Blackstock, J., Byass, P., Cai, W., et al. (2015). Health and climate change: Policy responses to protect public health. *The Lancet, 386*, 1861–1914.

Watts, N., Adger, W. N., Ayeb-Karlsson, S., Bai, Y., Byass, P., Campbell-Lendrum, D., et al. (2016). The Lancet countdown: Tracking progress on health and climate change. *Lancet, 389*, 1151–1164.

Wellcome Trust (2017a). *Understanding the links between human health and the environment.* Retrieved October 13, 2017, from Wellcome Trust website https://wellcome.ac.uk/news/understanding-links-between-human-health-and-environment

Wellcome Trust (2017b). *Our planet, our health: responding to a changing world.* Retrieved October 13, 2017, from Wellcome Trust website https://wellcome.ac.uk/what-we-do/our-work/our-planet-our-health

WHO (2017). *Climate change and health.* Retrieved October 13, 2017, from WHO website http://www.who.int/mediacentre/factsheets/fs266/en/

Part VIII
Call to Action

CHAPTER 29
Governor of California

Edmund G. Brown Jr.

It is a privilege to participate in one of the crucial steps in the process of making the turn to a climate-sustainable world, which is to call attention to the catastrophic health impacts of climate change. Much has been said about California, and that is appropriate because California, in fact, has been doing quite a lot.

But before I write about that, I want to mention the Pontifical Academy of Sciences meeting held in November of 2017: *Health of People, Health of Planet and Our Responsibility – Climate Change, Air Pollution and Health*. This book on climate and health resulted from that meeting. It really was all said right there and very clearly in the document which was prepared in connection with the meeting. The same is true of the declaration which resulted from that meeting, reproduced in this book. The declaration, which I signed along with astrophysicist Stephen Hawking, several mayors, Nobel laureates, and prominent faith leaders, fully captures the essence of what we are up against and what we need to do.

The Pontifical Academy meeting and this book that resulted from it are uniquely important. Of course, there are the annual conferences of the parties and many other meetings around the world dealing with climate change, but until the religious sensibility is engaged, until religious leaders from every part of the globe and from every denomination take up the challenge, we are not going to move aside the huge rock of indifference, complacency, and inertia. So I want to pay my respect for everything that the Pontifical Academy has already put in place, and will put in place, to further this cause.

The connection of health to climate change is central because climate change is an abstraction. Very few people can actually grasp it—or certainly be moved by it—but the air pollution from carbon emissions is visible and affects people directly. They get that. At the Vatican meeting, people were asking me: Why California? You might add, why is China getting so aggressive all of a sudden? It is called pollution.

E. G. Brown Jr. (✉)
Honorable Former Governor of California, Williams, CA, USA

W. K. Al-Delaimy, V. Ramanathan, M. Sánchez Sorondo (eds.), *Health of People, Health of Planet and Our Responsibility*, https://doi.org/10.1007/978-3-030-31125-4_29

It is the impact. It is the experience of the health effects of very serious air pollution. California had perhaps the worst air pollution, which was called smog, and this smog affected people in a very important part of America: Los Angeles, California. And that smog then led to the creation of the California Air Resources Board during the governorship of Ronald Reagan and the presidency of Richard Nixon. But it was the problem that created the solution, and the pollution in Shanghai and Beijing have created a more heightened and ambitious response on the part of China—different from just a few years ago.

So I believe the authors of this book and the World Health Organization itself are right on track in focusing on health and disease, on what air pollution is doing to people. And of course, there is the element of injustice that the people who have done the least to cause the injury are the ones who suffer the most. It is all laid out in this book. We know what the world faces, and going forward, we are going to find the pathway to awaken people, to get done what needs to be done.

So, in the declaration, you lay out four points. First, we have got to get to carbon zero by 2040 or 2045. That is number one. Secondly, we get there in the most expeditious and efficient way by getting rid of short-lived climate pollutants. We do that and we will gain a few more years to get to zero carbon. Thirdly, after we get to zero, which hopefully comes sooner rather than later, then we have to extract a tremendous amount of carbon from the environment itself. So that is an utterly daunting undertaking, and we are not going to get there with just science and technology. Tragically, there is no technical fix adequate to the challenge we face. That is why in the declaration there is a call for transformation, which is both the most important and the most difficult.

I know about the idea of transformation because I entered the Jesuit seminary in the world of 1956, pre-Vatican II. We spoke Latin, we were told we had embarked on the practice of perfection. We meditated, we underwent self-discipline, we mortified ourselves. We tried our best, and I can tell you, I did not achieve perfection. I was not transformed. In fact, some of my bad habits, which I will not reveal, are the same today as they were when I entered the Jesuit seminary when Pius XII was pope. So it is a word that is easy to use but a state that is profoundly difficult to achieve.

The image that comes to my mind is Saul of Tarsus, riding to Damascus and being thrown off his horse. A lot of powerful people have to get thrown off their horse—the horse of affluence, the horse of power, the horse of inertia. So we do need an illumination. Taking a slightly different metaphor, we could use a 12-step program because our task cannot be achieved by mere self-will.

There is a power greater than ourselves, and in understanding that, and in orienting ourselves to that greater power, we will find the power to get done what we need to do. So much of this is technical fixes, but the big fix is really not a fix, it is a transformation. It is a reorientation, a paradigm shift. That transformation—or the context of that transformation—is best captured in the words of Pope Francis that are placed right on the early pages of this document. And I want to quote just these few words: "Seeing the mysterious network of relations between things." So it is the network of things—the connection, the connectivity.

In connection with the Vatican, I could mention the term "mystical body"—the Mystical Body of Christ. Everyone is part of the body whatever, wherever, and whoever you are. It is a connectivity. The biosphere is connected, there is not a separation. We can never be separated from, we are always connected to. We are a part and so our individualistic assumptions that we are going to conquer nature, we are going to become the greatest power on Earth, become the greatest church, the greatest university, the greatest politician, that all is a delusional binary framework. In point of fact, we live in nature, that mysterious network of connections. Our very being is embedded in and dependent on other beings; the network is the interactions that sustain life.

We solve one problem yet create another one unless we reach an understanding about our relationship to nature, to creation. How are we going to do that? Not merely by "cost–benefit" analysis and not just with "evidence-based." We do not want a no-evidence-based approach, we do not want superstition, we do not want ignorance, but we cannot rely on data and evidence alone.

Ultimately, it gets down to something like belief or faith. In the approach to Zen practice, three aspects are relevant here: great doubt, great faith, and great commitment.

If you want to be enlightened, you need great doubt, you need great faith, and you need great commitment. If we want to get the job done, we have to have great doubt. Doubt about what we are doing, where we are, to stand back and see clearly. But that is not enough. We need faith, faith to move forward, faith to take action based on insufficient evidence. But faith is useless without works, so you have to have commitment. You have to do things. It is that three-part approach that we have to have. It is true of all of life, but it is profoundly true with respect to the challenge before us. We first have to see what it is—and it is not a technical fix. It's a transformation of the relationship of human beings to all the mysterious network of things.

But let me say something concretely, because it is not only this larger idea of our place in the world and in creation, but also concrete steps. In California, we have done seven things that are replicable and could be used throughout the world.

First, we have a clear commitment to promoting zero-emission vehicles. We have a commitment not only to zero-emission vehicles but also to reducing the emissions of fossil fuel cars. In that respect, the car companies and the federal government are all increasing the pressure on California to change the regime that we have and the standards that are now in place. But we will resist that, politically and in the courts.

As part of the transportation challenge, we also have to deal with land use and how people live together, and, in fact, our whole relationship to things, to packaging, to moving stuff around. All that is tied in.

I was in Vladivostok recently for an East Asian economic conference, and we heard speeches by the presidents of Russia, Mongolia, and South Korea and the prime minister of Japan, and they talked about trade and building pipelines and bridges and all sorts of things, but not one word was mentioned about climate change. Yet to talk of trade and not talk about the carbon impacts misses a fundamental point. It indicates to me that in the highest circles, people still do not get it. At least they do not get it at the level that we have to get it if we are really to get to carbon neutrality and stabilize our climate. So it is not just a light rinse, but we need a total, I might say brainwashing. We need to wash our brains out and see a very different kind of world.

The second thing we have in California is energy efficiency, including building and appliance standards. We are doing that, we have the strictest rules in the United States—I would say probably in the Western hemisphere. That can be done, it is just making better buildings. There is a problem, of course; it costs money, it is complicated. Our building code, which was relatively short in length, is now very, very long. So it is not a free good here. Everything we do has a price and requires a tremendous overcoming of inertia.

Third, we have a low-carbon fuel standard. It is what the oil companies have defeated in the state of Washington. They fight it here in California as well. So far, we have it. But this is real combat, and to maintain any real standard that bites means you are going to have fierce opposition.

Fourth, we have our cap-and-trade program that was recently extended to 2030. This program, begun under Governor Schwarzenegger, puts a price on carbon and generates billions of dollars for climate-related investments.

Fifth, we have a renewable portfolio standard. We are now about 30% renewable electricity, not counting hydro or nuclear. The San Diego electric utility is over 40% renewable electricity. I can remember just a few years ago the private utilities—and the public ones as well—said they could not get to 20% renewable electricity by 2020. Soon they are going to be at 50%, so we are making it. And by the way, the utilities, from being major opponents of what I am talking about, are now major cheerleaders, and they are very much engaged in expanding the electrification of the California economy. In fact, there is now competition between renewable electricity and oil, between the conventional vehicle and the zero-emission vehicle. The role of utilities, charging stations, and other businesses makes it all possible.

Sixth, we are cutting short-term pollutants such as methane, black carbon, and chlorofluorocarbons, which will see more rapid reductions in the warming over the next 20–30 years. If the whole world follows California's super pollutants bill (SB1383), the rate of warming over the next 25 years can be cut by half.

Seventh, California and Baden-Württemberg have started the Under2 Coalition to keep global warming under 2 °C. That initial partnership has now grown to over 200 states and regions throughout the world.

America itself is very much in line with the Paris goals—the American people are, as are cities, corporations, nonprofits, and universities. The American people are more than their government. I can tell you that the majority of Americans are very much in support of serious climate action, following the Paris accords.

The federal government put out a report in November 2017, a report mandated by federal law, that laid out very clearly the human causation behind global warming and all the dangers. In the United States we do not have just a top-down structure. There are many elements in our federal system. We are a "country of laws, not men," as stated by U.S. President Adams, and we have states, we have cities, we have a very vibrant private sector, we have a vibrant civil society.

In many respects, if you try to measure President Trump's contribution, I would have to say that his attempted withdrawal from Paris has actually put the matter much more at the forefront. It is now more salient, because of the contrast between what he is saying, what the laws of America allow, and what other states are doing. So I would say over time, given the commitments that we saw at the 2017 Vatican meeting and that we are seeing all around the world, the Trump factor is very small, very small indeed.

But that is nothing to cheer about because if it was only Trump that was our problem, we would have it solved. But that is not the only problem. The problem is us, it is our whole way of life, it is our comfort, it is the greed, it is the indulgence, it is the very pattern of modern civilization. It is the inertia. And that is why the transformative solutions have to be present alongside, and informing, all the technical points and the political agreements that we make.

Let me conclude by saying that there is nothing more important than mobilizing the religious and theological spirit, but even more important than that is the prophetic spirit. It was not a pope who did the work of St. Francis. I recall being at the Vatican when Mother Teresa opened up one of her homes, right in the Vatican, and all the important people, the Italian politicians were there in the Piazza and the pope walked by, and no one paid much attention. Then this little woman in blue and white walked by, and all eyes turned to Mother Teresa.

So the power here is prophecy, the power here is faith, and that is what the Pontifical Academy is supposed to be about. So let us be about it—the technical, the scientific, the political—but never let us forget that our challenge is transformation, transformation in the profoundest way possible.

CHAPTER 30
United States Congressman

Scott Peters

I will discuss the imperative of addressing the world climate from my perspective as a member of the United States Congress.

You are familiar with the warnings of scientists, including my friend Dr. Ramanathan of Scripps Institution of Oceanography in San Diego, California, which I represent. I need not add to those insights. Instead, I will offer observations on how Washington, DC is dealing with climate change and ways in which I think we can do better.

Climate action used to be bipartisan. In his 2008 campaign for president, Republican Senator John McCain ran advertisements in which he "sounded the alarm on global warming" and said his credentials on climate were stronger than those of his opponent, Barack Obama ("How G.O.P. Leaders Came to View Climate Change as Fake Science," *New York Times*, June 3, 2017).

Soon after that election, however, there began a well-funded campaign to change this. Many conservative groups, including one called the Cooler Heads Coalition, have waged campaigns to cast doubt on climate science and to intimidate Republican candidates from supporting climate action. According to reports, the Cooler Heads Coalition has received more than $11 million in donations from oil and gas companies and from foundations controlled by wealthy families, including the Kochs ("A two-decade crusade by conservative charities fueled Trump's exit from Paris climate accord," *The Washington Post*, September 5, 2017). The television advertis-

Scott Peters represents California's 52nd Congressional District, including the cities of Poway, Coronado and most of San Diego. He is a member of the House Energy and Commerce Committee, and is a former environmental attorney, San Diego City Council President and Port Commission Chairman.

S. Peters (✉)
La Jolla, CA, USA
e-mail: scott@scottpeters.com

© The Author(s) 2020
W. K. Al-Delaimy, V. Ramanathan, M. Sánchez Sorondo (eds.), *Health of People, Health of Planet and Our Responsibility*, https://doi.org/10.1007/978-3-030-31125-4_30

ing, social media campaigns, and political events across the country raised doubts in the minds of voters and fear in the hearts of politicians. For representatives of coal regions, it became easier to accept alternative facts than it was to accept a climate action policy that would phase out carbon emissions.

About 74,000 Americans work in the production and distribution of coal, and more than 812,000 work in oil and gas (U.S. Department of Energy, 2017, see Table 1, p. 29). (To be fair, that workforce barely exists in California.) If you visit Midland, Texas, as I did, you see an entire community and culture invested in and proud of their oil and gas production. For them, many climate action policies mean rethinking the economic system that has supported them, their families, and their town for decades.

In 2008, anti-climate-change groups began asking Republicans to sign a "No Climate Tax" pledge (Mayer, 2013). In 2009, cap-and-trade legislation died in Congress. The climate denial lobby went after Democrats who had voted for cap and trade, and then they took on Republicans who were sympathetic to green energy or climate action.

Today, according to Republican strategist Whit Ayres, "most Republicans still do not regard climate change as a hoax. But the entire climate change debate has now been caught up in the broader polarization of American politics. A denial of the human contribution to climate change has become yet another litmus test issue that determines whether you're "a good Republican" (Mayer, 2013).

When I have a beer with my Republican colleagues, they will generally acknowledge that they are aware of climate change and that human activities are driving that change. But they fear losing their elections.

Yet their voters are not necessarily climate deniers. In a study conducted by the Yale Program on Climate Change Communication, about 70% of Republicans think that climate change is happening and 53% think it is caused by human activity ("How G.O.P. Leaders Came to View Climate Change as Fake Science," *New York Times*, June 3, 2017).

So why does America not elect more Republicans who are as open-minded as their Republican voters on climate? Let me illustrate how moderation is penalized by the American electoral process in most places.

In most American Congressional districts, there is no competition between Republicans and Democrats. At least 350 seats are so-called safe seats, where voters will only elect a Democrat, as in San Francisco, New York, or Boston, for instance, or a Republican, as in most of Texas or Alabama or Kansas. In a safe Republican district, there is no chance that a Democrat will ever win the seat. Thus, the real contest is over which Republican will win the Republican Party election and make the run-off against the Democrat, who will certainly lose. This has led to a race to the right—on guns, taxes, the environment, and even science—so that only the most right-wing Republicans can get elected from these safe Republican seats. In this safe seat competition, moderate stances on climate hurt a Republican's chances.

The remaining few congressional districts are known as swing districts, where the number of Republican and Democratic voters is relatively even, and the seat can "swing" from Republican to Democrat from election to election. These competitive general elections, between a Democrat and a Republican, can cost tens of millions of dollars. This money comes from private donations and from outside spending by political action groups. I represent a competitive swing district; as an example, my own 2014 reelection was the fifth most expensive race that year. My campaign raised $4 million from private donors, as did my opponent's. Outside groups spent about $4 million trying to reelect me, and other groups spent about $4 million trying to defeat me. All told, $16 million was spent on television ads, mail pieces, and staff salaries for one 2-year term in one of 435 Congressional seats. My friend Brad Schneider was defeated that year in suburban Chicago, in a race that cost over $23 million in all.

In competitive races, candidates need both to energize their base voters and to raise money. For Republicans, the base voters are usually the most conservative. Often, the donors who want to support Republican candidates are similarly conservative. In a swing race, a moderate stance by a Republican on climate may limit the ability to raise campaign money.

So there is a lot of pressure on Republican office holders—who must fend off other Republican challengers—not to stray from the most conservative positions. A case in point recently was Representative Eric Cantor, the number 2 Republican in Congress, who offered a mildly conciliatory position on immigration. He was defeated by a more rabidly anti-immigrant Republican in central Virginia's 7th Congressional District. Every Republican Congress member took note, and most of them took cover.

Thus, despite the overwhelming science and public interest, the road to climate action for a Republican office holder is a precarious one.

Washington Republicans are finally acknowledging that climate is changing. Water is rising from the ground in Florida; hurricane after hurricane is bashing Texas, and the deep red Republican South. When Miami streets are wet on sunny days and when "500-year storms" are happening every 3 years, people take notice. As of October 2017, 60 members of the House of Representatives have joined a bipartisan Climate Solutions Caucus, with an equal number of Democrats and Republicans, "to educate members on economically-viable options to reduce climate risk and to explore bipartisan policy options that address the impacts, causes, and challenges of our changing climate" (Derby, 2017). Secretary of Defense James Mattis continues to warn that climate can be a driver of instability. That is why in this year's defense spending bill, adopted by a vote of 344–81, the House of Representatives declared that climate change is a "direct threat to national security." The bill also calls for a report on how climate change affects the military, including identifying the 10 bases most susceptible to flooding from sea level rise (Lardner, 2017).

Last month, Energy Secretary Rick Perry (who is from oil-producing Texas) testified to my Energy Subcommittee of the Energy and Commerce Committee and asserted that there was insufficient evidence to support the case that humans were the cause of climate change. However, he did state that he agrees the climate is warming. That is a start.

These events together at least offer some hope that the USA can agree on some steps toward climate adaptation. If we acknowledge that seas are rising, temperatures are warming, storms are more frequent and violent, and fires are happening more often and are brutal, we at least agree on a common set of facts that can help us prepare. We may consider new building standards that defend better against flood or fire or zoning laws that will keep buildings from expanding flood zones. Our Navy can adjust its strategy to reflect melting ice. We can reform government flood and fire insurance programs to incentivize investments in resiliency and discourage bad decisions.

It is great that the effects of climate change are sinking in. However, the refusal by Secretary Perry, the administration, and the vast majority of elected Republicans to acknowledge the basic scientific consensus that this change is human-driven remains an obstacle to the steps we need to take now.

What voices can move America to act on climate—not just to gird ourselves as the planet dies—but move us to save the planet?

Who Are the Trusted Voices That Can Turn the American Public and Our Government to Action?

First, Scientists

Americans rely on science every day, whether they are taking medications prescribed by a doctor or following a recipe for baking a cake. It must be discouraging for scientists to encounter the current irrationality, but rigorous climate science is still the most important input for our policymakers.

Second, Our State and Local Governments

They continue to demonstrate to Washington, DC, that the choice between a prosperous economy and a healthy environment is a false choice. More likely, we cannot have one without the other.

Third, the United States Armed Forces

While the United States military may be affected by domestic politics, it remains on mission. We are fortunate to have military leaders who are patriotic, brilliant, and honorable. It is within the Department of Defense where climate policy has been the least politicized and where our federal government has made the most progress.

In fact, the 2014 Quadrennial Defense Review identified climate change as a significant security risk:

> As greenhouse gas emissions increase, sea levels are rising, average global temperatures are increasing and severe weather patterns are accelerating. Climate change may exacerbate water scarcity and lead to sharp increases in food costs. The pressures caused by climate change will influence resource competition while placing additional burdens on economies, societies, and governance institutions around the world. These effects are threat multipliers that will aggravate stressors abroad such as poverty, environmental degradation, political instability, and social tensions – conditions that can enable terrorist activity and other forms of violence (Department of Defense, 2014, Chapter 1, Future Security Environment, p. 8).

The Department of Defense is one of the largest fossil-fuel users on earth, and the United States Navy has always been an innovator in energy—from sails to coal, from coal to oil, to nuclear, and now to alternative energy. In 2009, Navy Secretary Ray Mabus set a goal that by 2020, 50% of Navy energy consumption would come from alternative sources, and 50% of Navy and Marine Corps installations would be net zero. Since then, the Navy has launched the "Great Green Fleet," powered in part by advanced biofuels derived from beef fat, purchased at a competitive price even when oil prices were low. And they have secured contracts for 1.1 GW of renewable energy. And, in partnership with the State of California, the Navy acquired 400 electric vehicles for its bases in California.

Even changing to LED light bulbs is saving the Navy 20,000 gallons of fuel per year in its destroyers, with longer lives, better light, and reduced maintenance costs.

Of course, as Secretary Mabus himself said, "The goals I set were not about making us 'greener,' though that is an added benefit – they were and remain about making us better warfighters. Optimizing our energy use is a force multiplier, increasing our capabilities, our impact and our endurance across platforms and disciplines."[1]

Fourth, American Businesses

Business is not waiting for Congress to say it is okay to compete in the renewable economy. For instance, now that China has called for one out of every five cars sold there to run on alternative fuel within 8 years ("China Hastens a Global Move to Electric Cars," New York Times, October 10, 2017), every worldwide automaker will have to compete in that space. American politicians will see that it is not just Tesla, but all US carmakers, and all their voting employees, who are part of the climate revolution.

Shareholders in American companies are themselves demanding climate responsiveness. This May, ExxonMobil shareholders, with a vote of 62% of its shares,

---

[1] "Biofuel blend powers Navy's Great Green Fleet deployment," Governors' Biofuels Coalition, January 21, 2016.

instructed the company to analyze and report on how the company would respond to global efforts to achieve the 2 °C target ("Financial firms lead shareholder rebellion against ExxonMobil climate change policies," *Washington Post*, May 31, 2017). The two largest shareholders in ExxonMobil are BlackRock and Vanguard, major Wall Street investment firms.

Fifth, Respected Republicans

Republican leaders who are no longer in office are important advocates for climate action. This year, the Climate Leadership Council (www.clcouncil.org) called for a tax on carbon, labeling it the "Conservative Case for Carbon Dividends." The plan would impose a gradually increasing tax on carbon, fully refunded to American taxpayers, combined with the repeal of regulations like the Clean Power Plan that would be redundant and less efficient at encouraging renewables than the carbon tax itself. The proposal is not novel, but the identity of its authors is. The council is led by James A. Baker III, George P. Schultz, Henry M. Paulson, Martin Feldstein, all former members of the Reagan and Bush administrations, and respected conservative economists and businessmen.[2]

Former US Representative Bob Inglis has formed RepublicEn.org, an effort to educate the country about free-enterprise solutions to climate change. His EnCourage Tour 2017, a series of town hall style meetings across the country, engaged voters across America in discussions of the "conservative case" for climate action.

Sixth, the News Media

A free press is a foundation of American democracy, and we need its help on climate. Our news media must be more honest about the severity of the climate crisis and the real facts. The media seems more concerned about being "balanced" than being accurate. This has led to needless confusion about the validity of undisputed science. And our media tend to focus on the issues, tweets, and scandals of today to the detriment of the long-term challenges, even existential threats like climate change.

---

[2]The Climate Leadership Council is sponsored by formidable businesses, including BP, ExxonMobil, General Motors, Johnson & Johnson, Procter & Gamble, PepsiCo, Santander, Schneider Electric, Shell, Total, and Unilever.

Finally, The Church

The church can show the moral imperative of action on climate. The Vatican's leadership is critical as we try to change minds and turn hearts in the USA. It is even more significant as we try to influence policy in the United States Congress. Here's why:

According to the Pew Research Center, while the number of adults in the USA "who describe themselves as Christian has been declining for decades, the U.S. Congress is as Christian today as it was in the early 1960s." Ninety-one percent of the Congress identifies as Christian, and one-third of the Christians in Congress identify as Catholic. And while the number of Christians in Congress has been virtually unchanged in the past 50 years, the number of Catholics in Congress has increased dramatically—from 19 to 31%.[3]

So nearly one-third of Congress has a direct connection to this church, which is why Pope Francis's commitment to making environmental stewardship a priority of his papacy has such potential to affect US climate policy.

Something else is happening, too. Evangelical Christians, who typically align with conservative Republicans, are weighing in on the need for climate action. If we understand the impact these voters have on the American electorate, we see that this is potentially transformative. As Rev. Mitch Hescox explains in his companion chapter:

> The religious landscape is changing in the United States. Today, only 43% of Americans identify as white and Christian compared to 81% in 1976. Yet even amid these shifting demographics, white evangelicals still comprise 35% of the Republican party (Cox and Jones, 2017). Together, evangelicals and Catholics represent approximately 28% of the American populace, yet they consistently vote in disproportionately high numbers. Over 80% of white evangelicals voted for Mr. Trump in 2016, representing 28% of actual voters. More broadly, the last three United States Presidential elections saw pro-life voters average 45% of the electorate (evangelicals, 26%, Catholics, 19%) and for many of these voters, their anti-abortion (pro-life) stance overrode all other electoral considerations.

Like this pope, who is named for the saint from Assisi who considered all living things his brothers and sisters, many leading evangelical Christians are taking up the cause of climate because they, too, believe we have a moral imperative to save the planet given to us by God and to preserve His creation for our children.

I want to thank the Vatican and Pope Francis for your moral clarity and commitment to climate action and for including me in this historic event. I am very grateful to have been selected to represent the United States Congress in this book.

---

[3] Pew Research Center. Jan. 3, 2017, "Faith on the Hill".

References

Cox, D., & Jones, R. P. (2017). *America's changing religious identity*. Washington, DC: Public Religion Research Institute. Retrieved on February 16, 2020 from https://www.prri.org/research/american-religious-landscape-christian-religiously-unaffiliated/

Department of Defense (2014). Quadrennial Defense Review 2014. Retrieved February 16, 2020 from https://www.google.com/url?sa=t&rct=j&q=&esrc=s&source=web&cd=1&ved=0ahUKEwiQ0OeXwfXWAhWIQiYKHULD1oQFggmMAA&url=http%3A%2F%2Farchive.defense.gov%2Fpubs%2F2014_Quadrennial_Defense_Review.pdf&usg=AOvVaw1b7Pyfby1OhUA3KggNaQwk

Derby, K. (2017, September 26). Stephanie Murphy Joins Carlos Curbelo's, Ted Deutch's climate solutions caucus. Retrieved on February 16, 2020 from http://sunshinestatenews.com/story/stephanie-murphy-joins-climate-solutions-caucus-group-nears-60-members

Mayer, J. (2013, June 30). Koch pledge tied to congressional climate inaction. The New Yorker. Retrieved on February 16, 2020 from https://www.newyorker.com/news/news-desk/koch-pledge-tied-to-congressional-climate-inaction

Lardner, R. (2017, July 14) Defense bill calls climate change a national security threat. Associated Press. Retrieved October 16, 2017 from https://apnews.com/c119aa0ef87446108ab-d2aa637e91b7f/Defense-bill-calls-climate-change-a-national-security-threat

U.S. Department of Energy (2017). 2017 U.S. Energy and Employment Report. Retrieved on February 16, 2020 from https://energy.gov/downloads/2017-us-energy-and-employment-report

CHAPTER 31
Challenges and Opportunities for a Sustainable Planet

Yuan T. Lee

Summary We know that population and consumption are the two primary drivers of dangerous environmental changes. And we already possess the technologies and capabilities required to address them. It is well past time for us to translate that into determined global action. Politically, we must shift from a nation-based approach to one of true global solutions. Socially, we must realize that the world is already over-developed and that averting crisis means reducing unsustainable consumption as well as improving the lives of the poor. Technologically, as we pursue innovations that will take us back to sunshine as the dominant source of energy, we must also develop ways of sharing and distributing renewable energy across borders. It is not too late for us to turn things around, but we do not have much time left.

Introduction

Some years ago, on a long-haul flight, I struck up a conversation with the business-woman sitting next to me. She was interested in my work on climate change and sustainability, so I explained to her that it really came down to population and consumption.

I said that between 1900 and 2000, the number of people in the world grew by four times, while the amount that each person consumed doubled. This meant, of course, that human consumption overall grew eightfold within a century.

This is dangerous on a planet of finite space and resources, I explained. The Earth might seem rather big to us, but humans really only inhabit a thin layer of it, like the skin of an apple. And our voracious appetite for resources and our dumping of waste has already put a tremendous burden on this thin layer.

Y. T. Lee (✉)
Institute of Atomic and Molecular Sciences, Academia Sinica, Taipei City, Taiwan
e-mail: ytlee@gate.sinica.edu.tw

Two Forms of Pollution

Today, this burden is manifested in two forms of pollution. One is visible: the air pollution that is literally choking our cities and killing people.

The Lancet Commission on Pollution and Health just released its landmark eport, which concluded that, despite being a neglected topic for decades, "pollution is the largest environmental cause of disease and death in the world today, responsible for an estimated 9 million premature deaths." The report found air pollution to be the deadliest of them all, having caused 6.5 million premature deaths in 2015. In my native Taiwan, it is estimated that over 9% of deaths can be attributed to air pollution.

The second form of pollution is less visible but no less dangerous: greenhouse gases that have contributed to an energy imbalance in the Earth system. More energy is coming in than going out. We have heard the incredible numbers from the likes of former US Vice President Al Gore—that the heat being trapped by greenhouse gases is equal to 400,000 Hiroshima-class atomic bombs going off every single day and is driving dangerous environmental changes.

An Unsustainable Trajectory

At the root of both forms of pollution is the rising number of people in the world and the rising amount of resources they are consuming. And this is by no means a new insight. In 1994, 58 academies of sciences from around the world had already published a statement pointing out that "If current predictions of population growth prove accurate and patterns of human activity on the planet remain unchanged, science and technology may not be able to prevent irreversible degradation of the natural environment and continued poverty for much of the world."

The trajectory we are on will take us past 9.5 billion people globally by 2050, over 470 ppm in atmospheric CO_2 concentration, and a ruinous level of consumption. The potential impacts have been well covered, including by the participants of this important book: mass coral reef and forest die-off, extinction of a substantial fraction of the species on Earth, ever more frequent and extreme weather events, sea-level rise, a sharp decline in food production, and so forth.

What Are We Doing About It?

As you might imagine, my conversation with that businesswoman on the plane was not the most cheerful. But she continued anyway. She asked me: "So, if it really comes down to population and consumption, then that must be what your organization is working on right?"

I thought for a moment, and I said, "No."

I was the president of the International Council for Science (ICSU), and we were doing a lot of important work, but we were not working directly on population and consumption.

She had a point. We already know what the key drivers of environmental change and unsustainable development are. We already possess technologies and capabilities to begin addressing them. But taken together, all of our efforts across the world are still insufficient to make our development sustainable. Collectively, all of the intended nationally determined contributions would still be insufficient for us to avert the climate crisis.

So how do we do better? I would like to offer five points for consideration.

Global Responses for Global Problems

First, climate change has regional and local manifestations, but at its heart it is a global problem, and it requires global solutions. The intended nationally determined contributions (INDC) are a prime example of this. Each country has determined for itself the level of its contribution to cutting emissions, based on factors like its domestic economic goals and politics. But it is ultimately the global-level result that will matter. If all the nationally determined efforts succeeded in taking us to 2.5 °C, we would still be in deep trouble. So our solutions must be global.

Another example is in energy. Getting to 100% renewable energy is quite a different challenge depending on where you live. In California, the population density is about 90 people per square kilometer, while in my native Taiwan, we have over 650 people per square kilometer. That is a whole lot of energy demand and not much land to deploy renewable energy supplies. And Taiwan is hardly alone in this regard.

Left to their own devices, densely populated places will have a hard time getting to a fully renewable future. So it would be wise for us to develop a cross-border system for trading renewable energy.

Back to Nature, Back to Sunshine

Second, whatever solutions we devise, they must take us back to nature and back to sunshine. Our societies and governments are used to putting the economy at the center of our decision-making. Even when sustainable development is on the agenda at the UN, and people talk about the economic, the social, and the environmental, everyone knows that the economy sits at the top. The first five sustainable development goals are clearly economic and social.

But all this must change. We must come around to the realization that without the environment, there would be neither society nor economy. We must put the environment at the center of our decision-making. This will require a deep decarbonization of our societies and economies by, for instance, establishing a carbon tax or fee.

And we have to go back to sunshine. Instead of digging down for more fossil-based energy, we must now reach up for solar energy. It is the most powerful fuel source we have. One hour of sunlight on Earth can meet one year of global energy needs! We must restore the direct relationship between people and the sun.

Live Better for Less

The third point is that while we need to improve the lives of billions of people, we can do so in a way that uses far fewer resources than the rich countries are using today. For instance, we can design our systems to be radically efficient in their use of resources. We can design our buildings to be net producers of water and energy. And we can design our neighborhoods so that all of our daily needs are easily met by walking or short public transit rides.

Innovators have proven again and again that human life can be improved by using fewer resources, not more.

Control the Population Explosion

Fourth, we need to reduce the explosive growth in human numbers. This has been a rather unpopular and neglected topic in sustainability discussions. Whenever we make projections for the coming decades, we almost take it as a given that global population will shoot past 9.5 billion and even 10 billion by 2050. And we calculate resource needs based on that. But is this really inevitable? Is it really fate? I would be inclined to say no.

We can bring human numbers to a more sustainable level by, for instance, improving the economic conditions of the poor and needy around the world. Some people argue that we should worry about consumption rather than population, but that is a false dichotomy. Humans are collectively having such an enormous impact on the environment that we must work on both population and consumption.

Improve Inequality Around the World

Fifth and finally, reducing inequality is a priority, but there are different ways of going about it. When we look at the discussions around sustainable development and the Sustainable Development Goals, the idea seems to be that we should reduce inequality by pulling the poor's living standards up to those of the rich.

But what would that do to the environment? And while we are busy getting the poor up to a higher level of consumption and income, the rich are getting richer still.

What about curtailing the unsustainable consumption of wealthy societies? If we wish to avert environmental catastrophe, then as we reduce inequality and improve the living standards of the poor, we must also curtail the unsustainable consumption of wealthy societies. That would be the proper way to reduce global inequality.

CHAPTER 32
Sustainable Development Goals and Health: Toward a Revolution in Values

Jeffrey D. Sachs

> All of this shows the urgent need for us to move forward in a *bold cultural revolution*. Science and technology are not neutral; from the beginning to the end of a process, various intentions and possibilities are in play and can take on distinct shapes. Nobody is suggesting a return to the Stone Age, but we do need to slow down and look at reality in a different way, to appropriate the positive and sustainable progress which has been made, but also to recover the values and the great goals swept away by our unrestrained delusions of grandeur.
>
> Pope Francis, *Laudato Si'*, 114

Pope Francis calls for a revolution in governance, to manage technologies for the common good. In my mind, this brings us back to politics in the tradition of Aristotle rather than the tradition of Machiavelli. For Aristotle, politics is about wellbeing (*eudaimonia*) of the *polis*, the political community. For Machiavelli, politics is about the struggle for power, little more. It should be clear that we are today living in Machiavelli's political world, not in Aristotle's.

To return to Aristotle's politics, we need a shared conception of the common good based on human nature. It seems to me that our great moral traditions, and in particular the Church's social teachings, offer us deep guidance. We should give priority to the poor; defend human dignity; promote the cardinal virtues of justice, fortitude, prudence, and practical wisdom; and dismantle the structures of sin, including pervasive corporate cheating, corporate lobbying, corporate campaign financing, and tax havens that facilitate global corporate and personal tax evasion of the rich on an unfathomable scale.

The 17 Sustainable Development Goals (SDGs), shown in Fig. 32.1, were adopted by all 193 UN member states in September 2015 to cover the period 2016–2030. The SDGs provide the best opportunity for a new cooperative global politics oriented around the common good. The SDGs aim for economies that are prosperous, socially inclusive, and environmentally sustainable. In the jargon of the

J. D. Sachs (✉)
Columbia University and Director of the UN Sustainable Development Solutions Network,
Columbia University, New York, NY, USA
e-mail: sachs@columbia.edu

W. K. Al-Delaimy, V. Ramanathan, M. Sánchez Sorondo (eds.), *Health of People,
Health of Planet and Our Responsibility*, https://doi.org/10.1007/978-3-030-31125-4_32

Fig. 32.1 The 17 sustainable development goals

UN, the SDGs should "leave no one behind," including future generations and indeed other species and Earth itself.

There are two intersecting themes to consider. The first theme is that we must move to a clean energy system, one that stops the emission of greenhouse gases and cleans the air for human health. The energy goals are covered by SDGs 7 and 13. The goals on human health are covered by SDG 3 (healthy lives for all at all ages).

The second theme is that we must move to a sustainable food system, one that promotes healthy diets for all while eliminating the massive environmental destruction caused by current farming practices. These food-sector objectives are covered especially by SDG 2 (sustainable agriculture), SDG 14 (sustainable marine ecosystems), and SDG 15 (sustainable terrestrial ecosystems). Again, one priority is that healthy diets reduce disease and premature mortality, especially by combatting the worldwide obesity and metabolic disease epidemics caused by "fast food." Both for energy and food, the biggest challenge we face is the structures of sin, namely, the power of corporate food and agricultural giants whose greed and irresponsibility undermine the common good.

When it comes to energy, the challenge is rather easily expressed. The world should decarbonize the global energy system by 2050 in order to achieve the Paris Climate Agreement goal of keeping global warming below 1.5 °C. This means, at the core, a shift from fossil fuels (coal, oil, and natural gas) to renewable energy, especially wind, solar, hydro, and geothermal energy. This shift is feasible and even economical if carried out rigorously and expeditiously over the period to 2050.

In a nutshell, all electricity should be produced by zero-carbon energy. All transport should be electrified (as with light-duty vehicles) or should transition to using synthetic fuels produced by zero-carbon electricity. Such fuels include hydrogen and synthetic liquid hydrocarbons. All buildings should be electrified for heating and cooking. All heavy industry should find alternatives to carbon emissions. Technological solutions exist or are within reach in each case.

These energy shifts will have two benefits: saving the climate and cleaning the air. Zero-carbon energy sources will reduce PM2.5 aerosols and other environmental pollutants, and the co-benefits of health are an enormously important reason to accelerate the shift to clean energy. India, for example, currently loses more than a year of life expectancy from massive air pollution, and in some states, more than 2 years of life expectancy.

Despite the urgent need for rapid decarbonization of energy, the process continues to be blocked by the corporate greed of companies such as ExxonMobil, Chevron, Gazprom, BHP, and the like. These companies will cause massive detriment to populations vulnerable to air pollution unless they are decisively brought under regulation, yet in the United States, Australia, Canada, Russia, and other countries, the oil industry rather than the common good directs politics. We are in the political world of Machiavelli rather than Aristotle.

The same tendency is clear in the food sector. Current global agricultural practices are wholly inadequate, indeed in crisis, for three reasons. First, the agricultural sector is a huge contributor to environmental catastrophe: deforestation, eutrophication, chemical pollutants, massive greenhouse gas emissions, invasive species, and others. Second, the agriculture sector is undermining its own long-term viability by destroying the land and soil and increasing its own vulnerability to climate change. Third, the diet produced by today's food industry is disastrous, especially through the spread of obesity and food addictions caused by sugar additives, saturated fats, and heavy food processing.

Once again, the solutions are rather easily described. We need to move from heavy beef consumption to diets that are based far more on plant proteins, notably nuts, legumes, and fruits. We need to enforce strong land-use regulations that protect biodiversity and stop the conversion of forests and grasslands into new pasturelands for beef.

As with energy, these changes are scientifically sound, morally correct, and technologically feasible, yet they are being stalled or slowed by the likes of Coca Cola, PepsiCo, McDonald's, Kentucky Fried Chicken, Kraft Heinz, and other food giants. These companies are massive lobbyists and major campaign contributors in the United States. They are part of the structures of sin, since they carry on with their destructive products even when the evidence is overwhelming that they are the creators of the global obesity epidemic and much else that is wrong with the world's diets.

What, then, will be the path of the bold cultural revolution called for by Pope Francis? It will have the following steps:

1. Putting the common good and human dignity in the highest purpose of politics;
2. Choosing the preferential option for the poor;
3. Adopting the SDGs and the Paris Climate Agreement as guiding principles for the world's governments and businesses;
4. Preparing plans to achieve the SDGs and the Paris Climate Agreement based on the best scientific information and in processes marked by transparency, justice, and public inclusion;
5. Adopting SDG and climate indicators to ensure accountability of governments and businesses;
6. Dismantling the structures of sin that have placed corporate greed above the common good.

CHAPTER 33
A Call to Action by Health Professionals

Edward Maibach, Mona Sarfaty, Rob Gould, Nitin Damle, and Fiona Armstrong

Summary We contend that the goal of the Paris Climate Agreement—limiting global warming to no more than 2 °C (and if possible 1.5°C)—is the world's most important public health goal. Failure to achieve the goal will lead to a sustained public health disaster, worldwide, lasting many generations. While the leaders of most nations have committed to the goal, with few exceptions, their actions thus far have been grossly insufficient to achieve it. In this chapter, we argue that the world's health professionals—doctors, nurses, midwives, and other allied health professionals—are uniquely suited to lead the effort to persuade the world's leaders to increase their commitments to the goal of the Paris Agreement and to rapidly initiate the actions necessary to fulfill their commitments.

If not us (health professionals), who? If not now, when? We are among the most trusted members of every society worldwide. We speak with not just scientific but also ethical and moral authority. We have—or can get—the attention of our nation's leaders. We are free of conflicts of interest with regard to climate change and health and are scrupulously evidence-based. And most importantly, we are powerful advocates for the very thing that people worldwide care about most and is most threatened by climate change: our health and well-being and that of our children and their children.

E. Maibach (✉) · R. Gould
George Mason University Center for Climate Change Communication, Fairfax, VA, USA
e-mail: emailbach@gmu.edu

M. Sarfaty
George Mason University Center for Climate Change Communication, Fairfax, VA, USA

Medical Society Consortium for Climate and Health, Washington, DC, USA

N. Damle
Brown University School of Medicine, Providence, RI, USA

F. Armstrong
Climate and Health Alliance (Australia), Melbourne, VIC, Australia

© The Author(s) 2020
W. K. Al-Delaimy, V. Ramanathan, M. Sánchez Sorondo (eds.), *Health of People, Health of Planet and Our Responsibility*, https://doi.org/10.1007/978-3-030-31125-4_33

Efforts are already under way in many nations to activate health professionals as advocates for climate and health. Each of these efforts is important, but they are not yet tightly focused on achieving the global goal of limiting warming to no more than 2 °C. We therefore propose an initiative to rapidly engage, organize, and empower a movement of health professionals and health organizations to convince the leaders of every nation to double their commitment to the Paris Agreement. By establishing a simple, clear worldwide objective—and a parallel objective for every nation—linked directly to the Paris Agreement, and by identifying an exact date by which these objectives must be met, concerned health professionals in every nation will become compelling guides for the difficult but necessary pathway to protect the health and well-being of all the world's people, especially those most vulnerable to climate change impacts, now and for generations to come.

To achieve the goal of the Paris Climate Agreement—limiting global warming to no more than 2 °C (and if possible 1.5 °C)—and thereby avoid the most harmful impacts of climate change on human health and well-being, the global community must dramatically reduce its use of fossil fuels in the very near future (Allen et al., 2018; Figueres et al., 2017). This goal can be achieved if global carbon emissions peak by 2020 and fall rapidly to near zero over the next several decades (Rockstrom et al., 2017; Rogelj et al., 2018). Approximately half of the necessary emissions reductions have already been pledged by the nations of the world in the Paris Agreement, assuming that all nations remain in the agreement. *If success is to be achieved, nations must collectively double current global commitments over the next few years.* Realizing the 1.5 °C goal will require even bolder action (Hoegh-Guldberg et al., 2018). Building the political will necessary for this to happen will not be easy, but doing so is imperative. Health professionals are in a unique position to help build the necessary political will.

The most important actions that any individual, any organization, any group of concerned citizens can take to protect the world's climate—and with it, the health and well-being of all people alive today and for countless generations to come—are those that will help build the political will necessary to persuade the world's leaders to double their nation's current carbon pollution reduction commitments and to fulfill those commitments as quickly as possible. Fulfilling those commitments will require nations to embrace policies, regulations, and technologies capable of greatly accelerating the inevitable transition to clean energy and rapidly eliminating the emission of short-lived climate pollutants (methane, HFCs, and black carbon).

For a myriad of reasons, the world's health professionals, including doctors, nurses, midwives, and other allied health professionals and public health professionals are uniquely suited to lead the effort to persuade nations of the world to double their emission reduction commitments. They are among the most trusted members of every society worldwide. They speak with not just scientific but also ethical and moral authority. They have—or can get—the attention of their nation's leaders. They are free of conflicts of interest with regard to climate change and health and are scrupulously evidence-based. And most importantly, they are powerful advocates for the very thing that people worldwide care about most and is most threatened by climate change: our health and well-being and that of our children and their children.

The world's health professionals have—or should have—ample motivation to play this role. The health harms of climate change are already significant contributors to many of the leading causes of morbidity and mortality worldwide, and as the planet continues to warm, these impacts are projected to become far worse (Field et al., 2014; Levy & Patz, 2015; McMichael, Woodward, & Muir, 2017; Woodward et al., 2014). Moreover, the single most potent driver of climate change—burning of fossil fuels—also creates air pollution, which is another of the leading causes of morbidity and mortality worldwide, distinct from climate change (Ciesielski, 2017; Perera, 2017). As was made clear by *The Lancet* Commission on Health and Climate Change in 2015, *"Responding to climate change could be the greatest global health opportunity of the 21st century"* (Watts et al., 2015). Each nation that doubles it commitment to reducing carbon pollution will accomplish two important public health goals: (1) dramatically and rapidly improving the health of its citizens by embracing clean energy and gaining cleaner air, soil, and water and (2) protecting its citizens—and all people—by doing its part to avert further dangerous climate disruption.

Efforts are already under way in many nations to activate health professionals as advocates for climate and health. These include the work of the Climate and Health Alliance (Australia), the Canadian Association of Physicians for the Environment, the UK Health Alliance on Climate Change, the US Climate and Health Alliance, Physicians for Social Responsibility (US), the American Public Health Association, the Alliance of Nurses for Healthy Environments (US), and the Medical Society Consortium for Climate and Health (US). Many of these organizations collectively represent a substantial proportion of their national health and medical workforces. At the regional or global level, the Health and Environment Alliance is engaging health actors around energy choices across the European Union, and the Healthy Energy Initiative is working with health professionals globally on shifting public funds away from fossil fuels and into health and climate action. Also at the global level, the Global Climate and Health Alliance is working to tackle climate change and protect and promote public health, and through its Global Green and Healthy Hospitals Network, Health Care Without Harm is working to transform hospitals and health systems worldwide so they reduce their environmental footprint and become anchors for sustainability and leaders in the global movement for environmental health and justice.

Each of these efforts is important, but they are not yet tightly focused on achieving the global goal of limiting warming to no more than 2 °C. We therefore propose an initiative to rapidly *engage, organize, and empower* a movement of medical, nursing, midwifery, allied health professional, and public health experts and organizations that will seek to convince the leaders of every nation to double their commitment to the Paris Agreement. By establishing a simple, clear worldwide objective and a parallel objective for every nation, linked directly to the Paris Agreement—i.e., doubling the current level of emission reduction commitments—and by identifying an exact date (to be determined) by which these objectives must be met, concerned health professionals in every nation will become compelling guides for the difficult but necessary pathway to protect the health and well-being of all the world's people, especially those most vulnerable to climate change impacts, now and for generations to come.

Strategy One: Engage

The first and most important action that must occur is to engage with and recruit the support of a broad cohort of health stakeholders to build on and amplify the current set of climate and health initiatives and activities around the world. This should include the following groups of professionals.

Doctors

As noted earlier many medical societies and medical society consortia already understand and focus on this issue. Some of these consortia and their participating medical societies have substantial global reach. For example, the American College of Physicians has chapters in 19 nations (including Brazil, Canada, India, Mexico, and Saudi Arabia), and the American Academy of Pediatrics, the American Academy of Allergy, Asthma, and Immunology, the American Thoracic Society, and the American College of Obstetrics and Gynecology also have chapters in many nations around the world. Other important medical organizations with global footprints—including the World Medical Association and the World Organization of National Colleges, Academies and Academic Associations of General Practitioners/Family Physicians—have strong climate change policies and can be brought into the movement. Other medical organizations with a substantial international presence, such as the Interacademy Medical Panel and the International Pediatric Association, are also excellent candidates to join the movement.

Nurses, Midwives, and Allied Health Professionals

Nurses and midwives constitute the largest group of health professionals globally, with approximately 19.3 million nurses and midwives worldwide. These professionals—who provide care to almost every person in the world at some point during their lives—offer the potential for influencing significant global change through their size and scale and through their capacity to reach a large and diverse proportion of the global population.

Along with other allied health professionals (e.g., psychologists, physiotherapists, social workers, occupational therapists), these professions offer the opportunity to bring large numbers of health professionals to the movement, people who are on the ground providing health care and who possess organizational clout in virtually every community worldwide.

Around the world, nurses are often at the top of the list of the public's most trusted sources on health issues. Nurses are taking the lead in supporting climate action in many nations, such as the Association of Nurses for a Healthy Environment in the

USA that is developing the Nursing Collaborative on Climate Change and Health, which aims to engage national nursing organizations and elevate nursing leadership in addressing climate change; the Australian Nursing and Midwifery Federation that offers educational and professional development training on climate change to thousands of nurses each year; and the Canadian Nurses Association Code of Ethics that supports and encourages nurses to promote climate adaptation and mitigation.

Globally, in 2018 the International Council of Nurses (ICN)—with members in 130 countries, representing nearly 20 million nurses around the world—issued a position statement calling for government action and nursing leadership on climate change; this followed a focus on the health impacts of climate change at its inaugural Health Policy Summit in 2017. This cohort, including organizations such as ICN, can add significant size, scope, and power to the global climate and health movement.

Public Health

In the USA, the American Public Health Association declared 2017 the year of climate and health. Other public health organizations, such as the World Federation of Public Health Associations (WFPHA)—with member associations in over 100 countries—and many of their national counterparts, have also been active on the issue. WFPHA's 2015 World Congress on Public Health had a strong focus on climate change, energy policy, and health. Also in 2015, WFPHA conducted the first ever global survey on national climate change and health policy, with a subsequent report documenting the actions of 38 respondent countries on addressing the health impacts of climate change.

Existing Climate Health Alliances and Initiatives

Several important organizations are already working at the intersection of climate and health worldwide—including Health Care Without Harm, Health and Environment Alliance (HEAL), Global Climate and Health Alliance, the Lancet Countdown, and the Global Consortium on Climate and Health Education. Each has established architectures and missions that are synergistic with this initiative and will provide it with a cadre of leaders and bridge builders.

While the aim is to quickly recruit medical, nursing and midwifery, allied health, and public health professionals and organizations worldwide to actively participate in the initiative, there are obvious advantages to initially targeting institutions and individuals in those nations that emit the highest levels of carbon pollution (e.g., the top 15 emitting nations: China, the USA, European Union, India, Russia, Japan, Brazil, Indonesia, Canada, Mexico, Iran, South Korea, Australia, Saudi Arabia, and South Africa).

Strategy Two: Organize

A team must be assembled to work directly with health professionals, organizations, and consortia to organize and empower health professionals and institutions to build this movement. They must also recruit and mobilize significant communication and advocacy assets for the initiative.

The organizing team—to be composed of the organizations that have the greatest strategic and operating capacity to build and lead the movement—should be guided by an advisory board of leading climate and health, strategy, and communication experts worldwide. The advisors will help provide guidance and insight on strategy, while the organizing team will help implement the strategy and tactics. With the help of these experts and collaborators, the organizing team will take the lead in planning, organizing, and implementing the movement's activities in each partici-pating country and in identifying health leaders in each country who will be sup-ported to become the public face of the movement locally and nationally. They will also work to ensure the visibility of these efforts and those of the broader movement and work to embed key messages on climate change and health in global and regional news media outlets.

Strategy Three: Empower

Our proposed initiative will work to empower health professionals to deliver key messages on climate change and health—repeatedly and forcefully—with the aim of reaching audiences in every nation of the world. Through educational meetings, materials development, research, information sharing, and media relations support, concerned health professionals will be empowered to become powerful advocates for the health and well-being of their patients and the public at large.

Key Messages and Engagement Strategies

1. *The health harms of climate change and air pollution are happening now.* Our health is already being harmed by climate change and air pollution, and it's almost certain to become dramatically worse unless we take action now. This is especially true of medically vulnerable people including pregnant women, chil-dren, the elderly, people with chronic illnesses, disadvantaged persons of color, and the poor.
2. *We must take immediate action to protect our communities.* Communities should be taking action to protect people's health from the impacts of climate change that are already with us. For example, communities can reduce the health harms associated with climate change by developing early warning systems for impend-

ing heat waves, providing resources to protect vulnerable populations, and educating their members about the health effects of climate change.

3. *Prevention offers the most powerful protection against future harms.* To prevent these harms from getting worse, the most important actions we must take—as individuals, communities, businesses, and nations—are those that will greatly accelerate the inevitable transition to clean energy and limit emission of toxic short-lived carbon pollutants.

4. *Preventive actions will create immediate health and economic benefits.* Taking these actions will not only protect the health of the population and our off-spring—and their offspring—from a dangerously unstable climate in the future, they will also help clean up our air, soil, and water almost immediately, thereby enabling everyone alive today to enjoy better health for the remainder of our days and ensuring the well-being of future generations.

5. *We must double our commitment to reducing carbon pollution—and then live up to it.* The most important action every country can take is to ensure it is committing to—and then delivering—its fair share of the global responsibility to emissions reductions in order to collectively achieve our shared goal under the Paris Climate Agreement. This is the surest path to ending air pollution, limiting climate change, promoting health and well-being, and ensuring a prosperous future for humanity.

In each targeted nation, the organizing team will work to break down the barriers to public engagement and build political will through four complementary national strategies:

1. Educating the public, other health professionals, the media, business leaders, and policymakers about the relevance of air pollution, climate change, and climate solutions to human health.
2. Advocating for a doubling of their nation's emissions reduction commitment.
3. Mobilizing individual health professionals to join the effort.
4. Building or joining coalitions of other national, regional, and local organizations that will support the advocacy goal and spread the initiative's key messages.

The Critically Important Role of Health Professionals

Historically, the potential for health professionals to convince the world's political leaders to redouble their climate stabilization efforts is clearly reflected in the pivotal role they have played with regard to preventing nuclear war. In 1980, American and Soviet physicians established the International Physicians for the Prevention of Nuclear War (IPPNW), which in 1985—a mere five years after its creation—was awarded the Nobel Peace Prize for its central role in helping to open arms control discussions between the USA and the USSR.

In his 1987 book *Perestroika*, Mikhail Gorbachev specifically pointed to the influence of IPPNW on his decision to stop testing nuclear weapons and to enter a nuclear arms reduction treaty with the USA: "Their work commands great respect. For what they say and what they do is prompted by accurate knowledge and a passionate desire to warn humanity about the danger looming over it. In light of their arguments and the strictly scientific data they possess, there seems to be no room left for politicking. And no serious politician has the right to disregard their conclusions." Later, he thanked and credited IPPNW even more directly: "I want to thank you for your great contribution to preventing nuclear war. Without it and other effective antinuclear initiatives this [INF] Treaty would probably have been impossible" (Gorbachev, 1987).

Obviously, humanity is not yet free from the existential threats associated with nuclear war. In 2007, adapting to the evolving global situation, IPPNW organized and launched an even more aggressive campaign—the International Campaign to Abolish Nuclear Weapons (ICAN)—to advocate for a nuclear weapons convention that would ban the development, possession, and use of nuclear weapons. Since then, more than 450 partner organizations in over 100 countries have become members of ICAN. In 2014, ICAN launched its Humanitarian Pledge, asking nations to commit to creating a nuclear ban treaty, a call that was subsequently endorsed in a UN resolution by more than 125 countries. This resolution created the foundation for the 2017 UN Treaty on the Prohibition of Nuclear Weapons, which 50 member states have now signed and 3 have ratified. In 2017, ICAN, like IPPNW before it, was awarded the Nobel Peace Prize for its work, "a driving force when it comes to disarmament issues."

Other examples of the influence of health professionals on global treaties include their role in successfully advocating for and securing the Minamata Convention—a global legally binding instrument to protect human health and the environment from the adverse effects of mercury. The Minamata Convention includes a ban on mercury mining, the phase-out of products containing mercury, the introduction of control measures on air emissions and releases of mercury to the environment, and regulation of artisanal and small-scale gold mining (WHO, 2015). Adopted in October 2013, the Minamata Convention has been signed by 128 countries and ratified by 119 of them (as of April 2020) (United Nations Environment Program, n.d.). The health sector played a crucial role in the development of the convention, identifying risks to health and sources of exposure to mercury, assessing the burden of disease, developing guidelines for food, air, and water, and developing tools and guidance for action (WHO-Europe, 2017).

In a recent article in the *AMA Journal of Ethics*, Macpherson and Wynia (2017) make a strong case that physicians as a group (and presumably other professionals), and many individual physicians—especially those with unique climate- and health-relevant expertise, such as pulmonary and infectious disease specialists, and those practicing in affected regions (which, as they state are "increasingly everywhere")—have a professional responsibility to speak out about the health impacts of climate change. More recently, Law, Duff, Saunders, Middleton, and McCoy (2018) asserted that medical organizations must divest from investments in fossil fuel companies—because the "(e)vidence that fossil fuels pose serious threats to public and planetary health is overwhelming" and "(t)he fossil fuel industry has a long record of undermining climate science and continues to spread misinformation and obstruct climate action." They praise the growing number of medical organizations who are leading by example in this way, including the Royal College of General Practitioners, the American Medical Association, the Royal Australasian College of Physicians, the Canadian Medical Association, the Australian Medical Students' Association, the UK Faculty of Public Health, and the British Psychological Society. (Other organizations that are leading in this manner include the American Public Health Association, the British Medical Association, the Chicago Medical Society, Gunderson Health System, Healthcare Without Harm, Mediabank, Practice Greenhealth, the Royal Australasian College of Physicians, Society for the Psychology Study of Social Issues, the UK Health Alliance on Climate Change, UNISON Cambridge Hospitals, and the World Medical Association.) Law and colleagues concluded, "We cannot afford to remain silent when the health of so many is at stake."

A current Australian initiative, led by the Climate and Health Alliance, nicely illustrates the potential of the approach we have proposed here. Through a 2-year consultation process, the initiative engaged, mobilized, and empowered health and medical experts, professional associations, and other groups in the development of a comprehensive national policy roadmap on climate change and health—called the Framework for a National Strategy on Climate, Health and Well-being for Australia. This policy framework, released in 2017, was designed to support the Australian government in fulfilling its international obligations under the Paris Climate Agreement.

The framework's clear demonstration of a substantive consensus on policy among key influential health and medical groups provided an important mandate for the Australian Labor Party to adopt the policy, which they did, pledging to implement it when elected to government. Public opinion polls currently (at the time of this writing) show that the Labor Government is very likely to be elected in 2019, and they are actively promoting this policy position.

The issue of climate change had long been politically toxic in Australia—a nation that is heavily dependent on fossil fuel—having contributed to the downfall of at least three prime ministers. We believe that the leadership recently provided by the health and medical community has created a unique and effective bridge to overcome the climate policy divide.

Conclusion

The voice of health professionals in putting an end to politics as usual regarding climate change is particularly important because, for many people, recognizing the need for—and appreciating the promise of—dramatically decarbonizing our energy future has been hampered by a perceived "distance" from climate change. It is seen as something that will happen far in the future, somewhere else, and not to people but rather to plants, polar bears, and penguins. In contrast, the fundamental message of our proposed initiative is that climate change and air pollution are happening now, everywhere, and are harming us in profound ways—and that we have an extraordinary opportunity, right now, to ensure a better, healthier future for ourselves, our children, and their children. Achieving the goal of the Paris Climate Agreement will lead to better health and more sustainable wealth.

By making clear what's at stake and what must be done about it, the world's health professionals can—to paraphrase Archimedes—create a lever long enough (using their trusted voices, en masse), and a fulcrum on which to place it (the Paris Agreement), to move the world.

Philanthropic foundations, corporations, and agencies for international development (to support work in developing nations) should step forward to provide the necessary funding for this vitally important initiative. The health and well-being (and the future) of humanity may depend on it.

Acknowledgments We wish to thank MSPH student Marc Trotochaud (Johns Hopkins University) for his research assistance with this chapter.

References

Allen, M., Dube, O. P., Solecki, W., Aragón-Durand, F., Cramer, W., Humphreys, S., et al. (2018). Framing and context. In V. Masson-Delmotte, P. Zhai, H. O. Pörtner, D. Roberts, J. Skea, et al. (Eds.), *Global warming of 1.5°C. An IPCC Special Report on the impacts of global warming of 1.5°C above pre-industrial levels and related global greenhouse gas emission pathways, in the context of strengthening the global response to the threat of climate change, sustainable development, and efforts to eradicate poverty.* Retrieved January 9, 2019, from https://www. ipcc.ch/sr15/chapter/1-0/

Ciesielski, T. (2017). Climate change and public health. *New Solutions, 27,* 8.

Field, C. B., Barros, V., Dokken, D. J., et al. (2014). *Climate change 2014: impacts, adaptation, and vulnerability. Volume I: Global and sectoral aspects. Contribution of Working Group II to the fifth assessment report of the Intergovernmental Panel on Climate Change.* Cambridge and New York: Cambridge University Press.

Figueres, C., Schellnhuber, H., Whiteman, G., Rockstrom, J., Hobley, A., et al. (2017). Three years to safeguard our climate. *Nature, 546,* 593–595.

Gorbachev, M. (1987). *Perestroika: New thinking for our country and the world.* New York: Harper & Row.

Hoegh-Guldberg, O., Jacob, D., Taylor, M., Bindi, M., Brown, S., et al. (2018). Impacts of 1.5°C global warming on natural and human systems. In V. Masson-Delmotte, P. Zhai, H. O. Pörtner, D. Roberts, J. Skea, et al. (Eds.), *Global warming of 1.5°C. An IPCC Special Report on the impacts of global warming of 1.5°C above pre-industrial levels and related global greenhouse gas emission pathways, in the context of strengthening the global response to the threat of climate change, sustainable development, and efforts to eradicate poverty.* Retrieved January 9, 2019 from https://www.ipcc.ch/sr15/chapter/chapter-3/

Law, A., Duff, D., Saunders, P., Middleton, J., & McCoy, D. (2018). Medical organisations must divest from fossil fuels. *British Medical Journal, 363,* k5163. https://doi.org/10.1136/bmj. k5163

Levy, B., & Patz, J. (2015). *Climate change and public health.* Oxford: Oxford University Press.

Macpherson, C., & Wynia, M. (2017). Should health professionals speak up to reduce the health risks of climate change? *AMA Journal of Ethics, 19,* 1202–1210.

McMichael, A., Woodward, A., & Muir, C. (2017). *Climate change and the health of nations.* New York: Oxford University Press.

Perera, F. P. (2017). Multiple threats to child health from fossil fuel combustion: Impacts of air pollution and climate change. *Environmental Health Perspectives, 125*, 141.

Rockstrom, J., Gaffney, O., Rogelj, J., Meinshausen, M., Nakicenovic, N., & Schellnhuber, J. (2017). A roadmap for rapid decarbonization. *Science, 355*, 1269–1271.

Rogelj, J., Shindell, D., Jiang, K., Fifita, S., Forster, P., et al. (2018). Mitigation pathways compatible with 1.5°C in the context of sustainable development. In V. Masson-Delmotte, P. Zhai, H. O. Pörtner, D. Roberts, J. Skea, et al. (Eds.), *Global warming of 1.5°C. An IPCC special report on the impacts of global warming of 1.5°C above pre-industrial levels and related global greenhouse gas emission pathways, in the context of strengthening the global response to the threat of climate change, sustainable development, and efforts to eradicate poverty.* Retrieved January 9, 2019, from https://www.ipcc.ch/sr15/chapter/2-0/

United Nations Environment Program. (n.d.) *Minamata Convention on Mercury.* Retrieved January 9, 2019, from http://www.mercuryconvention.org/Countries

Watts, N., Adger, W. N., Agnolucci, P., Blackstock, J., Byass, P., Cai, W., et al. (2015). Health and climate change: policy responses to protect public health. *The Lancet, 386*, 1861–1914.

Woodward, A., Smith, K. R., Campbell-Lendrum, D., Chadee, D. D., Honda, Y., Liu, Q., et al. (2014). Climate change and health: on the latest IPCC report. *The Lancet, 383*, 1185–1189.

World Health Organization—Regional Office for Europe (2015). *Health sector involvement in the implementation of the Minamata Convention: assessment and prevention of mercury exposure.* Report of a meeting Bonn, Germany 24-25 June 2015. Retrieved January 9, 2019, from http://www.euro.who.int/__data/assets/pdf_file/0018/303642/Minamata-Convention_Meeting-Report.pdf?ua=1

World Health Organization—Regional Office for Europe (2017). Implementation of the Minamata Convention in the health sector: challenges and opportunities. *Information note.* Retrieved January 9, 2019, from http://www.euro.who.int/__data/assets/pdf_file/0006/340269/Minamata_WHO_web2.pdf?ua=1

Index

Printed in the United States
by Baker & Taylor Publisher Services